Hi-tech for industrial development

This book is about the role of government in hi-tech modernisation of developing countries. Should such countries concentrate on using the new technology or also produce it? What are the costs and benefits associated with each strategy? How feasible are they politically?

One of the most controversial experiences in producing and using electronics-based technology is that of Brazil. Its policy régime of reserving the internal market for national producers has attracted international attention for many years. This policy is due to be revised in 1992, so this book provides a timely assessment.

The contributors to this book investigate both the economics and politics of this experience and examine what lessons other countries can learn from it. The three main axes of the investigation are: the producer–user connection, the changing balance between national and foreign firms and the need for learning in industry and government.

The book is addressed to the students of industrialisation and technological development, those who specialise in hi-tech modernisation, as well as those concerned with policy-making.

The Editors

Hubert Schmitz is a Fellow of the Institute of Development Studies at the University of Sussex, where he has specialised in questions of industrialisation and technological development in the Third World. He is the author of *Manufacturing in the Backyard* and *Technology and Employment Practices in Developing Countries*. **José Cassiolato** has been a Research Fellow at universities in Brazil, USA and Britain, working on technology and development. He was Director of a Government Research Centre on Science and Technology Policy and from 1985 to 1988 Secretary for Planning of the Ministry of Science and Technology in Brazil.

Hi-tech for industrial development

Lessons from the Brazilian experience in electronics and automation

Edited by

Hubert Schmitz and José Cassiolato

Institute of Development Studies
University of Sussex

ROUTLEDGE

London and New York

First published 1992
by Routledge
11 New Fetter Lane, London EC4P 4EE

Simultaneously published in the USA and Canada
by Routledge
a division of Routledge, Chapman and Hall, Inc.
29 West 35th Street, New York, NY 10001

Typeset from authors' disk by J&L Composition Ltd, Filey, North Yorkshire

Printed and bound in Great Britain by
Mackays of Chatham PLC, Chatham, Kent

British Library Cataloguing in Publication Data

Hi-tech for industrial development: lessons from
the Brazilian experience in electronics & automation.
I. Schmitz, Hubert II. Cassiolato, J. (José)
338.091724

ISBN 0-415-07161-5

Library of Congress Cataloging in Publication Data

Hi-tech for industrial development: lessons from the Brazilian
experience in electronics and automation / edited by Hubert Schmitz,
José Cassiolato.
p. cm.
Includes bibliographical references and index.
ISBN 0-415-07161-5
1. High technology industries—Government policy—Brazil.
2. Automation—Economic aspects—Brazil. I. Schmitz, Hubert.
II. Cassiolato, José Eduardo.
HC190.H53H5 1992
338.981—dc20 91-25435
CIP

Contents

Figures

Tables

List of contributors

Maria Ines Bastos, Senate Commission for Science and Technology, Brasilia.

Ruy de Quadros Carvalho, Institute of Applied Economic Research, Ministry of Economy, Brasilia.

José Cassiolato, National Council for Scientific and Technological Development, Brasilia, and University of Sussex.

Fátima Gaio, Engineering Faculty (COPPE), Federal University of Rio de Janeiro.

Tom Hewitt, Development Policy and Practice Research Group, Faculty of Technology, Open University, Milton Keynes.

Mike Porteous, Department of Trade and Industry, London.

Hubert Schmitz, Institute of Development Studies, University of Sussex.

Elizabeth Bortolaia Silva, Department of Science and Technology Policy, University of Campinas.

Foreword

by Christopher Freeman

This is an important book for several reasons. Despite the deep and continuing financial crisis, the immense burdens of debt, the economic slow-down and the profound social problems, Brazil remains a country of immense vitality and potential. On any showing it is one of the countries which is so large and influential that its future will deeply affect not only the whole of Latin America but the entire world economy.

For some time now it has been the scene of interesting and innovative attempts to promote the development and application of new technologies, in particular the 'market reserve' policy designed to foster the growth of the Brazilian computer industry and related sectors. This book contains a thorough review of these policies (see especially Chapter 2 and Chapter 9) and their influence on several key sectors of the economy. The coverage is impressive and does not fall into the trap of looking at information technology as relevant only to a few sectors based on electronics. The editors have recognised that computer technology is so important because it affects every sector of the economy.

For this reason Chapter 3 on the banking industry is one of the most important. The justification for special measures to promote the development of the computer industry in any country (such as those which were used in Japan for a long time) lies not so much in the industry itself as in the linkages with the software industry and other sectors of the economy which can result. Banking automation is a major Brazilian success story but industrial automation (Chapters 5, 6 and 8) has been less successful.

The analysis of the strong and weak points of Brazilian technology policy is a feature of the whole book but Chapters 1, 2 and 10 are particularly valuable. They attempt the difficult task of an overall assessment of the market reserve policy. Whilst giving serious consideration to the neo-liberal arguments which are pressed so strongly today by international advisers in Brazil and other Third World countries, the authors' main conclusion is qualified approval of the policies which have now been scrapped because of these pressures.

This is an extremely important conclusion for other countries as well as for Brazil, despite the strength of the present neo-liberal tide. Arguments related to externalities in skill formation (Chapter 7), user–producer relationships, opportunity costs and technology accumulation are largely or entirely neglected in the neo-liberal analysis. This book shows that to ignore the role of learning in technology is to court disasters even more severe than those presently affecting Brazil and most other Latin American countries. The book is an impressive contribution to the theory and practice of industrial development.

Professor Christopher Freeman
Science Policy Research Unit
University of Sussex

Acknowledgements

This book is the culmination of a collaborative project. It started at the University of Sussex where a group of researchers from the Institute of Development Studies and the Science Policy Research Unit met regularly to discuss their work. Most were engaged in doctoral research on the economics or politics of producing or using hi-tech in Brazil. The initial idea was merely to comment on each other's ideas and papers. Indeed, the group provided critical feed-back, stimulation and discipline to individual research efforts. Then came the recognition that, by drawing together our findings, there was a project bigger than the sum of its parts. This involved setting up a framework for pooling our investigations, additional fieldwork to cover gaps, as well as sharpening and redirecting the analyses so that the policy implications could be made explicit. This book is the result of that project.

As coordinators of the work we wish to thank all the contributors to this book. Their chapters have provided the indispensable building blocks for our joint endeavour. We owe special thanks to Tom Hewitt who helped us in many ways in the course of the project, in particular by co-authoring several chapters and assisting with the editing of the book. We also benefited from the exchange of research materials and ideas with Jörg Meyer-Stamer and other colleagues of the German Development Institute in Berlin.

An important step in our project was the international workshop held at the Institute of Development Studies in June 1990. It provided a forum for discussing the findings with researchers, consultants and policy makers. We are grateful to the workshop participants for their constructive criticism.

There are many others who have helped with selected parts of the book. Their advice and support is acknowledged in the chapters concerned. The book as a whole had three key inputs which deserve special mention. Katherine Orme and Susan Moore copy-edited the entire text, Marion Huxley and her team transformed separate drafts and disks into a finished product, and Marguerite Cooke provided administrative and secretarial support throughout the project.

None of this work would have been possible without financial support. The field research for various chapters was carried out with grants from CNPq and Capes of Brazil and from the ESRC in Britain. The project as a whole, including complementary fieldwork, the international workshop, the preparation of all chapters and all work concerned with transforming them into a publishable product, was financed by a grant from the Volkswagen Foundation.

Hubert Schmitz
José Cassiolato
Institute of Development Studies
University of Sussex

Abbreviations

AAGR	average annual growth rate
ABES	Brazilian Association of Software Houses
ABICOMP	Brazilian Association for Computers and Peripheral Equipment Industries
ABIMAQ	Brazilian Machinery Producers Trade Association
ABINEE	Brazilian Association for the Electric and Electronic Industry
ABIQUIM	Brazilian Association of Chemical Industries
ABNT	Brazilian Association for Technical Standards
APPD	Brazilian Association of Data Processing Professionals
ASSESPRO	Brazilian Association of Processing Services Enterprises
BNDES	Brazilian National Bank for Economic and Social Development
CAPES	Brazilian Support Programme for Postgraduate Education
CAPRE	Brazilian Commission for the Co-ordination of Electronic Data Processing Activities
CDI	Brazilian Council for Industrial Development
CNC	computerised numerically controlled
CNPq	National Council for Scientific and Technological Development
CSN	Brazilian National Security Council
DCS	digital control system
ECLA	Economic Commission for Latin America
ESRC	Economic and Social Research Council
FIESP	Brazilian Industrial Employers Association
FINAME	Agency for Industrial Finance of BNDES
FINEP	Brazilian Funding Agency for Studies and Projects

FMC	Ford Motor Company
IACs	industrially advanced countries
IDS	Institute of Development Studies
INPI	Brazilian National Institute for Industrial Property
IPEA	Institute of Applied Economic Research
IT	information technology
ITC	indigenous technological capabilities
LDCs	less developed countries
MICs	middle income countries
MBI	Brazilian Movement for Informatics
MCT	Ministry of Science and Technology
MFZ	Manaus Free Zone
MNC	multinational company
MTBF	mean time between failures
NC	numerically controlled
NICs	newly industrialising countries
NIEs	newly industrialising economies
PC	personal computer
PLCs	programmable logic controls
PMDB	Party of the Brazilian Democratic Movement
RAIS	Brazilian Annual Register of Social Information
R&D	research and development
SBC	Brazilian Computing Society
SBPC	Brazilian Society for the Progress of Science
SCSS	Special Commission for Software and Services
SECT	Brazilian Special Secretariat for Science and Technology
SEI	Brazilian Special Secretariat for Informatics
SERPRO	Brazilian Federal Data Processing Agency
SNI	Brazilian National Intelligence Service
SPC	statistical process control
UNICAMP	State University of Campinas

1 Fostering hi-tech industries in developing countries: introduction

Hubert Schmitz and José Cassiolato[1]

This book is about the role of government in managing hi-tech modernisation in developing countries. One of the most interesting experiences in producing and using electronics-based technology is that of Brazil. The aim of this book is to investigate this experience and to examine what lessons other countries can learn from it.

We are not the first to undertake this task. The protected development of hi-tech industry in Brazil has attracted a great deal of attention internationally. The upshot is that it is one of the most controversial experiences in current industrial and technology policy; conclusions vary from condemnation to celebration of success. One reason for such differences is that researchers come from different schools of thought. Indeed, how reality is interpreted and what lessons are drawn varies with the theory which guides the researchers. This is why Section 1 of this introduction contrasts two approaches which have been particularly influential in the recent debate. It is our concern to be explicit about theoretical underpinnings both in the review of other researchers' conclusions and in our own analysis of the Brazilian experience.

Our objective is not merely to clean the lenses through which one observes reality. Equally, if not more important, is the empirical contribution which this book seeks to make by presenting new material. Most of this comes from recently completed doctoral theses whose authors have distilled their findings for the purpose of this book. Section 2 of this introduction presents a preview of what their chapters are about, why they are important beyond the confines of Brazil and how they relate to the two paradigms which are set out in Section 1.

A few pointers are in order to guide our readers. First, the research and policy issues addressed are more relevant for semi-industrialised than for least-industrialised countries. Second, 'hi-tech' in this book refers primarily to electronics-based technology. Third, in Brazil, the state intervened to foster a national capability to develop and produce such equipment by establishing a market reserve for local firms. In other words, neither imports of final products nor local production by foreign firms was allowed. However, this policy did not extend to all sub-sectors of the electronics industry.[2] One of the main reasons for its adoption was the reluctance of foreign firms either to enter joint ventures or to license technology to national firms. The reader who seeks a brief account of the history, mechanisms and coverage of this policy regime is referred to Chapter 2 which concentrates on the computer industry.

Fourth, a key feature of this book is that it goes beyond examining the production of hardware (Chapter 2) and software (Chapter 4). One of our main concerns is to explore the experience of user industries. Therefore studies on automation in banking (Chapter 3), petrochemicals (Chapter 5) and car production (Chapter 6) are included.

Fifth, our focus on hi-tech does not imply a view that the blind operation of technological forces determines industrial development. Far from it. Chapter 6 explicitly argues against such technological dictate and highlights the critical role of work organisation for competitiveness. Chapter 8 underlines the fragility of technological efforts in the recession and Chapter 9 shows how technological development is shaped by political forces.

Indeed, throughout this book we are concerned with the opportunities rather than the imperatives of hi-tech. How developing countries can best avail themselves of these opportunities is controversial. The following section contrasts the main positions that emerge from the recent literature.

1 TWO APPROACHES TO HI-TECH POLICY IN DEVELOPING COUNTRIES

In the course of this book, two paradigms are addressed which tend to lead to different conclusions with regard to hi-tech policy for LDCs. In a nutshell they go as follows: one of them suggests that both production and use of hi-tech in LDCs should be left to market forces. Market imperfections exist but are less serious than government failures. Moreover, LDCs are more likely to do well in industries which use mature technologies. Government intervention seeking to foster a local capability in the production of hi-tech tends to be a waste of resources. This line of argument has grown out of the neo-liberal school of thought and the product cycle theory (for an example, see Corsepius and Schipke 1989).

Contrast this with the conclusion which can be derived from the neo-Schumpeterian school of thought and its work on the present wave of technological change. Put schematically, it suggests that 'windows of opportunity' exist for LDCs, particularly at the early stage of the development of new technologies. In order to make use of these opportunities strategic state intervention is required so that synergies are created between the various hi-tech sub-sectors and between producers and users of hi-tech (for an example, see Perez 1989). In the following pages we set out both approaches in more detail.

Before doing so, we need to explain what purpose they serve in the book. They are not used to explain how the Brazilian hi-tech policy emerged and why it took the form it did. To the extent that theories of industrial development had any impact on the design of the policy, it came from the import substitution debate. In fact, the policy is a case of anticipated import substitution. Our objective is to assess this experience and to draw lessons for other countries. It is for this purpose that we draw on both the neo-liberal and neo-Schumpeterian school of thought. We focus on these two because they have effectively set the agenda of this debate. And, as said before, contrasting them helps to understand how and why analysts often differ in their conclusions.

We should also make it clear that, in elaborating both approaches, the intention is not to provide a literature survey.[3] There are numerous contributions which do not fall into either school of thought and sometimes draw on the arguments of both (see, for example, Pack and Westphal 1986, Ernst and O'Connor 1990).[4] Our intention is merely to represent what we consider to be the most important and coherent reference points in the current debate.

The neo-Schumpeterian approach

The starting point for this school of thought is the idea that economic development proceeds in long waves, the pivot of which lies in technological revolutions. Thus it builds upon Schumpeter's theory of long cycles in economic development and his exploration of the 'creative gales of destruction'. This idea has been developed further, initially for advanced countries (Freeman *et al*. 1982, Freeman 1984, 1987; Perez 1983; Freeman and Perez 1988) and more recently has been extended to LDCs (Perez 1985, 1988, 1989;

Perez and Soete 1988).[5] Here we reiterate the main points of the argument.

Essential to neo-Schumpeterian theory is the distinction between different types of technical change:

(1) incremental innovations: these occur more or less continuously, although at differing rates in different industries, but they are concerned only with improvements in the existing array of products and processes of production.
(2) radical innovations: these are discontinuous events which cannot be construed as the cumulative addition of small modifications to existing processes or products. For example, there is no way that nylon could have emerged from improvements to natural fibres.
(3) new 'technological systems': these consist of combinations of innovation which have ramifications for several branches of the economy as well as giving rise to entirely new branches. Examples are, in the past the cluster of synthetic materials and petrochemical innovations, or currently new composite materials and plant varieties based on genetic engineering.
(4) changes in techno-economic paradigm: some generic technologies or new technology systems have such a wide range of applications that they affect the conditions of production in all sectors of the economy. The introduction of steam power, the electric motor and micro-electronics are examples of such deep-going transformations.

It is such changes in techno-economic paradigm which are essential to how neo-Schumpeterians:

(1) explain periods of economic growth and crises in advanced countries;
(2) assess the constraints and opportunities for industrial growth in LDCs.

Both are dealt with in turn here but the former merely briefly so, since it is less central to our concerns.

A new techno-economic paradigm alters the best-practice frontier and sets new efficiency standards. While fuelled by revolutionary technological opportunities it takes time for a paradigm to crystallise (a decade or more) and even longer for it to diffuse right through the economy. Indeed, it is often difficult for companies (or countries) to adapt to a change of paradigm. Economic crises are seen to arise when there is a mismatch between the emerging new paradigm and the old institutional framework. Economic booms represent periods when there is a good match between a new techno-economic paradigm and social institutions. Whereas technological change is often rapid, there is usually a great deal of inertia in social institutions buttressed by interest groups as well as by slow response times of individuals or groups. In this perspective, the structural crisis of the 1970s and 80s is seen as a period of social adaptation to the new paradigm.

This theory is also used to explain the previous periods of long booms and protracted crisis. Indeed, the past two hundred years are seen as a succession of five long waves each associated with the rise and decline of a techno-economic paradigm.

Inevitably, such ambitious attempts at theorising are controversial,[6] but the critiques need not concern us here. First, because the debate is primarily about advanced countries, second, because – in our view – inadequacies of long-wave theory do not necessarily undermine the relevance of what neo-Schumpeterians have to say about constraints and opportunities for LDCs under the techno-economic paradigm that has come with micro-electronics. It is the latter question we now turn to.

The central proposition is that the early stage of a new techno-economic paradigm opens a window of opportunity for LDCs. This stems from two conditions which are characteristic to the early phase: the discontinuity in technical progress and lengthy duration of the adaptation period for the previous leaders (those who have made the fullest commitment to the previous paradigm). The first means that the race is in a new direction and at least part of the previous gap can be by-passed; the second means there is time to learn how to run along the new route.

A special feature of the early phase of the new paradigm is that knowledge is relatively freely available. The new scientific and technical principles of product and process innovation can be learned in universities and research institutes. After the early phase it is more difficult to enter in an innovative way because the accumulation of know-how is increasingly private and expertise is of the tacit sort.

Relatively easy and cheap access to knowledge in the early phase does not necessarily mean that new technological paths are explored or established. For this to happen, complementary assets are necessary. These include a pool of scientific and technological skills and an R&D infrastructure that fulfils minimum requirements of efficiency and reliability. Close interaction between suppliers and users of technology is also regarded as essential. The proximity of users to suppliers facilitates the design and de-bugging of new products. Hence the environment where the technology is developed shapes the pattern of technical evolution. And the more sophisticated the users, the more they will influence the direction of technical change. As a result, the outcome of the innovation process embodies country, region or firm specificity.

These locational advantages or externalities which are crucial in the early stages diminish as technology matures. The more standardised a product or a technology the less important a local supply network becomes.

Developing a 'national system of innovation' with country-specific priorities is regarded as necessary to take full advantage of the opportunities of the early phase. For this to happen government intervention is seen as essential. In particular the structuring of infant sectors is thought important through both the support for R&D infrastructure and the creation of space in which producers and users can engage in intensive search processes characteristic of the early phase of a new techno-economic paradigm. So much for a summary of the neo-Schumpeterian position.

The neo-liberal approach

There is an asymmetry in contrasting the neo-Schumpeterian and neo-liberal approaches. The former has a specific agenda, namely the connection between technological change and economic growth, while the latter is about general principles of allocating resources and maximising economic growth. These general principles have, however, been applied to many specific areas, one of which is the production and use of hi-tech. Here we delineate briefly what neo-liberalism stands for in general and what its position is on the issues which are central to this book.

In development theory and policy, the 1980s was the decade of neo-liberalism. This may seem an odd statement, since some industrial advanced countries, post-war Germany in particular, have had an influential neo-liberal school of thought for several decades. In the debate on developing countries, however, neo-liberalism has been a reaction to structuralism, the dominant intellectual paradigm of the years 1950–80. There are many differences within structuralism, but – as pointed out by Colclough (1991) – its contributors share the view that the structures of developing countries are significantly different from those of industrial advanced countries. In particular, LDC markets tend to be less socially efficient in allocating resources;

hence governments need to intervene more substantially on grounds of both efficiency and equity.

Neo-liberalism broke with this consensus. The leading proponents of what Toye (1987) has called the 'counter-revolution in development theory and policy' are Balassa, Bauer, Krueger, Lal and Little. In his succinct assessment of their work, Colclough (1991) distils their two central theses:

- Long-run growth is maximised through the pursuit of short-run allocative efficiency as determined by market prices.
- Market failures exist, but imperfect markets are better than imperfect states.

The validity of these and other neo-liberal theses is examined for the case of the Brazilian computer industry in Chapter 2 of this book. Here it is important to comment on the validity of the neo-liberal position more generally.

First, neo-liberals claim more for the long-run correlates of short-run optimisation than classical and even careful neo-classical writers would presume.[7] Variables such as technology and human resources which have an enormous influence upon long-run outcomes (and which are central to the investigation in this book) are relegated to a category which will look after itself. Such variables are not judged unimportant, but thought to need no attention other than that signalled by short-run prices.

Second, while neo-liberals have greatly contributed to the analysis of government failures, in particular the costs of intervention, much less serious attention has been given to the analysis of market failures. Moreover, neo-liberals tend to ignore that the most celebrated successes of fast industrial and technological development, namely Japan, South Korea and Taiwan, are showcases of hands-on rather than hands-off economic policy (White 1988; Amsden 1989; Wade 1990). This is not just a point of diagnosis, it also underlines the problem of prescriptions. Generalisations on how much, where and how governments should (or should not) intervene are unlikely to be helpful, since the executive capacity of governments varies a great deal (Toye 1987). This latter point is important in view of the ease with which neo-liberals often move from general analysis to specific policy advice. With regard to hi-tech, there is no neo-liberal approach specific to it. The hi-tech sectors receive the same treatment as any other. Let us elaborate.

LDC governments seeking to promote local production (as opposed to importation) of hi-tech face, almost inevitably, the well-known problems of protection. Neo-liberals have contributed a great deal to the assessment of the protectionist policies (for example, Little, Scitovsky and Scott 1970; Balassa 1982). The objective here is to highlight why a neo-liberal analysis is likely to take a different route and lead to different conclusions as compared with the neo-Schumpeterian enquiry.

Most neo-liberals would accept – even if grudgingly – the infant-industry argument. In their analysis of whether infant-industry protection is justified, they tend to pay greatest attention to the cost which arises to the user of the product made by the protected firms (by comparing domestic with international prices or value-added). The view behind this is that what matters is the price and quality of the product, not its origin. The latter matters to neo-Schumpeterians, because close interaction between supplier and user is seen as essential for the effective use of the new technology. More than that, it is seen as essential to the very development of technology. Such close interaction is unlikely if producer and user operate in different parts of the world.

The difference in approach comes out equally clearly when we ask what sectors are most likely to achieve indigenous technological capability. The immediate neo-liberal answer would be that governments are inept at picking winners and that markets, if allowed to operate freely, will produce the most efficient solution. Since, however, the principle of infant-industry protection is accepted, the question of which sectors to protect cannot be escaped altogether. Here, neo-liberals seek refuge in the product-cycle theory; in other words, LDCs are thought to have their best chances of achieving international competitiveness in sectors which use mature technology, while hi-tech sectors are the least promising candidates.

This seems to be a most plausible position. It says that if you try to catch up, choose a target that stands still and not one that keeps surging ahead. What is happening in the hi-tech sectors, however, is precisely the latter. Therefore, the neo-Schumpeterian view that LDCs should enter the hi-tech sectors at an *early* stage seems a mad recipe.

In contrasting the two approaches, we may have overdrawn the differences. Neo-Schumpeterians do not entirely discard product-cycle theory. They recognise that the production of mature products, if serious learning efforts are undertaken, generates a pool of managerial and technical skills and increases the externalities. In this sense, the choice of mature products and technologies as point of entry can help to initiate industrial development and establish a platform for further growth. However, in their view, catching up or leap-frogging requires entering the early phase of a new technological paradigm, as explained above. Thus, while neo-Schumpeterians emphasise the limitations of product-cycle theory, that theory is an essential part of the neo-liberal approach towards hi-tech in LDCs.

At the risk of drawing up caricatures, the main issues on which the two approaches diverge are pulled together in Table 1.1. Such juxtapositions can rarely do justice to the richness of what the proponents of either approach contribute.[8] Hence it should be borne in mind that the table shows major differences in emphasis, not necessarily blindness to the concerns of the other side.

Table 1.1 Neo-liberal versus neo-Schumpeterian approach towards hi-tech in LDCs

	Neo-liberal	*Neo-Schumpeterian*
Focus of analysis	Short-run cost	Interactive learning
Concept of technology	Technology as a tradeable recipe or input	Technology as tacit knowledge not easily transferred
Innovation process	Separateness of production and diffusion of technology	Synergy of producer and user
Technical progress	Continuous and cumulative (single-track race)	Continuous and discontinuous (opening windows of opportunity)
Concept of labour	Labour as a cost	Labour as a resource
Concept of competitiveness	Product (or price) competitiveness	Structural competitiveness
Role of the state	Reliance on market forces	Strategic state intervention
Point of entry	Late stage	Early stage

As will be seen later in this book both approaches have strengths and weaknesses. The main point to be made here is that, if two analysts, one from the neo-liberal and one from the neo-Schumpeterian camp, were to investigate the same case, they would probably not come to the same conclusion on whether, how and when government intervention was justified. It is a familiar problem in social science but one that often goes under in debates on technology.

2 TOWARDS AN ASSESSMENT

One of our main objectives is to ask what other countries can learn from the Brazilian experience. Drawing lessons requires an assessment of that experience. This second section of the Introduction shows in which way the subsequent chapters contribute to such an assessment – even though it should be emphasised that this is not their only purpose.

The economic assessment of a policy regime requires some form of cost–benefit analysis. This is a difficult undertaking. In continuation of the previous section it is useful to contrast the different ways in which neo-liberals and neo-Schumpeterians would steer us in such an evaluation. Our main argument is that neo-liberals tend to focus on the costs while neo-Schumpeterian analysis favours the exploration of the benefits. A related argument is that the tools used by neo-liberals direct attention to current costs and internal economies whereas neo-Schumpeterians are beavering away at dynamic efficiency and external economies. Proponents of neither school of thought are likely to portray themselves in this way but we believe that this is a simple and useful way of bringing out their differences as well as their respective strengths and weaknesses. Let us elaborate.

The core piece of the Brazilian Informatics Policy is the market reserve for national producers in a number of sub-sectors of the electronics complex. Thus it is a classical case of infant-industry protection except that not only are imports kept out through quantitative controls but neither are foreign firms allowed to produce behind the protective barrier. In measuring the cost or benefit of an infant industry's development it is useful to start with the difference between the cost of domestic production and the cost of equivalent imported products. This difference changes over time, hence the cost or benefit of developing an infant industrial activity is equal to the cumulative discounted value of the difference between domestic and import costs. This measurement does not however take account of externalities generated by the infant industry.

The neo-liberal examination of infant industry generally centres on the cost of protection measured by the above mentioned difference between domestic and international prices (or value-added). There are good reasons for this, particularly the availability of operational tools which make it possible to measure this cost with varying degrees of sophistication, such as nominal rate of protection, the effective rate of protection or domestic resource cost. In the case of the Brazilian electronics industry only the simple comparison of domestic and international prices has been carried out and has led neo-liberals to dismiss the infant industry as a failure (Cline 1987; Corsepius and Schipke 1989).

This issue of price competitiveness is addressed explicitly in this book. Chapters 2 and 10 confirm that the protected sub-sectors of the Brazilian electronics industry are high-cost. In spite of rapid cost reductions the prices of micro-computers, printers and digital control systems are approximately double the world market levels. However, these chapters emphasise that (*a*) in countries without import restrictions users also pay well above world market prices and (*b*) the non-protected sub-sectors of the Brazilian electronics industry (in which foreign firms dominate) produce at above world market prices. Furthermore, Ruy Carvalho, in Chapter 5 on the use of digital control systems in the Brazilian petrochemical industry, suggests that the higher cost of equipment produced by national firms is partly offset by advantages they offer in maintaining this equipment. None of this is to deny that the prices of items produced by national firms are above world market prices but to argue that the cost to the user is a great deal less than what the price differential suggests. This, however, is not sufficient to justify the fostering of a national hi-tech industry.

Whether it is or not depends on the extent to which external economies are generated by the infant industry. The problem is that 'the concept of external economies is one of the most elusive in economic literature' (Scitovsky 1954:143). While some advances have been made, the problems of measuring externalities remain unresolved. This is one reason why in evaluations of infant-industry protection their existence is acknowledged but rarely given due weight. In what follows we try to unbundle the externality issue. In the course of doing so we suggest that much of the neo-Schumpeterian work is essentially about improving our understanding of external economies, and also indicate what contribution some of the subsequent chapters of this book make in this respect.

Externalities occur where market-priced transactions do not fully incorporate all the costs and benefits to economic agents. Externalities thus constitute market failures and provide reasons for government intervention.[9] Meade (1952) and Scitovsky (1954) developed the important distinction between technological and pecuniary externalities. The former affect the actual conditions of production while the latter affect the monetary terms on which they take place (the prices of inputs and outputs). Our concern is primarily with technological externalities.

The importance of such externalities has been underlined by

Stewart and Ghani (1988). Their review suggests that they are more widely prevalent in developing countries than is generally acknowledged. They also suggest a typology which we will use to show what externalities were generated by the Brazilian informatics industry.

(a) Attitudes towards innovation. Industrialisation requires embracing the idea of continuous technical change. Attitudes and awareness of electronics-based innovations are particularly interesting in the case of Brazil. The informatics industry and its products have received extraordinary attention in the Brazilian public debate. This is partly due to fascination with what the new products and processes can do. In this respect the Brazilian situation is no different from anywhere else. There is, however, another reason which is specific to the country. The market reserve for national firms was controversial and had outspoken external and internal adversaries (for details, see Chapter 9 by Maria Ines Bastos). Hence the policy, the industry and its products were in the media headlines for most of the 1980s. A consciously defined awareness programme could hardly have done more to give the electronics industry a higher profile. The industry was able to deliver well-functioning equipment adjusted to local conditions (for example, the Portuguese language). Thus, the policy and the industry contributed to a positive attitude towards electronics-based technical change. Such 'influences are a form of externality – that is, for any agent they depend on developments resulting from the actions of many other agents, in adopting new technologies, selling new products, etc.' (Stewart and Ghani, 1988:19). It is not possible to measure such externalities. However, their existence is underlined by the diffusion rates of microcomputers in Brazil. According to the US Department of Commerce's Office of Computers and Business Equipment, in the mid-1980s these rates were among the highest in the world (see Table 3.1 in Chapter 3).

(b) Human capital formation. In comparison with the slippery issue of attitudes, the formation of human capital is more widely acknowledged to be a source of technological externalities. In the Brazilian case such externalities were indeed generated by the protected electronic infants. Chapter 7 by Tom Hewitt compares the utilisation of labour in the national computer industry with that in other subsectors which are not protected. He shows that, proportionately, the national industry devoted more human resources to local research and development. The benefits were only partially captured by the firms which trained such technicians and engineers. Even though no systematic tracer study exists, there is known to be a significant movement of such labour to competing producers or – less often – to users. Sometimes they leave to set up their own firms. Indeed, most of the new firms, especially those producing specialised equipment and inputs were 'spin-offs' of the first generation of Brazilian informatics firms (see Chapters 2 and 10).

(c) Technological interaction between firms. Many aspects of innovation occur through transactions between firms that are partially or wholly outside the market. The importance of such interactions is brought out in Cassiolato's chapter on user–producer relationships in banking automation. He concludes that Brazil's banking system is technologically very advanced, that it has developed and adopted technological solutions suitable to local needs and conditions, that some of the locally developed systems are internationally competitive in price and quality, that these systems were developed through close interaction between users and suppliers, and that this was made possible by the protection of the local informatics policy.[10] Carvalho in Chapter 5 also emphasises the importance of learning by interaction between producers and users. However, his message is that such externalities do not necessarily develop as a result of protecting local producers (in his case digital control producers). Demanding users are equally important. The fragility of users – in his case petrochemical firms – led to the under-utilisation of the new technology.

The kind of externalities discussed above have been at the centre of the neo-Schumpeterian work on the 'national system of innovation' (Anderson and Lundvall 1988; Johnson and Lundvall 1988; Edquist and Lundvall 1990). This is a broad concept which seeks to capture 'the network of institutions in the public and private sectors whose activities and interactions initiate, import, modify and diffuse new technologies' (Freeman 1987:1). The main idea is that no matter how good the technology and no matter how aggressive the entrepreneur, technological dynamism in the economy will not come about without infrastructure, incentive structure, and networks which support innovative activity and stimulate diffusion of new technologies.

While the concept of national system of innovation remains slippery, the research carried out under this heading is essentially about what are conventionally called 'externalities'.[11] In this sense this work contributes to the evaluation of policy experiences (such as the one in Brazil which is the focus of this book).

Taking a systemic view of innovation does not, however, require that one accepts the entire neo-Schumpeterian package. Indeed, some chapters in this book draw on it only partially. Other authors concerned with industrial policy and technical change, coming from different intellectual traditions (for example, Pack and Westphal 1986), stress equally the learning and externality arguments put forward above. The neo-Schumpeterian research helps to take such work further by disentangling the question of interactive learning and bringing out that the gains from such interactive learning are greatest at the early stage in the development of new technologies.

We have emphasised these issues of learning and externalities because they are central to our assessment of the Brazilian informatics policy. The chapters in this book contribute in different ways to such an assessment highlighting both deficiencies and achievements. The main ones are as follows:

(1) By the end of the 1980s, Brazil had built up a significant platform in informatics particularly in small computers, peripherals and various types of automation equipment for banks and industry due to the market reserve for national producers. Precisely because the policy was restrictive in the 1980s, the 1990s offer genuine policy choices about the degree of openness to imports and of collaboration with foreign firms (Chapters 2 and 10).
(2) This build-up entailed short-run costs but these are outweighed by substantial external economies that arise from the learning efforts of national producers and users (Chapters 2 and 3).
(3) Relying on internally developed products was particularly efficient where strong user–producer relations were established as in banking automation (Chapter 3).
(4) A comparable build-up was not achieved in software. This is mainly because it never received effective protection. Hence internal producer–user loops, essential for application software, rarely developed (Chapter 4 by Fátima Gaio).
(5) The platform in hardware was built by relying essentially on two simple mechanisms: quantitative import restrictions and exclusion of foreign firms in some sub-sectors. While such measures are contrary to the neo-liberal prescription, the shortcomings of the policy did not lie in the choice of instruments (Chapter 2).
(6) There is time and space for a national hi-tech policy. Users who, as a result of such a policy, do not have immediate access to the latest foreign technology, do not thus become uncompetitive in international markets (see Chapter 6 by Elizabeth Silva).

(7) The main defect of the Brazilian informatics policy was that it did not feed into an overall industrial policy (Chapters 5 and 10). Even within the electronics complex there were different and conflicting regimes for telecommunications and consumer electronics, contributing also to the relative failure in semi-conductors. (The experience within these latter sub-sectors is only referred to in this book and is not a focus of analysis.)[12]

These are some of the main conclusions which come out of the later chapters. In this introductory chapter we have sought to show how, in spite of major differences between neo-liberal and neo-Schumpeterian analysis, the two can complement each other in a policy assessment. There are however two issues which are not addressed by either and which are important, particularly if one wants to draw lessons from such an assessment. The first relates to politics and the political room for manoeuvre for a national hi-tech policy. The second concerns the implications of short-run crisis management for long-run technology policy. In this book these two issues are taken up as follows.

In practice, policy decisions rarely result from rational evaluation but from political and economic pressure. The informatics policy is not an exception. On the contrary, as shown in Chapter 9, it has been one of the most severely contested cases of recent industrial policy. Maria Ines Bastos shows how and why the policy was politically viable in spite of growing hostile forces. These included the US government threat to Brazil of trade war. Such pressure eroded the supportive internal alliance, but did not lead to a dismantling of the policy in the 1980s. Compromises were made but the policy remained essentially intact. Bastos links this achievement to the build-up of a negotiating capacity in government. We would go a step further and argue (in Chapters 2 and 10) that the heavy critique of the policy contributed to the learning which occurred in both industry and government.

In spite of the enormous accumulation of know-how and skills, the achievements of the 1980s remain fragile because of the country's general economic crises. This is why Chapter 8 by Mike Porteous is of especial interest. It investigates the differential impact which economic recession has for national and foreign firms' ability to produce computerised numerically controlled machine tools. Even though his study concentrates on the recession which hit the sector in the first half of the 1980s, some of the conclusions are of undiminished interest. At the beginning of the 1990s it is precisely the crisis which threatens to undermine the continued pursuit of long-run technological objectives.

Crisis management leaves little room for strategic industrial policy. The informatics policy with all its limitations was an effective subsectoral strategy. It needed revising, but it is now (1990) in danger of being abandoned. If earlier on we stressed learning in industry and government, we also need to address the problem of unlearning which occurs in time of crisis. This is one of the issues taken up in the final chapter.

3 LEARNING FROM THE BRAZILIAN EXPERIENCE

In this introductory chapter we have sought to show how, in the course of this book, the Brazilian informatics industry comes under scrutiny from different intellectual traditions and from the point of view of producers and users. While contributors do not agree on all issues, a general conclusion is that there are a good number of positive lessons which can be drawn from this case for other LDCs. The final chapter seeks to pull these together, drawing on both the experience of Brazil and that of other countries. This policy discussion concentrates on the unifying theme which runs through the various chapters of this book which is *learning*. This is particularly relevant since policy issues change dynamically through time.

The most frequent argument against policies for the production of hi-tech in LDCs is that users are the victims of such policy. The Brazilian experience shows that this is not the case. More than that, it reinforces conclusions from small European countries that juxtaposing producer and user policies is a false dichotomy and also that the process of interactive learning between producers and users, which is fundamental for the mastery of new technologies, is facilitated when early entry occurs.

In most countries such entry needs necessarily to be selective. Criteria for such selection are discussed. The message of the final chapter however is not that there is a blueprint for picking winners but that government capacity to take informed decisions needs to be built up. Indeed, we extend the concern with learning from industry to government. The clearest indication of the growing capacity of both industry and government in Brazil is the turnaround in the willingness of foreign capital to negotiate over technology transfer to local firms. The final chapter draws out lessons on the sequencing of measures to restrict foreign firms, on the relative effectiveness of explicit and implicit measures to engender learning, on the political and economic conditions under which such learning occurs and, finally, and on how preoccupation with short-term crisis management conflicts with strategic industrial policy.

NOTES

1 Helpful comments on a previous draft of this chapter were provided by Ruy de Quadros Carvalho, Afonso Fleury, Chris Freeman, Martin Godfrey, Tom Hewitt, John Humphrey and Carlota Perez.
2 Sub-sectors which operate under different rules are mainframe computers, consumer electronics and telecommunications. The policy for the first two is explained briefly in Chapter 7. For an excellent study of the telecommunication industry, see Hobday (1990). For comprehensive recent coverage of the entire electronics complex see BNDES (1990), Porto *et al*. (1990) and Meyer-Stamer (1990). The 'classical' works on the computer industry are Tigre (1983) and Piragibe (1985).
3 For surveys on the literature of technical change in Third World industry, see Bell (1984) and Fransman (1985).
4 This applies also to various chapters in this book.
5 Other contributors to neo-Schumpeterian work include Nelson and Winter (1982), Rosenberg (1976) and Dosi (1982). While sceptical of long-wave theory they also emphasise the role of technology and institutions for long-run growth patterns.
6 It is criticised in particular for technological determinism (Rosenberg and Frischtak 1986).
7 While neo-liberals rely on the tools of orthodox neo-classical theory, many neo-classical authors would disagree with some of the main neo-liberal theses. For details, see Colclough (1991).
8 Again, it must be emphasised that this contrast of two schools of thought does not capture the entire range of thinking. For example, on the question of whether markets or state should guide development one cannot ignore the influence of traditional structuralism. Even if traditional structuralists and neo-Schumpeterians (the neo-structuralists!) agree on the need for state intervention, there are marked differences in the type of advocated intervention. A threefold classification of approaches to economic policy which is useful (though not specifically concerned with hi-tech) can be found in CEPAL (1990).
9 This is the consensus in neo-classical economics. Neo-liberals, while using neo-classical tools of analysis, do not necessarily agree with this. They claim that states are rarely good at correcting market failures and that imperfect markets are often better than imperfect states (see Section 1 of this chapter).
10 The benefits from the learning by interaction between users and suppliers do not, in this case, fall neatly into the category of 'external economies'. Since the users (banks) have a major stake in some of the supplying firms, some of the benefits could be categorised as internal economies (internal to the conglomerate). The essential point however is not whether they are internal or external, but whether they arise or not. It is unlikely that they would have emerged without an informatics policy.
11 Of course, in highly integrated firms, some such economies are internal to the firm.
12 For a study of telecommunications in Brazil, see Hobday (1990), for consumer electronics see Baptista (1987, 1988) and for semi-conductors Rosenthal (1987), Mendes and Borelli (1988) and Mammana (1989). On

the institutional and policy conflict, see Erber (1990), and Meyer-Stamer (1990).

BIBLIOGRAPHY

Amsden, A.H. (1989) *Asia's Next Giant – South Korea and Late Industrialisation*, Oxford: Oxford University Press.

Anderson, E.S. and Lundvall, B.Å. (1988) 'Small national systems of innovation facing technological revolutions: an analytical framework', in C. Freeman and B.Å. Lundvall (eds) *Small Countries Facing the Technological Revolution*, London and New York: Pinter Publishing.

Balassa, B. and associates (1982) *Development Strategies in Semi-Industrial Economies*, Baltimore: Johns Hopkins University Press for the World Bank.

Bell, M. (1984) 'Learning and accumulation of industrial technological capability in developing countries', in K. King and M. Fransman (eds) *Technological Capability in the Third World*, London: Macmillan.

Baptista, M. (1987) 'A industria electrônica de consumo a nivel internacional e no Brasil: padrões de concorrência, inovação tecnológica e carater de intervenção do Estado', M.Sc. thesis, Instituto de Economia, Unicamp, Campinas.

—— (1988) 'The consumer electronics industry in Brazil: current situation and outlook', in C. Piragibe (ed.) *Electronics Industry in Brazil: Current Status, Perspectives and Policy Options*, Brasilia: Centre of Studies in Scientific and Technological Policy, Ministry of Science and Technology.

BNDES (1990) *Microeletrônica e Informática: Uma Abordagem sob o Enfoque do Complexo Eletrônico*, Rio de Janeiro: Banco Nacional de Desenvolvimento Econômico e Social.

CEPAL (1990) *Transformacion Productiva con Equidad*, Comision Económica Para America Latina y el Caribe, Naciones Unidas, Santiago de Chile.

Cline, W.R. (1987) *Informatics and Development – Trade and Industrial Policy in Argentina, Brazil and Mexico*, Washington, DC: Economics International.

Colclough, C. (1991) 'Structuralism versus neo-liberalism: an introduction', in C. Colclough and J. Manor (eds) *States or Markets? Neo-liberalism and the Development Policy Debate*, Oxford: Oxford University Press.

Corsepius, U. and Schipke, A. (1989) 'Die Computerindustrie in Schwellenländern – der Fall Brasilien', *Die Weltwirtschaft*, Heft 1, Institut für Weltwirtschaft, Universität Kiel.

Dosi, G. (1982) 'Technological paradigms and technological trajectories: a suggested interpretation of the determinants and directions of technical change', *Research Policy*, Vol. 11, No. 3.

Edquist, C. and Lundvall, B.Å. (1990) 'Comparing the Danish and Swedish systems of innovation', mimeo, Institute of Production, Aalborg University.

Erber, F.S., (1990) 'Política de informática e política industrial', mimeo, Rio de Janeiro: IPEA.

Ernst, D. and O'Connor, D. (1990) *New Technologies, Global Competition and Latecomer Industrialisation – An Agenda for the 1990s*, Paris: OECD Development Centre.

Fransman, M. (1985) 'Conceptualising technical change in the Third World in the 1980s: an interpretative survey', *Journal of Development Studies*, Vol. 21, No. 4.

Freeman, C. (1984) *Long Waves in the World Economy*, London: Pinter Publishing.

—— (1987) *Technology Policy and Economic Performance – Lessons from Japan*, London: Pinter Publishing.

Freeman, C. and Lundvall, B.Å. (eds) (1988) *Small Countries Facing the Technological Revolution*, London and New York: Pinter Publishing.

Freeman, C. and Perez, C. (1988) 'Structural crises of adjustment, business cycles and investment behaviour', in G. Dosi *et al.* (eds) *Technical Change and Economic Theory*, London: Pinter Publishing.

Freeman, C., Clark, J. and Soete, L. (1982) *Unemployment and Technical Innovation: A Study of Long Waves in Economic Development*, London: Pinter Publishing.

Hobday, M. (1990), *Telecommunications in Developing Countries – The Challenge from Brazil*, Routledge: London and New York.

Johnson, B. and Lundvall, B.Å. (1988) 'Institutional learning and national systems of innovation', mimeo, Institute of Production, Aalborg University.

Little, I., Scitovsky, T. and Scott, M. (1970) *Industry and Trade in Some Developing Countries: A Comparative Analysis*, Oxford: Oxford University Press.

Mammana, C. (1989) 'Considerações sobre a microeletrônica no Brasil', Research Report to CNPq, Instituto de Economia, Unicamp, Campinas.

Meade, J.E. (1952) 'External economies and diseconomies in a competitive situation', *Economic Journal*, Vol. 62, March.

Mendes, M. and Borelli, M. (1988) 'The semiconductor policy and IC design in Brazil', in C. Piragibe (ed.) *Electronics Industry in Brazil: Current Status, Perspectives and Policy Options*, Centre of Studies in Scientific and Technological Policy, Brasilia: Ministry of Science and Technology.

Meyer-Stamer, J. (1990) *From Import Substitution to International Competitiveness – Brazil's Informatics Industry at the Crossroads*, Berlin: German Development Institute.

Nelson, R. and Winter, S. (1982) *An Evolutionary Theory of Economic Change*, Cambridge, Mass.: Harvard University Press.

Pack, H. and Westphal, L. (1986) 'Industrial strategy and technological change', *Journal of Development Economics*, Vol. 22, No. 1.

Perez, C. (1983) 'Structural change and the assimilation of new technologies in the economic and social systems', *Futures*, Vol. 15, No. 5.

—— (1985) 'Microelectronics, long waves and world structural change: new perspectives for developing countries', *World Development*, Vol. 21, No. 3.

—— (1988) 'New technologies and development', in C. Freeman and B. Lundvall (eds) *Small Countries Facing the Technological Revolution*, London: Pinter Publishing.

Perez, C. (1989) 'The present wave of technical change: implications for competitive restructuring and for international reform in developing countries', paper prepared for the Strategic Planning Department of the World Bank, Washington, DC, World Bank.

Perez, C. and Soete, L. (1988) 'Catching up technologies: entry barriers and windows of opportunity', in G. Dosi *et al.* (eds) *Technical Change and Economic Theory*, London: Pinter Publishing.

Piragibe, C. (1985) *Industria da informática: desenvolvimento brasileiro e mundial*, Rio de Janeiro: Editora Campus.

Porto, J., Silva, A. and Laplane, M. (1990) *Avaliação da política nacional de informática*, Research Report to CNPq, Instituto de Economia, Unicamp, Campinas.

Rosenberg, N. (1976) *Perspectives on Technology*, Cambridge: Cambridge University Press.

Rosenberg, N. and Frischtak, C. (1986) 'Technological innovation and long waves', in C. Freeman (ed.) *Design, Innovation and Long Cycles in Economic Development*, London: Pinter Publishing.

Rosenthal, D. (1987) 'Microelectronics and industrial policies in developing countries: the case of the semiconductor industry in Brazil', Ph.D. thesis, University College, London.

Scitovsky, T. (1954) 'Two concepts of external economies', *Journal of Political Economy*, Vol. 62, No. 2, April.

Stewart, F. and Ghani, E. (1988) 'Externalities and trade', mimeo, Oxford: Queen Elizabeth House.

Tigre, P. (1983) *Technology and Competition in the Brazilian Computer Industry*, London: Frances Pinter.

Toye, J. (1987) *Dilemmas of Development – Reflections on the Counter Revolution in Development Theory and Policy*, Oxford: Blackwell.

Wade, R. (1990) *Governing the Market – Economic Theory and the Role of Government in East Asian Industrialisation*, Princeton: Princeton University Press.

White, G. (ed.) (1988) *Developmental States in East Asia*, London: Macmillan.

2 An assessment of the market reserve for the Brazilian computer industry

Hubert Schmitz and Tom Hewitt

1 INTRODUCTION[1]

The question of whether and how LDCs can successfully foster their own hi-tech industries is a key issue in the industrial policy debate. One of the most interesting experiences in this respect is that of Brazil. At its core is the market reserve for national producers in a number of sub-sectors of the electronics complex. The purpose of this chapter is to assess this experience, focusing mainly on small computers and peripherals.

We are not the first to undertake this task. The protected development of the hi-tech industry in Brazil has attracted a great deal of attention internationally. The upshot is that it is one of the most controversial experiences in current industrial and technology policy debate; conclusions vary from condemnation to celebration of success.

One reason for such differences is that researchers come from different schools of thought. How reality is interpreted and what lessons are drawn varies with the theory which guides the researchers. This chapter is concerned with the neo-liberal approach. Its underpinnings are set out more fully in the introduction to this volume. Here it suffices to repeat that neo-liberal thinking has been particularly influential in the industrialisation debate of the 1980s and has been applied to many countries and sectors including the hi-tech sectors.

The neo-liberal answer to the opening question of this chapter can be summed up as follows: both the production and use of hi-tech in LDCs should be left to *market forces*. Market imperfections exist but are less serious than government failures. Moreover, LDCs are more likely to do well in *mature* technologies. Government intervention seeking to foster a local capability in the production of hi-tech tends to be a waste of available resources and holds back the competitiveness of user firms. The market reserve for the Brazilian computer industry is presented as a case which verifies this line of argument by Corsepius and Schipke (1989). Similar conclusions are obtained by Cline (1987) and Hieronymi (1987) from their analyses of the Brazilian experience.

It is an explicit objective of this chapter to engage with the neo-liberal agenda. The Brazilian informatics policy is of particular interest in this respect, because it has been heavily contested by a political opposition which has made frequent use of neo-liberal arguments. The chapter proceeds as follows: the purpose of Section 2 is to set the case study in the context of the wider industrialisation debate. Sections 3 and 4 give a brief overview of the policy regime and of the political setting which gave rise to the policy. Sections 5 to 11 contain the assessment; we try to answer the questions which neo-liberals would ask about the effectiveness of intervention but also raise some essential issues which they would tend to ignore.

2 CRITICAL ISSUES IN THE INDUSTRIALISATION DEBATE

The debate on industrialisation has always pivoted around two interrelated issues:

- state versus market
- self-reliance versus integration into the world economy.

Over the last decade LDCs have come under unprecedented pressure to integrate into the world economy and reduce state intervention. This pressure comes both from academic and institutional sources, including advice and actions from the international finance institutions. In neo-liberal circles such policies are justified by reference to the history of industrialisation in LDCs over the last four decades. It is argued that protectionist policies in pursuit of import substitution have not worked (except in their early stages) and export-oriented industrialisation under liberal policies has been so successful that LDCs in general should follow this route (Balassa 1981; Krueger 1981).

In an earlier survey of industrialisation strategies (Schmitz 1984), we concluded that import substitution has indeed led to (static) inefficiency and foreign exchange problems, but that from a dynamic perspective (learning and externalities) many existing analyses of protection are unsatisfactory – for both conceptual and empirical reasons. Second, we concluded that the alleged superiority of export orientation is not so much due to the adoption of more 'rational' market oriented policies, as to a favourable combination of historical factors and substantial state intervention.

The latter point has since been developed and clearly documented. Luedde-Neurath (1986), for example, has shown that South Korea's export success was neither preceded nor accompanied by significant across-the-board import liberalisation and that market forces were not given a free reign to allocate resources. Instead there was a targeted two-pronged import policy: liberal towards inputs for export manufacturing and highly restrictive towards the domestic market by making access to the domestic market conditional upon satisfactory export performance. The overall conclusion was that the Korean import regime was highly managed.

In the wake of this and other empirical studies, agreement is emerging that the fast East Asian industrialisers (with the exception of the city states) are showpieces of hands-on rather than hands-off industrial policy (see, particularly, White 1988). However, many neo-liberal economists remain unconvinced. A currently pervasive attitude in the profession is that most LDC governments are incapable of pursuing strategic industrial policies.

Of course scepticism about the difficulties of successful state intervention is justifiable, given that many industrial failures have been policy-induced. As stressed in the introduction to this book, neo-liberals have contributed a great deal to the critical analysis of state intervention, particularly with their work on policies of protection. However, sight is often lost of the fact that governments need to *learn* how to promote industrial development. Moreover, aphorisms such as 'imperfect markets are better than imperfect states' seem to introduce an unwarranted passivism – if not defeatism – into the debate on industrial policy.

In what follows we present a case study of active industrial policy: the promotion of national firms in the Brazilian computer industry. The chances of this selective state intervention succeeding seemed very unlikely at the time because it occurred in a period of increasing internationalisation of the Brazilian economy and in a sector in which the technological frontier was (and still is) advancing rapidly.

3 POLICY REGIME

Until the late 1970s, Brazil's computer industry was in the hands of foreign firms which either imported finished products or carried out the final assembly of goods locally. By the late 1980s, this situation had changed almost out of all recognition. By 1986, the number of firms operating in the *nationally-owned* computer and peripherals market increased to 310 from four in 1977 (Piragibe 1987). Employment in these firms grew from just 4,000 in 1979 to over 50,000 in 1988. Total sales of national firms in 1988 were in the region of US$3,000 million, accounting for 66 per cent of the computer market (SEI 1989a).

This rapid growth resulted from the creation of a reserved market. Foreign corporations were not excluded from the Brazilian computer market. Rather, they were limited to specific market segments. Other segments were reserved for Brazilian private capital.

The principal tool of policy has been the creation of a reserved market for mini- and microcomputers and their peripherals. This policy is carried out by the Special Secretariat for Informatics (SEI) which is a specialised government agency working under the Ministry (now Secretariat) for Science and Technology. The main policy instruments used by SEI are quantitative import restrictions and the concession of manufacturing licences to national firms. Foreign firms are limited to the production of mainframe computers. They are also controlled by SEI to the extent that the granting and withholding of import licences can force these firms to have increasing indices of nationally produced inputs in their final products and also to show positive export balances.[2]

These policies have their origin in the Commission for the Coordination of Electronic Processing Activities (CAPRE) which was created in 1972 and had a regulatory role over information technology. Control over computer imports began in 1975, at which time a number of complementary policy measures were also set in motion.[3] The beginning of the market reserve dates from 1977, when the production of minicomputers was put to tender for *national* firms.[4]

In 1979, CAPRE was transformed into SEI with a wider mandate. The protected market for national firms under SEI's guidance gradually spread to other areas, such as industrial automation equipment, micro-electronic components, digital instruments, superminicomputers and others.[5] In 1984, the market reserve and complementary measures were enshrined in the 'Informatics Law'.[6]

4 THE NATIONAL ALLIANCE

The passing of the Informatics Law was the result of an intense political struggle. It is through understanding the origin of the policy for the computer industry that we can make sense of its political sustainability in the face of internal and external opposition. CAPRE emerged from a particular alliance of political forces, which included the military, technologists and economists. The glue was *nationalism*. At the time, this proved a strong enough link to resist pressures for an open door policy.

It was an unlikely alliance because it occurred at a time when the Brazilian economy was in a phase of internationalisation. In this developmental coalition there was, first, a small caucus of highly skilled engineers trained in technology institutes and universities in the US and in Brazil whom Evans (1986:792) has called 'frustrated nationalist technicians'. They were neither satisfied with working as salespeople for foreign computer manufacturers nor did they want to remain in universities. Their ambition was to find a commercial outlet (and, therefore, real test) for their prototype computer designs.

These engineers found political and financial support from three groups in the state apparatus. In the national development bank (BNDE) there was a group of economists who were keen to promote a national capital goods industry. The Federal data-processing agency, SERPRO, had been doing their own hardware and software development to suit their particular needs and supported the creation of a national industry which could supply equipment to their specifications. Finally, the navy was keen to see the development of a national computer technology to supply its frigates and thus avoid technological dependency in what was considered to be a strategic and sensitive military area (*ibid.*:793).

The convergence of these forces produced a particular breed of nationalism based on a combination of security, developmental, pecuniary and technological considerations (Evans 1985). The presence of private national capital was minimal in the *initial* decision-making process. With the installation and then rapid expansion of the industry, national capital played an increasingly important role and as a consequence has become a more powerful pressure group. In particular, the national banking sector now has a strong presence in the computer industry.

Nationalism is still a strong cohesive force but internal and external pressures to change computer policy have been growing.[7] The opposition comes primarily from two groups: the excluded foreign computer firms who are backed by their governments and from the user firms who demand quicker access to the latest international technology. The outcome of these conflicts will be determined not just by ideological factors but also by the record of the industry to date.

The latter will be examined in the sections which follow through the neo-liberal eye. While neo-liberals are not against infant-industry protection in principle, they would favour neither the use of quantitative controls, nor giving preferential status to national firms.

5 GROWTH AND COMPETITION

The computer industry was responsible for 40 per cent of the US$8 billion total sales of electronic equipment in Brazil in 1986. This represents the largest electronics market segment in Brazil (exceeding that of consumer electronics and telecommunications). Within computer production, it is the national firms which have undergone the fastest growth in output of 300 per cent from 1981 to 1986. Indeed, by 1986, national firms had passed a watershed by attaining 51 per cent of the total computer and peripherals market (Piragibe 1987).

Although impressive, rapid growth on its own is not a sufficient indicator of success. In our assessment we proceed as follows: first, we examine the questions which neo-liberals would ask about intervention in the computer industry. This is not merely an academic exercise since there has been powerful opposition to the policy using neo-liberal arguments. Accordingly, we examine the issues of internal competition, lack of international competitiveness, detrimental effects on users, opportunity costs and intrinsic problems of state intervention. We then examine the questions to which neo-liberals assign less importance, in particular the process of learning in both industry and government.

The widespread disillusion with protectionist policy is in many cases not the result of protection *per se* but of monopoly positions built up behind protective barriers.[8] An undoubtedly positive feature of the reserved market for small computers and peripherals is that there is *strong internal competition in most product lines*. The number of national firms increased from four in 1977 to 310 firms in 1986 (excluding software houses[9]). The number of foreign-owned computer firms increased from four to thirty-seven in the same period. However, IBM still held more than 50 per cent of the mainframe market in 1986.

The large number of firms operating at the lower end of the computer and peripherals market has meant that there is considerable competition between firms in most product groups, particularly the general-purpose computer-systems market;[10] in some specialised markets, however, competition is less intense. Developments in the computer industry since 1986 show that internal competition is forcing producers to become more competitive despite protective barriers. There are now signs that the growth of the national computer industry is slowing down for the first time in its (short) history. Even during the recession of 1981–83, the national industry grew at annual rates of 30 per cent to 35 per cent, but by 1988 growth had slowed down to 10 per cent. Competition has increased further and what was once perceived as a huge market now appears more limited. The need to become more internationally competitive is now perceived as a priority within and outside the industry.

6 INTERNATIONAL COMPETITIVENESS

This section examines to what extent there is a lag in prices and in technology between Brazilian products and international standards. First, the price comparison. Table 2.1 gives a summary of the Brazil/US price ratios for a range of market reserve products from 1982 to 1989.

Table 2.1 Brazil/US price ratios of computer products, 1982–9

Date	*Product*	*Brazil/US price ratio*	*Source*
	Micro-computers		
1982	Apple II clone	2.2	1
1984	Apple II clone	1.0	1
1982	Tandy 80 clone	2.79	1
1984	Tandy 80 clone	1.18	1
1985	IBM PC-XT compatible	2.49	3
1986	IBM PC-XT compatible	2.75	4
1989	IBM PC-XT compatible	2.1	5
1989	IBM PC-AT compatible	1.8	5
	Peripherals		
1984	Printer 80col/100cps	2.2	2
1985	Printer 80col/100cps	4.7	3
1989	Printer 80col	2.0	5
1984	Printer 132col/100cps	3.25	2
1985	Printer 132col/160cps	5.22	3
1985	Monitor 12inch	4.63	3
1989	Monitor 12inch	1.8	5
	Others		
1983	CNC unit	1.94	6
1987	CNC unit	1.63	6

Sources: 1. Tigre and Perine (1984); 2. Piragibe (1984); 3. Cline (1987); 4. Piragibe (1988); 5. Corsepius and Schipke (1989); 6. SEI (1988)

While these price comparisons have to be treated with caution,[11] some general conclusions can be drawn. In 1989, Brazilian products were approximately double the US price. Interviews carried out with leading manufacturers in Brazil in early 1990 broadly confirm this 1:2 ratio.[12]

So as to elaborate and trace changes over time it is useful to distinguish between products. As can be seen from Table 2.1, the price ratio of less sophisticated 8-bit microcomputers (Apple II and Tandy RS80 clones) was reduced from between two and three to virtual parity with the US level over the period 1982–84.

Since then, the IBM Personal Computer (PC) of 16 (and more) bits has become the *de facto* international standard in small systems. IBM PC compatibles now form the bulk of microcomputer production in Brazil and internationally. As can be seen from Table 2.2, Brazilian prices for such computers fell by about half between 1986 and 1989. The US/Brazil price ratio also declined over that period, but relatively little, remaining at a level of 1:2 (see Table 2.1).

As regards peripherals (monitors, printers, disk drives) price reductions in Brazil have also been significant (see Table 2.2). Analyses carried out in the mid-1980s concluded that the international price lag in peripherals was greater than in microcomputers (for example, *Info*, August 1985). Table 2.1 confirms this, but also suggests that this international price difference has since been reduced considerably. In the case of both simple monitors and printers, the price ratio is now the same as that in microcomputers, approximately 1:2. Note, however, that Table 2.1 does not contain data on flexible disk drives and hard disk units. It seems that their price development has been less favourable, particularly that of 'Winchesters' (hard disk units), which are approximately three times above the US price.[13]

Finally, the relative performance in CNC (computerised numerically controlled) units for machine tools is worth mentioning. Table

Table 2.2 Average reductions in prices of Brazilian microcomputers and peripherals, 1986–9

Equipment	*Price reduction (%)*
PC-XT with 2 drives	42.3
PC-XT with 1 drive and Winchester 20/30MB	52.3
Monitor	18.3
Printer	44.3
Flexible disk drive	30.3

Source: Tigre (1989)

2.1 suggests that their international price lag has been gradually reduced and is below the 1:2 'norm' found for most other informatics products.[14] The most favourable performance of Brazilian informatics producers is not reflected in the above tables. It has occurred in banking automation systems which are now less expensive than comparable internationally available systems.[15]

To sum up, in the electronics industry, the quest for competitiveness means chasing a fast moving target. Even though international computer prices fell dramatically, Brazilian producers of microcomputers and peripherals were able to reduce the price gap over the 1980s. This has not, however, occurred in a linear way. As can be seen most clearly in the case of microcomputers, when Brazilian firms have almost caught up, international competitors make a new leap forward and a new round of catching up begins. In spite of these achievements, Brazil's industry remains high-cost in international terms. By the end of the 1980s, its prices tended to be double those in the US.

This conclusion requires further clarification: US prices are an important yardstick for international competitiveness, but must not be equated with world market prices. In view of substantial differences between, say, US and European prices (see Section 7, Table 2.3), the very notion of a world market price must be questioned. This has ramifications for the assessment of the Brazilian performance. According to Tigre (1989) price gaps between Europe (France/Britain) and Brazil were: 22 per cent for microcomputers XT, 11 per cent for monitors, 14.5 per cent for printers and 60.5 per cent for flexible disk drives.[16] Similarly, Meyer-Stamer (1990) suggests that Brazilian microcomputers are between 15 and 25 per cent more expensive than in Germany. Thus, Brazilian prices are above European prices, but the size of the gap is smaller than that with the US.

The question remains: what are the internal reasons for the higher cost of Brazilian products? This issue was raised with a number of the leading producers during interviews carried out in early 1990.[17] The causes for the higher cost and their relative weight vary with the firm and the product, but some general conclusions can be drawn from the interviews and complementary case material.

The main cause for the higher cost of computers and peripherals made in Brazil seems to lie in sourcing. This is a consensus of national and foreign producers who operate in different markets but frequently rely on the same suppliers. The cost of inputs acquired *nationally* are between two and five times the international price. It seems that a major reason for this differential is that suppliers, many of which are multinational companies, abuse their monopoly or cartelised positions in their pricing policy. A case study carried out by Elebra calculates that high-cost inputs made in Brazil account for between 22 and 40 per cent of the international/domestic price differential of the *end* product (Elebra 1988).

Importing components (where allowed) does not necessarily reduce costs. Brazil's weak bargaining position in negotiating favourable input prices and payment conditions accounts for approximately 23 per cent in the international/domestic price differential of the *end* product (Elebra 1988). The effective cost of imports is further raised by government-imposed duties: tariffs on imports ranged between 30 per cent and 80 per cent in the mid-1980s (Tigre 1985). By the end of 1989 they had been lowered to between 10 per cent and 40 per cent; also a former 25 per cent tax on foreign exchange for imports had been withdrawn, but other taxes on imports (ICM and IPI) still come to 29.8 per cent. The relative weight of such costs varies with the import intensity of the product, but – as said before – both imported and local inputs are high-cost.

Putting together information from various sources, it seems safe to conclude that where the international/domestic price differential of computers and peripherals is in the order of 1:2, more than half of this differential is due to sourcing. Correspondingly, less than half is due to inefficiency internal to the firm, be it in product development or production.[18]

While only providing orders of magnitude, this conclusion differs from that of some other analyses of the market reserve which put more emphasis on inefficiencies internal to the firms. Notably Frischtak (1986) and Cline (1987) primarily blame the lack of scale economies for the high cost of Brazilian microcomputers and peripherals. In their work, the threshold for efficient plant size of microcomputers (16-bit) is set at 50,000 per year. The leading producers of microcomputers in Brazil were approaching that output figure by the end of 1989. These firms' strategies, however, do not seem to be governed by scale economies, since such economies are relatively small under their technological conditions. Automation is as yet of minor interest given the relatively high cost of capital and low cost of labour. In any case, automation is only feasible in some parts of the production process.

None of this is to invalidate a concern with plant size and internal efficiency in the analysis of Brazilian computer firms, but there is a case for putting greater emphasis on inefficiencies outside the computer industry, most notably the suppliers of inputs. Scale economies matter more in the production of inputs than in the assembly of microcomputers. However, little is known about plant efficiency of suppliers. Equal, if not more, attention needs to be given to domestic markets for inputs. There are indications (already mentioned) that monopolistic price practices of suppliers are one of the main factors holding back the competitiveness of local computers and peripherals *vis-à-vis* equivalent international products.

The assessment of competitiveness has so far concentrated on price performance. This needs to be complemented with an indication of whether Brazilian products are – by international standards – technologically up-to-date. Frischtak (1986) suggests that in the mid-1980s the technological lag was in the range of two to four years. It seems that by the end of the decade this gap has narrowed. Notably in professional microcomputers and interface equipment, manufacturers claim to have reduced the gap to below two years.[19]

Initially, the output of the industry was not sufficiently different from what was available internationally to enable exports of niche market products. This situation had changed by the end of the 1980s. National firms were exporting locally-designed automation systems for banks and commerce to Europe and Latin America (Botelho 1987; Tigre 1988). While such exports are small compared to those of foreign computer producers, there seems to be a future for Brazilian exports in some niche markets where independent design strategies are a real possibility (see Cassiolato in this book). To a great extent, this will depend on the continued development of technological capability by national firms.

7 IMPLICATIONS FOR USERS

In spite of the advances made by the national computer industry, opponents claim that the informatics law dooms the user firms to obsolescent and expensive equipment and undermines the broader competitiveness of Brazilian industry in the world market (Campos 1985; Moad 1988; Corsepius and Schipke 1989).

In examining such claims, five points need to be made. First, the diffusion of computers made by national firms has been rapid by any standards. The average annual growth rate of the Brazilian microcomputer market between 1984 and 1987 was the highest in the capitalist world at 74 per cent.[20] This exceptionally fast growth does not in itself answer the case made by the opposition, but it gives some testimony to the quality and relative cost performance of the national firms. It certainly contradicts Hieronymi's claim (1987:10) that the market reserve threatens the modernisation of the Brazilian economy.

Second, the difference between Brazil and US prices does not measure the cost to the Brazilian user.[21] There is little evidence to suggest that the prices of computer products imported into developing countries reflect US prices. For example, according to Piragibe (1988b) the price of imported microcomputers in Argentina was two or more times the US price (after excluding tariffs from the comparison). The fact that European countries also pay above the US price was mentioned in the previous section. This is further underlined by Table 2.3 which shows that British prices for microcomputers are between 28 and 67 per cent above US prices.

Third, opening up the production of all computers to foreign corporations would not necessarily reduce the cost to the user. Their LDC subsidiaries often sell at above world market prices and/or produce equipment that is less up-to-date than that produced for advanced countries. For example, the prices which IBM charges in the Brazilian market are higher than in the US market. A case in point is the IBM 4341 system which was in 1984 between 2.4 (according to IBM) and 3.02 (according to the association of national computer producers) times more expensive in Brazil (Frischtak 1986:29). Prices of another mainframe producer in Brazil were in 1989 between 1.7 and 1.8 times above US prices. This was due, it was stated, to two reasons: the higher cost of inputs in Brazil and higher profit margins used by the firm in Brazil compared with the US.[22]

Fourth, there are indications that even openness to foreign investment plus tariff-free imports of components would not guarantee lower prices for users. Consumer electronics production in Brazil, most of which is located in the Manaus Free Zone, operates under the above conditions, yet prices are substantially higher than in the US. Tigre (1989) shows video cassette players to be 2.08 times the US price, CD players 1.57 times and even colour televisions, a more mature product, to be 1.38 times US prices.[23]

Fifth, higher prices or less up-to-date equipment does not necessarily mean a loss of competitiveness for users. Five years ago (a long time in the history of this young industry!) a Brazilian researcher concluded that 'there is little evidence that the technological gap creates a serious productivity bottleneck for their users' (Erber 1985:303). Does this remain true?

In answering this question one may have to distinguish between those user firms who incorporate computers into their own *products*

Table 2.3 UK/US price ratios of microcomputers, 1989

	Average US	*Average UK*	*Ratio UK/US*
IBM-AT compatible[1, 2]	US$1,285	US$1,638	1.28
IBM-XT compatible[3, 4]	US$ 849	US$1,420	1.67
Apple MacPlus[3]	US$1,295	US$2,121	1.64
Apple Mac SE20	US$2,495	US$3,476	1.39
Apple Mac SE30	US$3,499	US$4,685	1.34

Notes: 1 US AT models Rosehill AT Rose 1900C, Epson Equity 11+ AT, Blue Star PC/AT, Kaypro KC-2
2 UK AT models Tandon 12sl-20, Ness AT286–10, Tandon 12sl-20, Epson PCAX2 AT
3 USXT models Zenith XT Turbo, Rose Hill Systems XT Rose Turbo
4 UKXT models Lucky Gold Star Logix, Ness XT-10, Epson PCeXT.

All models include 20 Mb hard disk.

Sources: *Personal Computing* (November 1989), *Personal Computer Magazine* (November 1989), *Personal Computer World* (November 1989), *Macuser (US)* (November 1989), *Macuser (UK)* (November 1989)

and those who use them in order to automate their production processes. The former are likely to suffer more than the latter. For example, the competitiveness of Brazilian microcomputer firms seems to be hampered by the higher prices of national hard disk units.[24] User firms that need the new technologies in order to automate their own production *processes* also complain, but one must question whether the market reserve seriously threatens their competitiveness. It is often forgotten that free access to international technology does not mean efficient use of that technology; especially when very advanced equipment is imported, maintenance and repair tends to bc a major problem (see Carvalho in this book). In fact, the indirect evidence that is available strengthens the position of the defenders of infant-industry protection. For example, a study based on a comparison of a British and Brazilian car plant argues against the 'technological imperative' (see Silva in this book). It shows that the Brazilian plant is less automated than its British equivalent but has not as a result lost competitiveness, either in price or quality of output.[25]

A similar conclusion can be drawn from Brazil's macro-economic record. In spite of the world economic recession, Brazilian industry stepped up its exports during the 1980s.[26] This does not suggest that industry suffered from outdated or overpriced technology. It could even be argued that user firms benefit from the establishment of a national computer industry. The international literature emphasises that the successful use of electronically controlled equipment depends on the close relationship between suppliers and users (Sayer and Morgan 1987; Cassiolato in this book). The high percentage of skilled personnel in national firms (see Section 9) provides a capability of adjusting equipment to local conditions. Probably the heightened internal competition will force national firms to put this capability to use.

None of this is to deny that some user firms encounter problems in their attempt to modernise as a result of the government policy. In many cases these are unavoidable learning costs which result from infant-industry protection. Of course this learning argument can be abused. However, the political opposition to the market reserve policy is so strong that this is unlikely to happen. 'There is a limit to the differences between local and international products, beyond which the combined pressure of customers and excluded competitors becomes irresistible' (Erber 1985:306).

8 OPPORTUNITY COSTS

The case of the opposition is driven by an interest group which feels it is losing out because of the market reserve. Its arguments are largely identical with the neo-liberal agenda. In one respect, however, the latter would go beyond the former. A neo-liberal enquiry would include the question of opportunity costs. Could labour and capital not have been employed more productively elsewhere in the economy?

By 1988, the national informatics industry employed over 50,000 people. The case that there is a social opportunity cost to this labour could only be made if there were full employment. This assumption does not hold in Brazil. The opportunity cost argument would be difficult to uphold even for highly trained personnel. Although the national computer industry has attracted some highly educated people from other productive sectors, there is little doubt that the industry has made a substantial net addition to 'human capital' (see Section 9).

There is a similar answer in the case of the opportunity cost of capital. One of Brazil's major problems in the 1970s and 1980s was that capital flowed into financial operations rather than into manufacturing. Profits in short-term financial markets were easier and higher than in industry.[27] Particularly noteworthy was the high profitability in the banking sector arising from short-term financial and speculative operations.[28] Unlike their counterparts in Japan and South Korea, Brazilian bankers were rarely strategic promoters of industry.

The only major exception is the computer industry. Banks are not only major users of computers,[29] but also invest heavily in their production (Fioravanti 1989). National banks own and/or control six of the ten largest national computer firms (SEI 1986). Thus, far from drawing capital away from more productive uses, the computer industry attracted capital which otherwise would have probably stayed in short-term financial and/or speculative operations.

9 LEARNING AND EXTERNALITIES

From a neo-liberal perspective, one might ask whether the build-up of the computer industry could not have been achieved more successfully if foreign capital had been granted entry into these sub-sectors. That is, if there had been no market reserve for national firms.

Ultimately, this counterfactual is impossible to answer. But we can suggest some of the likely trajectories of the industry. This can be done by way of a comparison with other sub-sectors of the electronics industry where the market reserve has not applied, namely the foreign-owned mainframe producers and the consumer electronics industry most of which is foreign-owned.

The Brazilian electronics market amounted to sales of US$8 billion in 1986. Foreign computer firms accounted for approximately 20 per cent of these sales and consumer electronics firms for a further 36 per cent. Neither of these market segments is subject to such rigorous control over imported inputs, although they have to abide by a generalised 'law of similars'.[30] In comparison with national computer firms, they rely to a greater extent on imported know-how. This is reflected in their labour force composition.

Piragibe *et al.* (1983) show that foreign computer firms employ 3 per cent of their workforce in R&D; by contrast, in national firms R&D accounts for 14.3 per cent. Our own research (Hewitt 1988) tends to confirm these findings: while a representative sample of national computer firms employ 12.5 per cent of their labour force in R&D, the share in foreign computer firms is only 3.7 per cent.[31] There is an even lower incidence of R&D employment in the largely foreign-owned consumer electronics sector: 1.9 per cent of the labour force.[32] The relative R&D intensity of national firms finds further confirmation in a recent survey (SEI 1989a:25) which shows the activities engaged in by employees who have completed higher education: R&D accounts for 25.3 per cent in national and 5.7 per cent in foreign computer firms. (So as to complement this comparison, it should be added that foreign firms put a greater emphasis on marketing where they employ 36.1 per cent of their personnel with higher education, compared with 19.7 per cent in national firms.)

The relatively high percentage of skilled engineering and technical labour engaged in R&D of national computer firms is striking. This build-up in human expertise in electronics has occurred as a result of the market reserve which has forced these firms to develop and accumulate their own technological expertise since they have had no parent companies to turn to.[33] The above data cannot prove the case, but it nevertheless suggests that such a fostering of talent and human resources would not have occurred if foreign firms had been allowed to enter this industry.[34]

While it is, then, beyond doubt that the national computer firms employ more of their professional human resources in R&D than foreign firms do, neo-liberals would query the usefulness of this. Some, indeed, have dismissed electronics R&D in Brazil as simply 'reinventing the wheel' and a waste of valuable resources (Ayres and Guanães 1985:58). Contrary to this view, we argue that building up an R&D capacity is a cumulative process which passes through various stages, and that what is occurring has to be viewed in a dynamic, longer-term sense.

Since neo-liberals tend to see long-run growth arising out of the pursuit of short run efficiency (see Chapter 1) they tend to ignore the benefits of cumulative learning. In what follows, its significance for the Brazilian computer industry is examined. The stages in the learning process which can be distinguished in accordance with Tigre (1986) are: imitation, modification, redesign and innovation.[35] The labour requirements for these activities differ. For example, imitation requires mainly manufacturing and organisational skills, but, with the progression towards innovation, a design capability becomes increasingly important.

Most national computer firms are engaged in imitation and modification of existing products. Indeed, since the major international microcomputer manufacturers have been able to establish *de facto* world standards in hardware and software, national firms have relied heavily on reverse engineering to develop a capacity for *creative imitation*. This requires not only copying and understanding[36] but also adaptations to local needs and available inputs. The complexity of this task should not be underestimated. Furthermore, the educational returns for the engineers and technicians involved in the process are substantial.

The *redesign* of existing products may be viewed as a form of 'minor innovation'. It appears that those firms which have, in the past, engaged in creative imitation are in a much stronger position to redesign products to enter niche markets. This, for example, is so for firms which have developed specialised applications for microcomputer hardware and software.

Innovation represents the highest level of R&D activity. In the *general-purpose* computer market, innovation is a tall order for all but th few leading firms in the world.[37] By contrast, creative imitation has been an effective way for Brazilian firms to enter the microcomputer market and to create a local technological capability.

The growth of the industry, of its competitiveness and of its skill base all indicate a degree of success which can, in good measure, be attributed to timely policy intervention. Intervention has given the industry time to learn (which includes making mistakes) and to build up a local technological capability. It is not clear how a policy based on short-run efficiency would have produced longer-term learning on the scale that has taken place in Brazil.

The process of learning has repercussions for the whole economy in the form of external economies. Judging from the experience of other sectors of electronics, computer products and know-how, if left to the market to provide, would have been imported. Instead, a capability was generated locally. Its benefits now stretch beyond the computer industry itself. Given the increasing use of electronics throughout the economy, the computer industry has a similar role to that normally attributed to the capital goods industry (Hoffman 1985:266). It has become an important general source of innovation and skills which are applied in other sectors.

In practice, externalities are difficult to measure. But the results of an assessment of the success of infant industries will often depend upon the ways in which externalities are considered in the analysis.[38] Whilst we have only indirect indications from our case study, we nevertheless conclude that externalities are greater under the present protected regime than they would have been in its absence. As shown before, national firms invest a higher percentage of their resources in R&D than the foreign-owned sub-sectors do. Equally significant is the fact that there is a high loss of R&D engineers and technicians by national firms (Hewitt 1988:126). This drift of engineers from firm to firm, lured by greater remuneration and/or new professional challenges may be damaging to individual firms, but it contributes to the technological capability of the computer industry as a whole and strengthens that of user firms.[39]

10 INTRINSIC PROBLEMS OF GOVERNMENT INTERFERENCE

In assessing the policy regime adopted in Brazil, we have so far concentrated on developments in industry. We now turn to the government itself and to its performance.

The widely observable poor record of governments in pursuing protectionist policies has contributed a great deal to the influence of neo-liberalism. It has strengthened the thesis that there are 'intrinsic problems of government interference' such as corruption or bureaucratic delays (Little, Scitovsky and Scott 1970). In this section and the next we argue that first this is only partially applicable to the protection of the Brazilian computer industry. Secondly, and more positively, a learning process within the government is observable, which would tend to escape the neo-liberal eye.

Let us begin with the question of corruption (a form of rent-seeking within government). SEI has a remarkable record. Our discussions with government officials and critics of SEI suggest that bribery within this organisation is unheard of.[40] Of course, one can never be totally sure that there are no clandestine deals; but it is noteworthy that the opponents of the informatics policy have not been able to come up with a single case of corruption. Since the policy was so fiercely contested, the agency and its officials could ill afford to be seen to be involved in corrupt practices.

As regards the question of administrative delays, SEI's record is more blemished. All imports and projects for internal production have to be approved by SEI. In addition, SEI administers the registering of software, the granting of fiscal incentives to firms and licences to technical service firms along with a number of other less frequent transactions. In the past, both producers and users of new technology often had to wait between six and twelve months to have their application dealt with. Such delays were excessive and brought the agency and the informatics policy into disrepute.

The pressure of its critics, however, has forced SEI to speed up the processing of requests. Since 1988 particularly, SEI has been undergoing an efficiency drive. While the law stipulates maximum times for the processing of projects, SEI has set its own targets well below this. In order to enforce these targets, an internal rule was introduced that projects which are not processed within the target period are approved automatically. Thus SEI achieved a significant overall reduction in processing times – the time it takes from receiving a request from a firm to providing the final response to that firm. The recent improvements are documented in SEI's own statistics (SEI 1989b).

- One-off requests for the import of hardware were being processed on average in forty days in 1986/7. In 1989 this had been reduced to an average ten days. Repeated requests for importing standard components and sub-assemblies (for already approved projects) were taking on average two to three days in 1989.
- The processing of projects for setting up new product lines fell from an average of sixty-five in 1988 to forty-three in mid-1989.
- The time for processing of internal software projects fell from forty-six days in the second half of 1988 to twenty-five days in 1989, while the waiting time for software import requests has stayed at around sixty days since 1988.

Linked to the growth in productivity of the agency is a change in its approach from 'policing' to providing a 'service'. Growing resentment from the private sector has forced it to speed up its work and become more accountable. It is interesting to note that a more efficient information and approval system was developed in liaison with national and foreign firms; SEI participated in a working group, consisting primarily of private sector firms, in which members explored ways of raising their productivity through technical and organisational innovations.

This experience of SEI runs counter to the neo-liberal view of the intrinsic inefficiency of government bureaucracy. It shows that government agencies can be responsive and become efficient. More important, it shows the process of learning in government at work. To explain, in those projects which have been longer established – for example, hardware imports, projects for internal hardware production – one can observe a steep decline in the learning curve. On the other hand, the regime in software has shown a slower decline because the rules governing imports and internal projects have only recently been established.

The wider significance of this learning process is taken up in detail in the following section.

11 LEARNING IN GOVERNMENT

Certainly, this case of protectionist policy does not conform to the following neo-liberal expectations: protection provides opportunities for both public and private actors to engage in rent-seeking; once established, these interest groups try to ensure the continuation of protection; if policy modifications are made, these are influenced by rent-seeking rather than developmental objectives. In contrast, we have observed a more positive trend: government agencies are learning and adjusting their policies in order to find an appropriate balance between, on the one hand, the domestic accumulation of capital, skills and know-how, and on the other, the international competitiveness of producers and users.

However, anti-statism is so deeply ingrained in neo-liberal thinking that government learning has little room on its agenda. In our case study, such learning is an essential part of the assessment. The previous section showed the increasing efficiency of government in implementing the rules of the market reserve, that is in administering import controls and new projects. The main argument of this section is that the learning extends beyond the speeding up of bureaucratic procedures into policy making.

In spite of the achievements made, government agencies in charge of information technology are caught up in the same race as the national firms. Neither can rest on their laurels. The political opposition to the market reserve and the fast technological developments in the world market do not allow them to do so. The 1984 informatics law itself sets a deadline of 1992 for the end of the protection.[41] In anticipation of that date, some members of the nationalist coalition have already indicated that the market reserve policy needs to be revised. The former chairman of the Special Secretariat for Informatics (SEI), Edison Dytz, suggested that the phase of stimulating the creation of new national firms is over.

> We need to launch a second phase, encouraging the association among firms, universities and research centres with well defined technological goals. After some time of supporting producers and researchers, we would have to assess what we have obtained and then select the most successful areas for continued support. Those areas in which our support has not led to adequate local capability will have to be supplied from abroad, even if this means the elimination of some national firms.
>
> (*Globo*, 7 June 1987)

Similarly, Veiga da Rocha, Chairman of SEI from 1987 to 1989, announced that where protection 'only results in higher product prices and no significant technological advance is made, there is no reason for keeping imports out'.[42] More generally, his successor, Kival Chaves Weber, emphasised that 'the consolidation of the Brazilian computer industry requires a greater concern with competitiveness'.[43] The implication is that for some items which are now in the market reserve the government might liberalise imports ahead of schedule or agree to joint ventures with foreign firms.

In our view this is not a retreat from the nationalist position. On the contrary, it strengthens it and makes it more viable. It is the clearest indication that the government is learning along with the national industry and its political opponents.[44] This learning argument is worth spinning out because there is more to it than the trivial point that one learns from mistakes. It is true that errors were made in the raising of Brazilian electronic infants, but the foster parents have also learnt from the successes. More importantly, the acquired knowledge and experience can be put to use because many of the infants have grown. Let us explain.

First, choices now exist because a national capability for producing computers has been created. The capacity to innovate still lags behind that of international competitors, but these latter firms can now be brought to the negotiating table. For selected product and process innovations it is important to have at least the option of collaborating with foreign electronics firms; in some cases it is the best way of finding a balance between strengthening local technological capability and achieving international competitiveness. This possibility now exists but did so less before.

In the late 1970s, for example, foreign computer firms were unwilling to make any concessions to Brazilian firms. The large foreign firms (IBM, Burroughs, DEC, Data General) refused to transfer technology for minicomputers without majority ownership (Piragibe 1985:142).[45] By 1985 they had reconsidered this position and had begun to forge links with national firms. For example, IBM embarked on a joint venture with a nationally-owned data processing company in order to commercialise its software, while other large North American firms granted licensing agreements for national firms to manufacture super minicomputers (Tigre 1986; Meyer-Stamer 1988). It should be stressed that these collaborative strategies by foreign electronics firms were a response to the policies adopted by the Brazilian government. Foreign firms which are unwilling to be excluded from the fastest growing market segments have been forced to cooperate with Brazilian capital. For other examples of collaboration, see Abicomp (1990:7).

Second, the government itself has learnt alongside its industry.[46] Even though SEI has rightly been criticised for slowness in administering import controls (mentioned before), the government has had a small cadre of officials who (1) keep abreast of national and international developments in the electronics industry, (2) are capable of participating in technical debates about the benefits of the market reserve, and (3) recognise the need for greater flexibility and selectivity in the application of policy to different segments of the market. To understand such government learning and the need for more selectivity, we need to look back briefly. In 1984, the informatics laws extended the market reserve. The following areas were added: semiconductors, software, automation equipment, superminicomputers, process control and instrumentation. SEI took on more than it could deal with. With hindsight, a more differentiated treatment would have been desirable. But that is a technocratic point to make. The preparation of the informatics law ignited ideological warfare between nationalist and internationalist forces. The issue received extraordinary attention in public debate and the media. Eventually a bout of nationalist fever saw the law through Congress. It was not a time for careful picking and selecting.[47]

It would, however, be wrong to see the reason for the ambitious scope of the informatics law in nationalist zeal alone. It was also influenced by the arguments of electronics experts. They emphasised the interdependence of the various electronics sub-sectors. The buzz word was synergy. While Japan has demonstrated that the technological and economic benefits which arise from having an 'electronics complex' are real, fascination with synergy clouds the reality of most countries.

The present Brazilian reality has led government agencies to move towards a more pragmatic position. There is a recognition that it is less necessary to maintain the protection for all sub-sectors covered by the law and also that it is more important to concentrate on selected sub-sectors and to complement import controls with more support for research and training.[48]

The informatics law has been in operation for only six years and some sub-sectors have only enjoyed protection from competition since then. It could be argued that for them a change in policy is premature, that they need more time. Perhaps. The fact is that the government agencies are under enormous pressure to re-examine the application of the law. This brings us to the final point in this section: the (involuntary) contribution which the political opposition made to learning in government.

The opposition has been both external and internal. In the initial years, it came mainly from the foreign computer firms operating within the country. Since then, the opposition has grown considerably.[49] External pressure has come from the American Electronics

Association, the US Computer and Business Equipment Manufacturers Association and the US government which has threatened a series of sanctions against Brazilian exports and has taken the Brazilian case to GATT. Growing internal opposition has come from users and industrialists (e.g. the industrial employers association, FIESP) who became dissatisfied with high product prices and administrative delays. Foreign electronics firms have maintained their opposition and have, in some cases, managed to obtain support from the Brazilian judiciary in the interpretation of the informatics law. Apart from helping to keep the nationalist alliance together, this unrelenting pressure has forced the government to review its policies, to become more selective and to improve implementation by speeding up the processing of import requests.[50]

In summary, the protectionist policy kept out the competition from the world market. Without state protection, most Brazilian infants would never have seen the light of day, let alone grown. While the international competitive pressure was not felt directly it was mediated through political pressure from internal and external opponents. The latter gave local producers and government little option but to adapt and learn quickly. The benefits of these pressures have thus been substantial.

It could be argued that neither industry nor government learned quickly enough. This is a matter of judgement. Nevertheless, one outcome which is undeniable is that Brazil now has choices which would not have existed without the market reserve. These are choices about the degree of openness to imports and of collaboration with foreign firms. The latter are now prepared to negotiate where previously they were not. What is more, they have to negotiate with Brazilian partners who know what they are after. In both industry and government a familiarity with complex technology and markets has been established. This is a necessary condition for successful negotiation.

Neo-liberals rightly force us to investigate the problems which arise in the execution of government policies. However, what tends to escape their analysis are the equally important questions of learning in government, and the conditions under which it occurs. If the frame of mind does not even allow for such learning to take place, there is indeed little alternative to believing in the wisdom of the market.

12 CONCLUSION

The record of the computer industry and of its regulation by government reveals both strengths and weaknesses. Indeed, a final assessment of this case of infant-industry protection is not yet possible. The market reserve has operated effectively for little more than a decade and the bulk of national computer firms is even younger. Is this sufficient time to grow up and achieve international competitiveness? Bell *et al.* (1984), in their review of infant-industry experiences, emphasise that there is no agreement on what duration of infancy is acceptable. Neo-liberals generally tend to be impatient in this respect. For example, Balassa (1975 and 1982) prescribes that protection for infant industries should be lifted after five to eight years. Bell *et al.* (1984) point out that there is often a large discrepancy between this expectation and the time that appears to be needed to become internationally competitive.[51]

While available data do not allow a rigorous quantitative analysis, we have tried here to make an *interim* assessment. We have shown that a verdict based only upon a comparison of domestic costs of production with the costs of equivalent imported products would be negative. This is not because the infants have been unable to reduce costs but because the technological frontier in this industry has been moving very quickly. Nevertheless, we have also shown that:

- an export capacity in the nationally-owned segment is beginning to emerge, particularly in specialised markets;
- there are substantial external economies arising from the learning efforts of national firms;
- without the market reserve, the creation of skills and know-how would have been lower than in fact happened;
- foreign firms are now willing to transfer technology on terms which they had previously refused.[52]

If these points are sufficient to justify a positive assessment of this experience – as we believe they are – what general lessons can we draw for other countries? Clearly there is a need for caution in advocating that other LDCs follow the same route because of the size of Brazil's internal market and its relatively advanced stage of industrialisation. However, in the 1970s, few would have thought that the Brazilian computer industry could have achieved as much.[53]

The case study has shown that opportunities for successful state intervention arise even under adverse political and technological circumstances. A rapidly expanding infant industry, in which national firms compete behind a temporary protective barrier, is growing up. This growth has been facilitated by government agencies which have pursued a sectoral strategy, and have themselves been positively affected by the experience of the infant industry and by its adversaries.

State efficiency is thereby increasing. There are, then, some short-run costs, but it seems likely that they will be more than outweighed by benefits over the longer run. Taking a more dynamic perspective than typical neo-liberals, then, it seems that selective state intervention along the above lines can work.

It is worth reiterating that most neo-liberals are not against infant-industry protection in principle, but they would insist that such protection be pursued by using price measures rather than quantitative restrictions. One cannot tell what would have occurred if tariffs or subsidies had been the main instrument. However, we do know that the combination of quantitative and administrative measures produced, in this case, a range of *national* benefits which the price instrument could not deliver. The latter would have stimulated production within the country, but by foreign rather than national capital. Thereby many benefits would have been lost.

To conclude, this chapter does not present a clear-cut success story for protection. An examination through the short-run lens reveals deficiencies. Taking the longer view, what impresses is the enormous learning which both infants and fostering parents underwent. It is a lesson in national industrial development and politics which is still under way. In our view it conveys a message of optimism for active industrial policy which contrasts with the neo-liberal gloom on state intervention.

If this optimism needs dampening it is not because of the performance of the computer industry or related government agencies. The two main problems lie outside their parameters of action. First, Brazil has no industrial strategy which supports initiatives at the subsectoral level.[54] Second, the greatest danger comes from the general crisis of the Brazilian economy which has its roots in both internal policy failures and failures of international capital markets.

NOTES

1 This paper is a revised and updated version of 'Learning to raise infants: a case study in industrial policy', in C. Colclough and J. Manor (eds), *States or Markets? Neo-liberalism and the Development Policy Debate*, Oxford University Press, 1991. It has benefited from the comments of many colleagues. We would like to acknowledge in particular the help of Chris Colclough and José Cassiolato.

2 See Piragibe (1985). For a recent and comprehensive analysis of the policy instruments, see also Paiva (1989).

3 These have been amply documented in a number of texts, for example, Piragibe (1985 and 1988a), Tapia (1984), Tigre (1983).

4 The *national ownership* of the computer firms has a strict definition: they are corporations established in Brazil, under permanent, exclusive and unconditional direct and indirect control (over managerial decision-making power, technological development and stock ownership) of individuals resident in Brazil or of domestic public entities (Frischtak 1986:8).

5 It is this wide range of activities related to data processing through machines which is termed 'informatics' in Brazil, and not just the computer industry itself.

6 For details see Piragibe (1988a) and Meyer-Stamer (1988).

7 For a detailed account, see Bastos in Chapter 9 of this book.

8 Monopoly is attributable often to foreign capital (e.g. the cathode ray tube production of Brazil) which can then afford to market products which, elsewhere, have become obsolete.

9 In 1986, there were an estimated 1,200 software 'producers' and 15,000 data-processing centres in Brazil (*Dados e Idéias*, August 1986:8).

10 For example, in 1985 there were an estimated 37 different firms producing copies of the IBM-PC family of computers alone (*Informática Hoje*, 15 October 1985).

11 The outcome varies with the choice of exchange rate. See Tigre (1989) for a brief discussion of the merits of different currency conversions. Price comparisons in Table 2.1 are based on the black market rate or proxies of the market-based rate of exchange.

12 Interviews carried out by Hubert Schmitz and José Cassiolato in March 1990 with five leading national and two multinational firms.

13 Based on interviews in March 1990. A case study on flexible disk drives (Elebra 1988) gives a Japan/Brazil price ratio of 1:2, but tariffs and taxes on imported inputs were excluded from the calculation.

14 Note that this study used international prices as reported by national producers rather than using independent sources.

15 According to Frischtak (1989) and information provided by José Cassiolato based on interviews with producers of banking automation equipment in May 1990. See also Chapter 3 of this book.

16 Owing to increasing currency speculation in 1989 and the difficulties of appropriate currency conversion, the above findings need to be taken with caution.

17 Earlier mentioned interviews carried out by Hubert Schmitz and José Cassiolato in March 1990.

18 Such a distinction between inefficiency external and internal to the firms is problematic because of changes in the degree of verticalisation. In the early stage, a good proportion of internal inefficiency was due to the fact that informatics firms had to verticalise production, since local supplies were inadequate (for example, plastic moulding or production of mechanical parts). Our interviews suggest that over time the industry has de-verticalised and become more efficient. However, there are also recent examples of verticalisation where new quality requirements cannot be met by existing suppliers.

19 Based on interviews in March 1990.

20 Estimated by the Office of Computers and Business Equipment, US

Department of Commerce. See Table 3.1 in Cassiolato's contribution to this book.

21 With some refinements, this is the method used by Cline (1987:4) to reach the conclusion that 'the informatics strategy in Brazil is probably costing user firms and the public as much as US$500 million annually'.

22 Based on interviews in March 1990.

23 Note that fiscal benefits in the Manaus Free Zone allow these items to be 20 per cent to 30 per cent cheaper than if produced elsewhere in Brazil (Baptista 1985).

24 Interviews in March 1990.

25 In this respect it is worth referring also to comparisons between Japanese and Western car makers. They show that in the early 1980s Japanese firms often had a lower degree of automation than their European and North American rivals and that Japan's competitive advantage stemmed above all from their superior organisational structure (Kaplinsky 1988). Similarly, Bessant and Haywood (1986) and Hoffman (1988) show that the benefits of advanced flexible manufacturing systems come more from organisational than technological innovation (even though the two cannot be strictly separated).

26 Of course, one would need to disaggregate these industrial exports and examine the reasons for their competitiveness, but the overall industrial export performance needs to be recalled in view of the allegations referred to earlier that the market reserve for informatics threatens the competitiveness of the rest of the Brazilian economy.

27 See, for example, *Gazeta Mercantil*, 4 October 1978.

28 See, for example, *Gazeta Mercantil*, Balanço Anual, various years.

29 Approximately 30 per cent of the national industry's output is used by the financial sector (SEI 1986:34).

30 Imports are only allowed if equivalent locally made products are not available.

31 The share of engineers and technicians in the labour force of national and foreign computer firms is very similar, namely 22.7 and 20.6 per cent respectively (Hewitt 1988:Table 4.5). However, engineers in the latter firms are more often involved in marketing, management and administration.

32 For a detailed account of occupational and skill composition in the various sub-sectors of the Brazilian electronics industry, see Hewitt in this book.

33 In absolute terms, 4,900 professionals were employed in R&D by national firms (in 1986) and their R&D expenditure reached US$154.1 million which amounted to 10.1 per cent of total sales (Piragibe 1987).

34 Apart from the example of the consumer electronics industry which has experienced two decades of the free market without showing any signs of a build-up of technological capabilities, the case of the Argentinian electronics industry is illustrative. What was once a burgeoning sector under national control is now virtually non-existent owing, it seems, to the opening up of the sector to uncontrolled imports of electronics goods (Tigre 1988).

35 Bell *et al.* (1984:34) employ a similar schema: replicative copying,

minor improvements and modifications, formalised overall redesign, and innovation through R&D.

36 Having know-how *and* know-why in Lall's terminology (Lall 1982).

37 Frischtak (1986:22) argues that 'The Brazilian market does not seem to be sufficiently large to support firms which, in isolation, would have the financial means to conduct R&D in the scale and depth necessary to introduce major innovations.'

38 Recovering the costs of coming down the learning curve (as in Bell *et al.* 1984), even if infants become internationally competitive, can probably only be achieved by including externalities in the calculation.

39 As Bell (1982:132) has pointed out: 'Perhaps the most important intra-industry externality is the direct acquisition of labour-embodied change-capacity that has been accumulated through expenditure by other firms . . . the inter-firm movement of experienced managers, engineers and technicians will carry not only knowledge and skill about how to effect change, but also information about what changes to effect – detailed information about specific changes that have been implemented or attempted in other situations.'

40 This is not true, of course, for Brazil as a whole.

41 The informatics law is open to different interpretations with regard to the duration of the market reserve for national firms and for protection from imports. Article 9 states explicitly that the adoption of restrictive measures in favour of *national* firms is temporary, without setting a time limit. Article 8 stipulates that SEI's mandate to control imports ceases 8 years after the introduction of the law, that is, 1992. The latter could be interpreted to mean that CACEX would take over this function from SEI. (CACEX is the government agency in charge of administering foreign trade regulations for most other sectors.)

42 *Jornal do Brasil*, 24 October 1988.

43 Interview in March 1990.

44 For a different interpretation, see Fleury (1987:31) who emphasises the 'bureaucratization of the Brazilian informatics policy formulation process'.

45 As a result national firms were forced to seek and obtain licences from smaller firms such as Sycor, Logabax, Nixdorf and Fujitsu.

46 National industrialists joined forces in ABICOMP, a sectoral association which is active in lobbying and producing information. The joint learning was helped by the fact that the government itself has a sectoral focus and a specialised administration for this industry.

47 This is based on direct observation. We were in the country at the time and participated in a number of open seminars and public debates.

48 This is a lesson also learnt in other Latin American countries, for example, Venezuela. See Perez (1985).

49 For a detailed account, see Bastos, Chapter 9 of this book.

50 This emphasis on learning in government must not hide the existence of inter-ministerial conflict. This has arisen mainly because the formulation and execution of policy for one sub-sector of the electronics complex, namely telecommunications, lies with the Ministry of Communication.

51 For example, the textiles industry took two or three decades to mature in Japan and four decades in South Korea. The Japanese auto industry took

anything between 3 and 6 decades to mature. By contrast, the South Korean steel industry appears to have matured in two years (Bell *et al.* 1984:23, 25).

52 We should add a factor which has not been examined in this chapter: the expatriation of profits. Between 1974 and 1983, foreign computer firms sent profits home from Brazil of almost twice the value of their total direct investment (Botelho 1987:44). While there are no comparable data for national firms, their retention of profits and dividends was almost certainly higher.

53 One dimension which could not be pursued in this chapter, because of lack of space, was a comparison with countries such as Argentina, Mexico, India or South Korea. However, brief mention should be made of a recent comparison between Brazil and South Korea, the most advanced Third World computer producer. Evans and Tigre (1989a and 1989b) show that, overall, Brazil had a larger computer industry than South Korea in 1986 and, more significantly, that Brazil has progressed further in designing and producing 'non-commodity' computers which can be exported to specialised markets. On the other hand, they show that South Korea is more competitive in the production of 'commodity computers' (PC clones). For a recent comparative study of Brazil, South Korea and India, see Schumacher and Wilkens (1990).

54 Erber (1990) and Meyer-Stamer (1990) discuss problems which arise from the fragmented nature of industrial policy.

BIBLIOGRAPHY

ABICOMP (1990) Associação Brasileira da Indústria de Computadores e Periféricos, 'Informática – a principal indústria deste fim de século', mimeo, Rio de Janeiro.

Ayres, R.U. and Guanães, V. (1985) 'Automation and competitiveness of multinationals in Brazil', paper prepared for UNDP/ILO Project Bra/82/024 , Brasília, June.

Balassa, B. (1975) 'Reforming the system of incentives in developing countries', *World Development*, No. 3.

—— (1981) 'The process of industrial development and alternative development strategies', in B. Balassa, *The Newly Industrializing Countries in the World Economy*, Oxford: Pergamon Press.

Balassa, B. and associates (1982) *Development Strategies in Semi-Industrial Economies*, Baltimore: Johns Hopkins University Press, for the World Bank.

Baptista, M.A.C. (1985) 'A indústria electrônica na Zona Franca de Manaus: considerações sobre seu impacto na política nacional de informática e no processo de desenvolvimento na Amazônia', report prepared for ABICOMP.

Bell M. (1982) 'Technical change in infant industries: a review of the empirical evidence', mimeo, Brighton: Science Policy Research Unit, University of Sussex.

Bell, M., Ross-Larson, B. and Westphal, L.E. (1984) 'Assessing the performance of infant industries', *World Bank Staff Working Paper*, No. 666, Washington, DC.

Bessant, J. and Haywood, B. (1986) 'The introduction of flexible manufacturing systems as an example of computer integrated manufacturing', *Operations Management Review*, Spring (Part 1), Summer (Part 2), Fall (Part 3).

Botelho, A.J.J. (1987) 'Brazil's independent computer strategy', *Technology Review*, Vol. 90, No. 4, May/June.

Campos, R. (1985) *Além do Cotidiano*, São Paulo: Record.

Cline, W.R. (1987) *Informatics and Development – Trade and Industrial Policy in Argentina, Brazil and Mexico*, Washington, DC: Economics International.

Corsepius, U. and Schipke, A. (1989) 'Die Computerindustrie in Schwellenländern – der Fall Brasilien', *Die Weltwirtschaft*, Heft 1, Institut für Weltwirtschaft, Universität Kiel.

Dados e Idéias, *Gazeta Mercantil*, São Paulo.

Elebra (1988) 'Custos industriais na indústria de informática – um estudo de caso', mimeo, São Paulo: Elebra Informática, May.

Erber, F.S. (1985) 'The development of the "Electronics Complex" and government policies in Brazil', *World Development*, Vol. 13, No. 3.

—— (1990) 'Política de informática e política industrial', mimeo, Instituto de Pesquisa Econômica Aplicada (IPEA), Rio de Janeiro.

Evans, P. (1985) 'Varieties of nationalism: the politics of the Brazilian computer industry', paper presented at the MIT Symposium on The Computer Question in Brazil, Cambridge, Mass., April.

—— (1986) 'State, capital and the transformation of dependence: the Brazilian computer case', *World Development*, Vol. 14, No. 7.

Evans, P.B. and Tigre, P.B. (1989a) 'Going beyond clones in Brazil and Korea: a comparative analysis of NIC strategies in the computer industry', *World Development*, Vol. 17, No. 11.

—— (1989b) 'Paths to participation in 'Hi-Tech' industry: a comparative analysis of computers in Brazil and Korea', *Asian Perspective*, Vol. 13, No. 1.

Fioravanti, C. (1989) 'Os donos da reserva', *Istoé Senhor*, 10 May.

Fleury, A. (1987) 'An institutional analysis of policy-making in Brazil: the case of microelectronics', mimeo (paper prepared for the United Nations University, New Technologies Centre Feasibility Study), Escola Politecnica, Universidade de São Paulo, February.

Frischtak, C. (1986) 'The informatics sector in Brazil: policies, institutions and the performance of the computer industry', mimeo, Industrial Strategy and Policy Division, the World Bank, March .

—— (1989) 'Specialization, technical change and competitiveness in the Brazilian electronics industry', paper prepared for OECD Workshop, Technological Change and the Electronics Sector – Perspectives and Policy Options for Newly Industrializing Countries, Paris.

Hewitt, T. (1988) 'Employment and skills in the electronics industry: the case of Brazil', D.Phil. thesis, Brighton: University of Sussex.

Hieronymi, O. (1987) *The Domestic and External Impact of National Industrial Policies: The Example of the Electronics Industry*, Geneva: Batelle Centre for Applied Economics.

Hoffman, K. (1985) 'Microelectronics, international competition and development strategies: the unavoidable issues – editor's introduction', *World Development*, Vol. 13, No. 3.

—— (1988) *Technological Advance and Organizational Innovation in the Engineering Industry: A New Perspective on the Problems and Possibilities for Developing Countries*, report submitted to the World Bank, Sussex Research Associates, Brighton.

Info, Rio de Janeiro: Editora Jornal do Brasil.

Informática Hoje, São Paulo: Plano Editorial.

Kaplinsky, R. (1988) 'Industrial restructuring in LDCs: the role of information technology', paper prepared for Conference on Technology Policy in the Americas, Stanford University, December.

Krueger, A.O. (1981) 'Export-led industrial growth reconsidered', in W. Hong and L.B. Krause (eds) *Trade and Growth of the Advanced Developed Countries in the Pacific Basin*, Seoul: Korea Development Institute.

Lall, S. (1982) 'Technological learning in the third world: some implications of technology exports', in F. Stewart and J. James (eds) *The Economics of New Technology in Developing Countries*, London: Frances Pinter.

Little, I. Scitovsky, T. and Scott, M. (1970) *Industry and Trade in Some Developing Countries: A Comparative Analysis*, Oxford: Oxford University Press.

Luedde-Neurath, R. (1986) *Import Controls and Export Oriented Development: A Reassessment of the South Korean Case*, Boulder and London: Westview Press.

Meyer-Stamer, J. (1988) *Informatik in Brasilien*, Hamburg: Institut für Iberoamerika-Kunde.

—— (1990) *From Import Substitution to International Competitiveness – Brazil's Informatics Industry at the Crossroads*, Berlin: German Development Institute.

Moad, J. (1988) 'The plight of the Brazilian user', Special Report, *Datamation*, 1 November.

Paiva, S.M.C. (1989) *Política Nacional de Informática: Intervenção do Estado, Resultados e Desafios*, Tese de Mestrado, Instituto de Economia Industrial, Universidade Federal do Rio de Janeiro.

Perez, C. (1985) 'Hacia una estrategia de desarollo integral del sector electronico en Venezuela', Proyecto CONDIBIECA – ONUDI VEN 80/003, July, mimeo.

Piragibe, C. (1984) 'Competitividade dos equipamentos periféricos fabricados no Brasil – impressoras', Texto para Discussão No. 61, Instituto de Economia Industrial, Universidade Federal do Rio de Janeiro, December.

—— (1985) *Indústria de Informática: Desenvolvimento Brasileiro e Mundial*, Rio de Janeiro: Editora Campus.

—— (1987) 'Policies towards the electronics complex in Brazil', Textos em Política Científica e Tecnológica, No. 28, CNPq, Brasília.

—— (1988a) 'Políticas para a indústria electrônica nos novos países industrializados: lições para o Brasil?', in H. Schmitz and R. Quadros Carvalho (eds).

—— (1988b) 'Electronics industry in Brazil and the role of the state: some aspects for a comparative analysis', in C. Piragibe (ed.) *Electronics Industry in Brazil: Current Status, Perspectives and Policy Options*, draft report for OECD, CPCT, CNPq, MCT, Brasilia, November.

Piragibe, C., Penna, M.V. and Tigre, P.B. (1983) 'Recursos humanos na indústria Brasileira de equipamentos de processamento de dados', Relatório de Pesquisa, Instituto de Economia Industrial, Universidade Federal do Rio de Janeiro, Rio de Janeiro.

Sayer, A. and Morgan, K. (1987) 'High technology industry and the international division of labour: the case of electronics', in M.J. Breheny and R. McQuaid (eds) *The Development of High Technology Industries: An International Survey*, London: Croom Helm.

Schmitz, H. (1984) 'Industrialisation strategies in less developed countries: some lessons of historical experience', *Journal of Development Studies*, Vol. 21, No 1.

Schmitz, H. and Quadros Carvalho, R. (eds) (1988) *Automação, Competitividade e Trabalho: A Experiência Internacional*, São Paulo: Editora Hucitec.

Schumacher, D. and Wilkens, H. (1990) *Marktabschottungspolitiken wichtiger Schwellenländer und wirtschaftliche Entwicklung: Empirische Untersuchungen am Beispiel der Elektronikindustrie und des Kraftfahrzeugbaus in Brasilien, Indien und Korea*, Heft 118, Deutsches Institut für Wirtschaftsforschung, Berlin: Duncker & Humblot.

SEI (1986) 'Panorama da indústria nacional – computadores e periféricos', Secretaria Especial de Informática, *Boletim Informativo*, Brasília, mimeo.

—— (1988) 'Comandos numéricos computadorizados (CNC) – estudo de avaliação dos preços no país', mimeo, Secretaria Especial de Informática, Brasília.

—— (1989a) *Panorama do Setor de Informática*, Secretaria Especial de Informática, Séries Estatísticas, Vol. 2, No. 1, Brasília.

—— (1989b) 'O processo de melhoria da produtividade e da qualidade dos serviços prestados', *Indicadores de Desempenho*, Vol. 1, No. 5, Secretaria Especial de Informática, Brasília, October.

Senhor, São Paulo: Editora Três.

Tapia, J.G.B. (1984) 'A política nacional de informática: 1970–1984', Núcleo de Política Científica e Tecnológica, Universidade de Campinas (UNICAMP), mimeo, November.

Tigre, P.B. (1983) *Technology and Competition in the Brazilian Computer Industry*, London: Frances Pinter.

—— (1985) 'Estrutura de custos e formação de preços na indústria Brasileira de computadores', estudo preparado para a Secretaria Especial de Informática, Instituto de Economia Industrial, Universidade Federal de Rio de Janeiro.

—— (1986) 'Perspectivas da indústria Brasileira de computadores na segunda metade dos anos 80', mimeo, Instituto de Economia Industrial, Universidade Federal do Rio de Janeiro.

—— (1988) 'How does Latin America fit high technology?', paper prepared for the International Symposium: Technology Policy in the Americas, Stanford, December 1–3.

—— (1989) 'Perseguindo o alvo movel: evolução dos preços e competitividade dos equipamentos eletrônicos nacionais', mimeo, Forum Informática, Rio de Janeiro, September.

Tigre, P.B. and Perine, L. (1984) 'Competitividade dos microcomputadores nacionais', Texto para Discussão No. 60, Instituto de Economia Industrial, Universidade Federal do Rio de Janeiro, November.

White, G. (ed.) (1988) *Developmental States in East Asia*, London: Macmillan.

3 The user–producer connection in hi-tech

A case-study of banking automation in Brazil[1]

José E. Cassiolato

1 INTRODUCTION

The issue of participation of less developed countries in the effort to develop new high technologies has been the subject of a fierce debate. Analysts who draw on neo-liberal thinking insist that it is a waste of time and scarce resources for LDCs to join the hi-tech race. They say it would be better for such countries to concentrate their efforts on acquiring mature technologies from advanced societies and then try to achieve competitiveness using their so-called traditional comparative advantages (see Chapter 1).

Contrary to this, this chapter argues that it is precisely the early stages of the new techno-economic paradigm that offer opportunities to LDCs, at least those which possess minimal requirements, particularly in industrial and technological infrastructures. Thus this chapter follows Perez (1986, 1988) who suggests that early entry enables LDCs to follow different technological paths to those established in advanced countries. Indeed 'the main objective of [these new] R&D systems should be the exploration of possible technological trajectories, in order to concentrate on those best adapted to the conditions of the region' (Herrera 1986).

A 'window of opportunity' (Perez 1986) is open to LDCs due to the specificities of the early stages of the new techno-economic paradigm when markets for products and technologies are not yet completely structured, and when inexpensive or even relatively free knowledge is still available. It is at this stage that LDCs can gear their strategies to internal or regional markets and follow technological trajectories different to those of advanced countries.

This specificity is related to the diffusion process of new technologies since this process is itself a key determinant of the direction in which technologies evolve. The mutual influence between diffusion and innovation processes usually leads to the establishment of very specific technological paths.

The literature on the simultaneous determination of the process of innovation and diffusion in the new techno-economic paradigm is relatively recent. It is discussed in Section 2 which emphasises, in particular, the importance of close supplier–user interaction in shaping specific technological paths. It also suggests that the existence of location advantages of the innovation–diffusion process of new technologies allows – given certain conditions – the implementation of policies which emphasise country-related specificities.

In Section 3 the process of banking automation in Brazil is presented as an example of how country-specific trajectories can be pursued and how, given the specificity of the innovation–diffusion process and the importance of the producer–user relationship in such a process, the participation of a developing country in the early stages of new technologies is essential.

2 THE SIMULTANEOUS PROCESS OF INNOVATION AND DIFFUSION IN THE NEW TECHNO-ECONOMIC PARADIGM

Technological paradigms are made up of a general set of search procedures which are broadly accepted by a technological community (Clark 1987). Such searching will result in basic designs which set the technological standards for a considerable period of time. The diffusion of such regimes is mediated by social acceptance (Perez 1986 and Freeman 1986). In capitalist economies firms, markets, governments and social relations play a considerable role in this acceptance as they act as a selection environment (Nelson and Winter 1977:63; Dosi 1982:155; Clark and Juma 1987:40). There are then both deterministic and voluntaristic elements in technological change for a given technological trajectory (Hagerdoorn 1989).

There are clearly obstacles that could delay or even impede a transition from the old technological trajectory to a new competing one. In his analysis of these obstacles Clark (1987) showed how the dominant technologies provide the heuristic for any subsequent technological change, how a particular technological practice is enshrined socially and how 'interests' of various types play an important role in the genesis (or not) of technological change. These tendencies help to delay the substitution of the new paradigm for the old one because

> for any given paradigm to be defeated, there has to be an alternative candidate which shows promise of dealing with the

> pitfalls and puzzles experienced in the operation of existing technological practice, since in the absence of such a candidate it is likely that attempts will continue to be made to prop up the status quo – to regard 'anomalies' as 'puzzles'. Indeed just as with many examples in science, revolutionary change with respect to any given technological paradigm often arises from a source outside that community.
>
> (Clark 1987:32)

At this stage it is useful to recall a systematic approach developed by Perez (1983), where in the transition from one techno-economic paradigm (defined as an ideal type of productive organisation or best technological common sense) to another, changes in the institutional and social framework are emphasised. Then, at the beginning, a great deal of inertia in social institutions and the various interest groups are bound to defer the actual economic use of promises brought about by a very rapid pace of technological change. In other words, the new core technologies do not always meet the appropriate socio-institutional framework and in fact their diffusion may indeed be held back by the social restraints of the past (Perez 1986). Changes in capital stock, skill profile, industrial structure and social organisation, which the introduction and diffusion of revolutionary new technologies require, are a matter of decades rather than years.

Radical technical change then is both influenced by and influences its surrounding environment. Contrary to the traditional wisdom the innovation process and the diffusion process are simultaneously determined. Metcalfe (1986) has shown how it is misleading to speak about the diffusion of a particular innovation during the early stages of the development of a particular technology. At the beginning of the diffusion process 'there exists a set of competing innovations whose technologies are changing systematically in response to experience and the incentives arising during diffusion' (Metcalfe 1986:62). Selection and inducement mechanisms linked to the diffusion process will actually shape the technology path. In other words, the supplier–user interaction of the technology and the existence of a relatively sophisticated pool of technical skills in the surrounding environment are important elements of the development process of a new technology.

The fact that technical problems arising during the innovation–diffusion process require collaboration between suppliers and users has two major implications. First, the collective learning which arises through inter-firm links during the early stages is more feasible when open systems are at play. (Consequently proprietary technology – to the firm – loses its relative importance.) Second, the supplier's competitiveness becomes structurally linked to the user's competitiveness (Perez 1988).

The importance of users in the development of new technologies in advanced economies during the early stages of the new technological cycle has already been shown. Von Hippel (1976), for instance, found that in the US 67 per cent of the new process machines employed in the manufacture of semiconductors and electronics sub-assemblies were developed by the users, not by machine manufacturers. Utterback and Kim (1985) suggest that users are important introducers of radical innovations since these cast aside the value of established positions amongst suppliers in materials, design, markets and skills. Moreover, users tend to play a significant role in the new paradigm since uniformity of consumption patterns is giving way to production for diversified demand, made possible by programmable equipment (Perez 1986).

This basic characteristic of the development of high technologies implies that being an effective user of them is not a sufficient condition for a country to compete at international level. Gaining access to available techniques can in itself only lead to efficiencies in the short term. To move from static to dynamic efficiency, a firm (or a country) must not only possess the knowledge that determines the basic functionality of various information-based systems, but also has to acquire knowledge through a long interactive process of learning-by-innovating where user–producer relations play a significant role.

This is precisely why the early stages of the development of an innovation are so important for Third World economies. The innovation literature (Freeman 1982) suggests that the proximity of the users to the suppliers of an innovation plays an important role in the design and de-bugging of new products, especially high-technology products. Indeed, many hi-tech innovations involve collaborative development between supplier and user. For this process to come about, the supplier needs ready access to a reasonably large constituency of potential users with the range of skills necessary to apply the innovation successfully or the expertise necessary to formulate problems for the innovator to solve (Walsh 1988).

Given that this process tends to occur in advanced countries, Third World requirements tend to be dealt with only when the new technologies mature. What is significant for the new high technologies is that as the process of technological innovation, particularly in its early stages, comprises a significant degree of uncertainty both at technical and economic levels, users may be in a position to shape the subsequent paths of technical development.

The diffusion to the Third World of technologies which have already gone through this lengthy and costly process may contain a considerable degree of 'inappropriateness' and deprive the local economies of this important 'learning by doing' process which obviously has important effects for the economy as a whole. Putting it another way, *different diffusion environments generate different patterns of technological evolution* (Georghiou *et al.* 1986). This is particularly important in understanding national differences in the establishment of a technological regime. But it is also important in that the crystallised designs coming out of such an evolution will also be location-specific.

The work of Lundvall (1983, 1985, 1986) on user–producer interaction emphasises the above mentioned points. He argues that the geographical proximity between groups of users and producers constitutes a comparative advantage, which increases with the complexity of messages to be sent between them and with the degree of non-standardisation of production, having a more significant effect the more radical the innovation and the earlier it is introduced in the life-cycle of the industry concerned. Indeed the significance of user–producer interaction acts as a counter-tendency to that of the traditional understanding of the international division of labour whereby segments of the research, development and production process may be located separately (Walsh 1988).

The initial stages of the development of the new technologies in the incoming paradigm are characterised by the fact that the necessary scientific knowledge is relatively freely available in universities and research institutes or could be bought from small hi-tech firms (Perez 1986). Obviously this is a necessary condition for entry, but not a sufficient one. As was recently argued:

> even in the pre-paradigmatic stage, very substantial stocks of complementary assets are likely to be critical to success . . . and such . . . capabilities have to be built up over extended periods of heavy investments in R&D, involve very significant elements of firm-specific know-how and are not simply a part of the stock of scientific knowledge in public domain.
>
> (Ernst and O'Connor 1989:40)

But what seems to be crucial is that only entry in the early stages of development of a new paradigm will provide the synergies related to user–producer relationships to Third World countries which possess the minimal requirements. It is precisely in these earlier stages that most of the learning required to design, produce, distribute and service these technology-intensive products is generated (Lundvall 1985 and Perez 1988). As innovation and diffusion in such stages are parts of the same process their specificities and the synergies they comprise are most likely to be obtained in such periods.

3 BANKING AUTOMATION IN BRAZIL

The experience of banking automation in Brazil provides an interesting example of how user–producer relationships are important in shaping specific innovation–diffusion paths during the early stages of the introduction of a new techno-economic paradigm in a developing country. It is also very revealing in so far as it points out that during such stages and when certain conditions are fulfilled, developing countries can follow different technology options which may be more suitable to their conditions.

One of the most important achievements of the Brazilian informatics policy is precisely to have opened up the possibility of an independent technological path. Technological development was achieved through a close relationship between user needs and

Table 3.1 Main microcomputer markets: size and growth, 1984–7

Countries	*1984 (US$ million)*	*1987 (US$ million)*	*Average rate of growth per year (%)*
Developed countries			
USA	13,124	19,953	15
Canada	208	274	10
West Germany	730	1,774	34
United Kingdom	544	1,319	34
France	510	1,206	33
Italy	334	680	27
Sweden	157	468	44
Japan	1,618	2,920	22
Developing countries			
Brazil	189	992	74
China	112	662	81
India	61	300	70
Australia	144	201	12
Mexico	74	179	34
Korea	54	108	26

Source: US Department of Commerce (1987)

suppliers. This fact lies behind the very rapid diffusion of microcomputers, particularly to the services sector, and make Brazil the world's second fastest growing microcomputer market during the period 1984/87 and the sixth largest in size (Table 3.1).

The importance of the diffusion process in shaping the technological development of the Brazilian computer industry is studied surprisingly little since, in the absence of explicit policies, international firms would supply national customers with 'packages' over which those customers have little influence. Consequently the packages are relatively inappropriate to local needs and relatively poor servicing may be expected.[2]

Service sectors in general (and banking in particular) have been traditionally considered in more advanced economies as non-performers of R&D and 'passive' recipients of technology developed elsewhere. The increasing relative importance of software developments and maintenance and the emergence of strong systems groups in banks and other service activities will probably change this picture. But until now it seems reasonable to characterise banks in advanced countries as 'supplier dominated firms',[3] as far as technology appropriation is concerned.

In Brazil, the introduction of advanced technology in this sector has followed a significantly different pattern. First of all the commercial banking system has been one of the most significant user sectors of computers in Brazil (Peritore 1988). From 1982 to 1985, approximately 29 per cent of all computer sales went to the financial sector, either private or public (SEI 1989). It is claimed that until 1986 banks were responsible for nearly 50 per cent of Brazilian IT firms' sales (Frischtak 1989). In 1988, of the ten largest users of microcomputers in Brazil, five were banks (*Dados e Idéias*, Nov./Dec. 1988).

The banking system in Brazil has gone through an accelerated process of automation in less than a decade. In the recent words of one of the most important Brazilian bankers,

> it was to reduce costs, improve the bank image, attract new clients and become more competitive that bank automation reached [in Brazil] heights never seen before, contributing to the development of our national informatics industry. Electronic money, through the use of magnetic cards in any kind of bank terminals, has gained the trust and preference of the Brazilian public and the banks; their clients and the country are the winners. Other services that only informatics can provide and are already being developed in

> Brazil are worth mentioning: terminals in customers' premises, terminals for the general public or installed with on-line access to the banks' mainframes for immediate checking of current accounts and portfolios – the audio-respondent supplies information on account balances, portfolios, shares and investments by phone.
> (Santos 1988:211)

These aspects of customer services in banking automation in Brazil were preceded and made viable by a rapid process of office automation through a strategy of relative self-reliance. Documenting and analysing this process is the aim of this case study.

The reasons for automation

Banks, together with government agencies, pioneered the use of computers in Brazil from the early 1960s. Following the same tendency as that observed internationally, the main Brazilian banks set up central data-processing services in that period which enabled office work to be centralised.[4] This phase of banking automation corresponded to a highly Taylorist organisation of labour which clearly had technical limits since business was still taken over the counter and cheque processing still required handling even if only to a limited extent (Petit 1984).

But it was only during the 1970s that the rapid process of centralisation and concentration of the Brazilian financial sector imposed and created both the conditions and means for relatively rapid automation of the sector. Historically, the banking system in Brazil developed through a large number of small and medium-sized regional banks which catered for regional specificities in a continent-size country.

One of the axes of government policies in Brazil during the late 1960s and the 1970s was to acquire better financial efficiency of such a system. As part of a broader reform of the financial system, the banking policy established that economies of scale were to be obtained via banking concentration. Later on, this policy made it possible for the main banks to become financial conglomerates. Financial, fiscal and credit incentives were the measures the government put into practice in order to foster banking mergers and acquisitions. Bigger and stronger banks were considered essential as searchers of foreign funds and as mediators of a growing public-sector debt. This caused the number of banks in Brazil to decrease from 336 in 1964 to 105 in 1986. In that year these banks together had

Table 3.2 Commercial banks in Brazil: total number and number of branches, 1983–6

	Number of banks				*Number of branches*			
	1983	*1984*	*1985*	*1986*	*1983*	*1984*	*1985*	*1986*
Private	85	83	78	76	8,629	8,902	9,062	8,375
Brazilian	66	64	60	57	8,570	8,843	9,000	8,312
Foreign	19	19	18	19	59	59	62	63
Fed. Government	4	4	5	5	2,694	2,711	3,187	3,150
Banco do Brasil	1	1	1	1	2,400	2,401	2,489	2,458
State Government	24	24	24	24	3,113	3,123	3,173	3,154
Total	113	111	107	105	14,436	14,736	15,422	14,679

Source: Banco Central do Brasil

approximately 15,000 branches spread throughout the country (Table 3.2).[5]

In 1965 the four major private banks (1.3 per cent of the total number of banks) held 15 per cent of total deposits, 14 per cent of total loans and 11 per cent of total net assets. By 1986 the four largest (3.7 per cent of total) held 34 per cent of total deposits, 23 per cent of total loans and 37 per cent of the total net assets of the financial system. Bradesco and Itaú, the two largest Brazilian private banks, acquired more than fifty smaller institutions over this period.

Such a process of concentration led to a situation where five of the ten largest banks in Latin America (in terms of stock-holding equity) are Brazilian. Bradesco and Itaú are the two largest private banks in Latin America. In 1988 their combined assets were US$23.1 billion for their banking activities alone (*The Banker*, August 1989).

Foreign competition was stopped by preventing the few foreign bank subsidiaries which operated in Brazil at that time from expanding their branches and thus from entering into some of the most lucrative niches of the market. In addition, new foreign banks were excluded from setting up as commercial banks in the Brazilian market. This situation still holds today. At the same time leading banks were encouraged to become financial conglomerates. In contrast to developed countries where anti-trust legislation impedes commercial banks from diversifying their activities beyond certain limits, large Brazilian banks work as commercial, clearing, investment banks, building societies, securities houses and so on.

This process of concentration through transforming regional banks into national ones and commercial banks into financial conglomerates created specific managerial needs. In the mid-1970s, when this process of concentration acquired maturity, the previous phase of automation appeared outdated. Until then, banks used to rely only on data-processing centres in which millions of documents were sent and processed overnight; in the morning updated lists were sent back to branches. With new branches being incorporated into the network from far away, these procedures no longer seemed viable.

Highly qualified and trained technicians at the banks' data-processing centres started to conceive automation strategies specific to their own particular needs. It is important to point out that the needs of the Brazilian banks in their search for automation were very singular and country-specific: how to integrate agencies nation-wide in a country with a vast territory. Even the US, the leading supplier of products and technology in the area and also a continent-size country, could not supply the appropriate solutions since banks do not operate nationally there.

This second phase of the automation of the banking sector can be characterised as one in which differences in the national organisation of banking systems played a significant role in shaping patterns of technical change and in explaining differences in rates of diffusion and so on. The universal banking system in West Germany, the segmentation of deposit-taking and intermediation functions in France, the enforced deconcentration in the USA and the considerable significance of foreign business for financial institutions in the United Kingdom all illustrate how varied the bases were from which the move towards more computerisation began in the 1970s (Petit 1984). Unlike the first phase which involved only a Taylorist organisation of work, the second phase of banking automation was centred on money management.

The next section analyses how during this phase users, that is the banks, participated in defining technologies and directly invested in technological development and the production of the hardware and software they needed.

The technology strategy of Brazilian banks

Decentralisation of data-processing at branch level was the general feature of banking automation strategy as devised by local technicians during the late 1970s. It was perceived that between 70 and 80 per cent of all operations processed in a particular branch were only relevant to the same agency. Then, why not decentralise data-processing and install a minicomputer in each branch? By then Olivetti was selling a minicomputer (Sycor) under licence in Brazil which was considered to be ideal for data transmission. But the specifications designed by Brazilian users (the banks) would require appropriate programs and some changes in the hardware which Olivetti was not interested in making (Dantas 1989).

Multinational affiliates with their strategies of selling standard products did not have the required technologies to provide the products Brazilian banks needed. That they were not interested in developing them when approached by the Brazilian banks' technicians is well documented (Dantas 1988). For instance, in 1978 Bradesco designed its own 'Bradesco Instantâneo' system so that teller machines could read magnetic characters printed into the cheque. This was before such readers were available worldwide. Multinational subsidiaries were not interested in developing such a product[6]

Table 3.3 Brazilian IT firms where BRADESCO has participation (via its holding DIGILAB)

Firm	*Share of voting capital (%)*	*Turnover, 1988 US$ thousands*	*Main products*
Scopus	70.0	72,432	Microcomputers, terminals
Digilab	100.0	57,539	Printers, teller terminals, disk drives
Sid Informática	11.5	114,750	Minicomputers, microcomputers, bank terminals
Sid Microeletrônica	14.6	NA	Components
CPM	50.0	29,326	Communication processors, mainframes[1]
Elebra Computadores	30.0	32,328	Supermicros, bank automation
Matel Tec. de Teleinf.	26.3[2]	70,593	Teleinformatics
Cobra SA	1.4	99,849	Supermicros, microcomputers, superminis
Victori Comunicações	34.0	NA	Telecommunication
PDV Informática	50.0	NA	NA
DNI Informática	100.0	NA	NA
Rima Impressoras[3]	17.0	NA	Printers

[1] CPM is starting in 1989 the production of IBM compatible mainframes with technology obtained from Hitachi; it was also the fastest growing IT firm in Brazil during 1988 – 249.2 per cent in real terms (*Exame Informática*, Vol. 4, No. 7, 12 July 1989)

[2] Digilab has also a 48 per cent stake in Matel Participações e Administração, the holding which controls Matel Tec. de Teleinformática.

[3] Joint venture with Itautec and Mecaf (Mecânica Fina SA)

Sources: *Exame Informática*, Vol. 4, No. 7, 12 July 1989, *Veja*, Vol. 22, No. 14, 5 April 1989 and ABICOMP, *Catálogo Brasileiro da Indústria de Informática*

so Bradesco created its own digital electronic laboratory headed by a former IBM engineer. In less than a year they created the first magnetic characters reader in the world (Dantas 1988). This and other specific products were designed by the Bank's laboratory to be later produced by some Brazilian firms (Sid and Digilab) where Bradesco held a minority stake.

As far as production of hardware and components was concerned Bradesco pursued this strategy of minority shareholding. In the beginning they participated in only three ventures (the above-mentioned two, plus COBRA, the government-controlled computer firm). In the early 1980s they acquired total control of Digilab and also used it as its holding company for IT investments (See Table 3.3).

Itaú, the other leading Brazilian bank, followed a different strategy. They decided to create their own totally controlled firm, Itautec, to develop and manufacture their systems, and also to supply other segments of the Brazilian market. The firm was created using technologies and human resources from the bank's data-processing centres. In the beginning it concentrated on the production of microcomputers and terminals for its own use. Since then Itaú's interests have increased and diversified into practically all areas of IT (Table 3.4).

Both of these two large financial conglomerates followed a very

Table 3.4 Composition of Itaú's interests in IT in Brazil

Firm	*Sub-sector of activity*	*Sales (US$ millions)*	
		1987	*1988*
Itautec	Computers and peripherals	114	158
Itaucom	Semiconductors	42	72
SESA[1]	Telecommunications	48	NA
Philco[2]	Consumer electronics	220	NA
Itaucam[3]	Components	NA	NA
Adiboard[4]	Components	—	—
Rima[5]	Printers	—	—

[1] Itaú has a 25 per cent stake in SESA.
[2] Itaú acquired Philco from Ford in December 1988.
[3] Itaucam produces components in the Free Production Zone at Manaus.
[4] Adiboard is a joint venture with Kollmorgen Co., of USA, to produce components from 1989, mainly to exports.
[5] This is an association of Itautec, Mecaf and Digilab (Bradesco Group) to produce printers.

Sources: Evans and Tigre (1989); data of 1987 sales from *Anuário Informática Hoje* 87/88 and of 1988 sales from *Exame Informática*, Vol. 4, No. 7, 12 July 1989

specific automation path. The other medium-sized banks, without the financial strength to pursue a similar strategy, decided to pool their efforts and created a firm – Tecnologia Bancária – with the aim of providing optimal scale for the large investments needed for banking automation. But instead of pursuing a strategy for hardware production, Tecnologia Bancária opted to concentrate on achieving a common pattern for automation of the medium-sized banks, buying similar products and establishing facilities (such as automatic teller machines) for joint use.

By that time there was fierce competition between the two leading financial conglomerates. Both saw technology and its rapid introduction to the financial system as a key to further growth. In the early 1980s, both banks were able to inaugurate their first fully automated branches.

The strategy of designing and developing the equipment according to need also proved important in minimising dependence upon a particular supplier. Itaú, for example, developed its own network management system, even though IBM was able to supply its CICS;[7] a world-wide success. Itaú's reasoning was that 'by developing their own network management system, communications control units and terminal concentrators, and using only IBM's central processing units, they would be in a position, if necessary, to acquire central processing units from other suppliers' (Dantas 1988:228).

Banks were able to free a significant share of their profits in order to implement such strategies. They have obtained high profits over the last twenty years as a result of concentration and government financial and monetary policies. It is worth mentioning that commercial banks in Brazil were in the past not traditionally directly involved in any productive investment but were known for timidity in financing risky long-run projects, even though Brazilian law does not impede moves in such directions.[8]

Environment-specific factors in banking automation

Five other related factors facilitated the speedy introduction of automation to the Brazilian banking system and its specific characteristics of producer of IT goods and R&D performer. These were: government IT policies, the availability of highly skilled human resources, the availability of a good telecommunications infrastructure and services, the lack of resistance of the labour force to automation and rapid changes in the banking legal regulations, some of which specifically aimed to promote automation.

The importance of government policies towards IT in Brazil has been widely discussed (see Erber 1985; Ernst 1986; Schmitz and Hewitt in this book). Here it suffices to say that in providing the market reserve for national firms precisely in market segments that were crucial to the automation process of the banking system – namely small computers and peripherals – the IT policy allowed time and the conditions for banks to invest in R&D capabilities.

The availability of highly skilled personnel was fundamental to banking automation. As a matter of fact, since the 1950s physics has been regarded as the most important scientific area in Brazil. In the beginning this was strongly linked to an idea of achieving scientific capability for research in nuclear energy but it also proved to be essential to R&D in computers. The Instituto Tecnológico da Aeronáutica, set up in 1947, one of the top engineering schools in Brazil and linked to the Air Force, was 'the training ground for an active generation of physicists and engineers, who later became principal actors in the sequence of events that enabled Brazil to become a major manufacturer of microcomputers' (Langer 1989). Also the University of São Paulo (through its physics and engineering departments), the Federal University of Rio de Janeiro, the Catholic University of Rio and later on the new University of Campinas provided highly qualified physicists and engineers and, through their research teams, participated in R&D efforts from their early stages (Adler 1986; Dantas 1988; Langer 1989). Many of these technicians either became entrepreneurs in the infant industry or were offered key positions in firms and data-processing laboratories of the main users. Thus, a minimum scientific infrastructure was available.

A further infrastructure requirement was telecommunications, not least because of the size of the country. Telecommunications services in Brazil are provided by EMBRATEL, a government-owned monopoly firm. The efficacy and success of Brazilian industrial and technological policies in telecommunications and the efficiency of the system is well documented (see for instance, Hobday 1990). The fact that EMBRATEL made available, at a reasonable cost, means of transmission such as TRANSDATA and the Package Communication Network (RENPAC) proved to be crucial for banking automation. Moreover, banks' headquarters can communicate with their agencies in any part of the country through the BRASISTAT satellite.[9]

Another facilitating factor was that bank workers did not obstruct automation, in spite of inevitable job losses. Although there is no comprehensive study of the impact on labour of the swift automation of the Brazilian banking system, some general features can be derived.

First of all, overall employment in the sector has not kept pace with the growth reflected in general performance indicators of the financial sector. T.A. Santos, President of the powerful Brazilian Federation of Banks, stated that 'between 1979 and 1985, employment in commercial banks grew 56 per cent . . . and . . . it must be underlined that, had it not been for the process of banking modernisation (automation), the number of jobs created would have been much higher' (Santos 1988:221). In fact, during the period 1981–85 the compound rate of growth of the product of the financial sector was close to 60 per cent in real terms (*Conjuntura Economica*, various issues).

A net increase in the number of jobs occurring during this period is related to the increasing diversification of banking activities. On the other hand, the number of employees per branch has fallen since the late 1970s suggesting a rationalisation process fostered by automation. There are also reports[10] of several divisions of the main banks, such as those related to paperwork, being simply scrapped with automation, with resulting losses of thousands of jobs.[11]

Lack of resistance from the unions to automation due to legislation which has clearly been biased against workers' participation in the decision process regarding automation was coupled with a feeling that modernisation with Brazilian technology was an important and legitimate goal both politically and economically.

This apparently has not been the case in commercial banks in more advanced economies. Even in the UK, where overall policies throughout the 1980s were also directed towards a reduction of labour union power, the introduction of new technologies in the banking system has been slower because of, among other things, some awareness (even if only small) of workers' reactions. Child *et al.* (1983) in their study of micro-electronics and the quality of employment in some service sectors in Britain found, for the banking system, that apart from some legal requirements which constrain the scope for automation, workers' response is noted and considered by the banks' senior management when automation decisions are to be undertaken.[12]

Finally, it is well known that regulatory policies can either stimulate or inhibit innovation and diffusion processes of new technologies (Beckler 1982). Brazilian banks have been continually subjected to changes in legal requirements specifically aimed to foster automation. In 1984, for instance, the Central Bank forced commercial banks to implement cheque magnetisation (Marques 1987). There are some claims that in most advanced countries legal requirements have, on the contrary, delayed the process of automation.[13] Moreover, as government in Brazil relies extensively on banks as intermediaries for its taxation and transfer activities, every citizen is a potential user of the system. Indeed, besides their traditional banking functions, banks are expected (by law and other regulations) to fulfil a range of tasks which increase the demand for data-processing equipment.[14]

Accelerating inflation further contributed to the drive towards automation. Banks perceived that inflation provided them with an additional source of revenue. As deposits did not pay interest, an efficient and rapid internal organisation of activities could mean the maximisation of revenues without outlay to the deposit holder.

All such factors explain the pace of banking automation in Brazil. Together with the peculiarities of a concentrated national banking system in a quasi-continental country they explain the need to pursue specific hardware production and technical solutions.

The suppliers of equipment

The purpose of this section is to show how the structure of hardware production for banking automation evolved in Brazil during the 1980s. It is argued that when the large banks (Itaú and Bradesco) decided to become directly involved in the production of equipment they rapidly dominated the market since their technical solutions were designed through user–producer relations. Independent equipment suppliers were set up during the second half of the decade, as almost all banks needed to become automated. These new ventures, 'spin-offs' of the first crop of banking automation firms, developed novel design strategies and increased competition in the area. At the end of the decade firms controlled by the major banks started a process of capital concentration in the IT sector as a whole through a series of acquisitions.

In 1979, the year that the decentralisation strategy of data-processing began, sales of Brazilian IT firms to banks amounted to US$95 million consisting mainly of minicomputers (US$48 million), data-entry terminals (US$41 million) and peripherals (US$6 million). Cobra, the pioneer state-owned computer firm, was responsible for the majority of sales: 1,628 of 1,638 data-entry terminals and 176 of 375 minicomputers.[15] Three other firms (Sid, Labo and Edisa) were responsible for supplying the remaining minicomputers.

As already mentioned, during the early 1980s automation as a competitive tool among banks meant automating several functions at branch level. Distributed and decentralised data processing was the key feature of banking automation in Brazil and such strategy increasingly demanded products designed to fulfil specific market needs.[16] For instance, distributed processing was achieved by installing minicomputers at branches and on-line electronic terminals in the bank teller. As these were rapidly introduced in 1980 and 1981[17] (when Itautec started to deliver its products), Cobra's share rapidly declined, since they continued to produce general purpose machines without attempting to fill the specificities of the market. More powerful microcomputers rapidly superseded minicomputers and together with financial terminals and concentrators,[18] both to be used at branch level, comprised the bulk of the market. Figure 3.1 shows that the Brazilian market for such products (i.e. excluding medium-sized computers and mainframes which are only supplied by foreign multinationals) rapidly grew in the 1980s from US$130 million in 1982 to US$243 million in 1985. The US$205 million of such banking automation equipment sold in 1984 is estimated to represent almost 7 per cent of the world market or 15 to 20 per cent of the non-US market for the same group of products.[19]

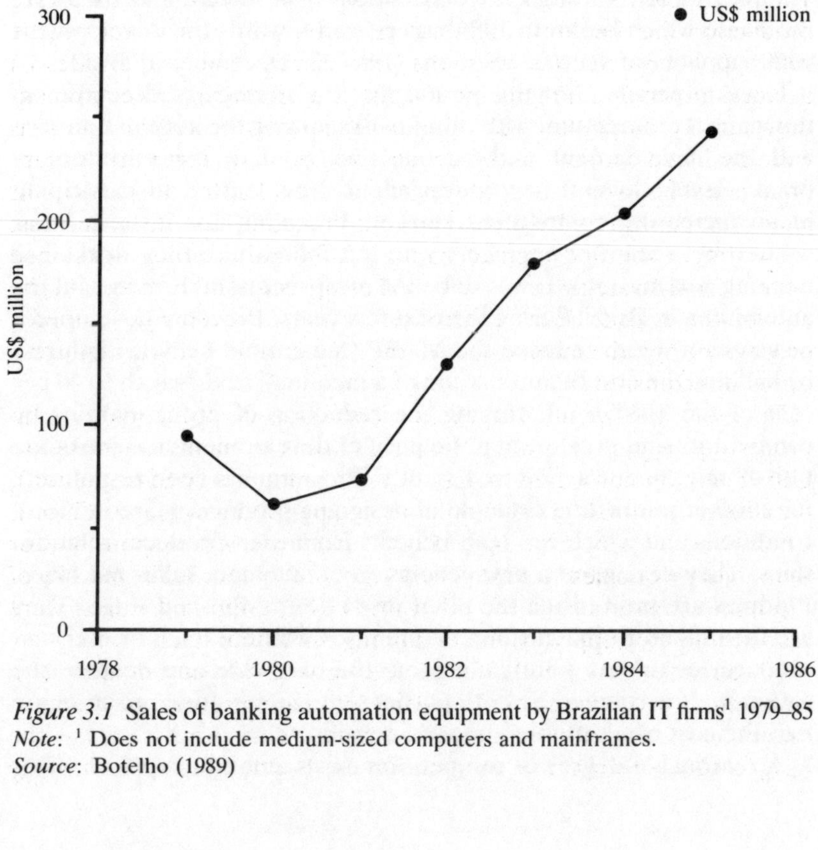

Figure 3.1 Sales of banking automation equipment by Brazilian IT firms[1] 1979–85
Note: [1] Does not include medium-sized computers and mainframes.
Source: Botelho (1989)

An interesting example of designing products according to needs occurred when Bradesco decided to install automatic teller machines in 1984. After importing eighty-five machines, Bradesco felt these to be expensive and inappropriate for its automation plans since they operated 'off line'. So, jointly with an affiliate (Digilab) they designed a simpler and cheaper machine which could work 'on line' (*Dados e Idéias*, April 1984:18–19).

The firms with stronger links to the banking system, such as Itautec and Sid, rapidly dominated the market and later on diversified into other IT areas. Banking automation products were responsible for 100 per cent of Itautec's sales during 1981 and 1982. Since then the proportion has fallen consistently to 36 per cent in 1987, when it was producing over 150 hardware and software products (Botelho 1989). Sid earned nearly 90 per cent of its revenue from banking automation in the early 1980s. By 1987 this market accounted for only 23 per cent of its sales.

After a period (up to 1985) when banking automation in Brazil was achieved mostly through the acquisition of hardware and software, the phase which began in 1986 was geared towards the development with suppliers of specific solutions (*Info* 83, December 1989:32).

It was precisely after this period that the strategy of decentralisation gained momentum, with other banks joining the automation race and the development and introduction of data concentrators at branch level. Several new independent firms started to participate in an increasing competitive market. Procomp, for instance, was created by a team of engineers who left Sid (where they worked in banking automation). It was to be the most successful firm in banking automation in Brazil during the next few years. Procomp developed a new system which reduced the MTBF (mean time between failures) by half and the cost of automation of a medium-sized branch to 40 per cent of the 1985 level, forcing the reduction of profit margins by competitors and accelerating the pace of development and introduction of new products.[20] More recently, Procomp has been responsible for another interesting example of designing products geared to local conditions and which can reap benefits from user–producer relationships. They developed a new generation of automatic teller machines. Products are built inside the plant up to a semi-finished stage. They are then taken to the customers' premises where a team of user and producer engineers jointly complete the hardware and develop the software. It is claimed by both parties that the resulting products are half the cost of alternative imported products.[21]

A reasonable degree of competition exists amongst suppliers. This suggests that, even if the move towards automation was conducted by the two main banks, learning and spin-off effects allowed other suppliers to enter into the competition (Table 3.5).[22] The pioneer hardware producers which belong to the main banks (Itautec and Sid/Digilab) saw their combined share of the market for concentrators and bank terminals decrease respectively from 72 per cent and 62.3 per cent in 1984 to 21.3 per cent and 47.8 per cent in 1988.

Table 3.5 Sales of concentrators and bank terminals by firm (%), 1984–8*

Firm	*Concentrators*					*Bank terminals*				
	1984	*1985*	*1986*	*1987*	*1988*	*1984*	*1985*	*1986*	*1987*	*1988*
Itautec	33.9	38.4	29.8	16.1	9.6	22.4	24.6	36.8	16.1	21.0
Sid/Digilab	38.1	34.7	21.1	19.3	11.7	39.9	41.5	30.7	43.5	27.1
Procomp	—	0.1	14.9	26.5	38.2	—	0.1	7.2	15.3	35.3
Digirede	28.0	24.5	27.4	11.5	24.8	28.0	24.0	18.1	NA	NA
Others	0.1	0.5	6.8	26.6	15.7	9.7	9.8	7.2	25.1	16.6

* The 1984 figures refer to the total of machines sold up to that year.

Source: SEI, *Panorama da Indústria de Informática*, August 1988 and *Séries Estatïsticas*, No. 1, August 1989

However, some recent developments show that both main banks have broadened their interests in the IT sector in Brazil, diversifying their investments and starting to promote a process of concentration in the area.

Itautec, for instance, set up a joint venture with IBM for the production of super-minicomputers. Itaú acquired Philco, a leading producer of consumer electronics, with the objective of securing a market with sufficient scale for its components production at Itaucom and Itaucam and of gaining learning economies.

In fact, one of the main criticisms of the development of the Brazilian IT industry is precisely the fact that unlike Japan and South Korea it has been growing without any forward linkage with the scale-important sector of consumer electronics (Evans and Tigre 1989:28). As they emphasise, this move towards consumer electronics 'cannot help but strengthen the prowess of Brazilian computer firms'.

Bradesco, through CPM, one of its subsidiaries, entered into the important market of mainframes via an agreement with the Japanese firm, Hitachi. They started producing the computer 80XX which is compatible with IBM 3090 but is more compact and consumes less energy, being air-refrigerated.[23] Bradesco also acquired 70 per cent of the shares of Scopus, one of the leading producers of micro-computers and terminals. Together with Digilab they will be the second largest Brazilian computer producer, after Itautec and ahead of Cobra, the pioneer, government-owned Brazilian computer producer. Sid, a firm in which Bradesco has a minority holding, set up a joint venture with IBM to produce personal computers. Finally, Itaú and Bradesco (through Digilab) have recently entered a joint venture (Rima Impressoras) with a small Brazilian fine mechanics firm (Mecaf) to produce high-performance printers. Until now this area has been considered to be one of the main bottlenecks of the Brazilian computer sector. In fact, these moves are part and parcel of recent changes which suggest that through mergers, acquisition programs and strategic alliances, larger corporations, particularly the ones linked to the banking sector, have increased their influence in the IT sector as a whole (Banhos 1989).

Summing up user–producer relations was crucial to the successful design and development of banking automation systems. Probably because such relations are behind a rapid rate of diffusion of IT in the sector, competition increased with the setting up of new ventures which 'spun off' from the old firms (the subsidiaries of the big banks). At the same time, the financial and technical strengths of these banks are promoting capital concentration in the whole IT sector and making possible very positive arrangements with foreign multinationals.

Results of the banking automation process

Some of the results of the automation strategies of the Brazilian banking system can be seen in Table 3.6. The five largest private

Table 3.6 Fully automated branches and suppliers of equipment for the 8 largest (in amount deposited) Brazilian commercial banks, 1988

Banks	*Number of branches*	*Automated branches*	*Percentage automated*	*Suppliers of equipment*
Private banks				
Bradesco	1,721	1,539	89.4	Digilab (Bradesco group), Sid (Sharp group)[1]
Itaú	1,680	920	54.8	Itautec (Itaú group)
Bamerindus	863	580	67.2	Digirede and Procomp
Unibanco	606	270	44.6	Digirede and Procomp
Real	545	35	6.4	Procomp and Sid
State banks				
B. do Brasil	4,665	290	6.2	Itautec, Sid and Procomp
Banespa	591	350	59.2	Sid, Digirede and Procomp

[1] *Bradesco has a minority stake in Sid.*

Source: *Veja*, 5 April 1989

banks have a total of 5,415 branches of which 3,344 were fully automated by 1988. Even public banks such as Banespa, the State of São Paulo's bank, has 59.2 per cent of its agencies fully automated and Banco do Brasil is slowly recovering because it had to restart its automation process after recognising the failure of a system it developed in the early 1980s.

In 1988, the twenty largest Brazilian banks had installed 133 mainframes, 1,454 minicomputers and 11,501 microcomputers. Bradesco, the main private bank had twenty-nine mainframes, with a total memory capacity of 660 Mbytes, coordinating a network which includes 552 minicomputers, 1,186 microcomputers and 41,000 teller terminals (Table 3.7 and *Dados e Idéias*, Nov./Dec. 1988:62–4).

The evolution of banking automation in Brazil, by type of equipment is shown in Table 3.8. In 1987, the installed base of four types of equipment – concentrators, bank terminals, automatic teller machines and cash dispensers – was estimated to value US$675

Table 3.7 Main Brazilian banks: number of mainframes, minicomputers and microcomputers, 1988

Banks	*Mainframes*	*Minicomputers*	*Microcomputers*
Private	101	876	7,307
Bradesco	29	552	1,186
Itaú	22	—	3,300
Citibank	5	2	1,030
Unibanco	4	—	807
Real	5	42	66
Bamerindus	9	48	316
Nacional	7	4	59
Econômico	5	55	210
Frances e Brasileiro	2	47	160
Safra	3	58	99
Sudameris	3	—	38
Chase Manhattan	2	64	167
Montreal Bank	1	1	39
Noroeste	4	3	110
Government	32	578	4,194
Banco do Brasil	15	323	2,167
Banco Central	4	7	772
Caixa Econômica Federal	2	92	645
Banespa	2	—	470
Banerj	4	118	58
Banestado (PR)	5	38	82
Total	133	1454	11,501

Source: *Dados e Idéias*, Nov./Dec. 1988:62–8

million. This represented 129.2 per cent more than the same figures for 1984, and an average rate of growth of approximately 32 per cent per year. There were more than 106,000 bank terminals in Brazil in 1987 which were valued at US$512 million. This compares with a projected figure of 343,000 in 1990 for the whole of the European market (Revell 1983).

But equipment such as automatic teller machines and cash dispensers is introduced less often (even if the rate of growth in recent years has been quite high), suggesting that over-the-counter automation was regarded as less important in Brazil during this phase of banking automation. Apparently the introduction of such equipment was deemed less crucial as it would represent a substitution of machines for cashiers. Furthermore, wages for this kind of work in Brazil are very low so there appeared to be less pressure for such substitution. In Europe, for instance, where cashiers earn much higher wages, the introduction of such standard machines has been happening more rapidly since the early 1980s (Revell 1983).[24] Another possible explanation could be that the widespread (and safer) use and acceptance of guaranteed cheques in Brazil did not press banks to introduce such machines quickly.

Table 3.8 Banking automation in Brazil by type of equipment, 1984–7 (number of installed pieces of equipment)*

Type	*1984*	*1985*	*1986*	*1987*	*1988*
Concentrator	3,160	5,265	6,563	7,427	11,276
Bank terminal	47,096	80,284	93,123	106,328	126,244
ATM	NA	NA	NA	862	NA
Cash dispenser	—	—	NA	356	1,168

* *The figures for CPUs are significantly underestimated because they do not include Cobra which did not provide SEI with detailed figures.*

Source: SEI, *Parque Brasileiro de Informática*, Vol. 1, No. 1, August 1988 and Vol. 2, No. 1, August 1989

In Brazil such equipment was supplied by nine different Brazilian-owned firms using locally designed technology[25] with only one multinational (UNISYS) supplying a very small proportion (3.4 per cent) of automatic teller machines in 1987.

Perhaps the clearest indicator of the success of the ‘on line’ strategy is that any Bradesco and Itaú cheque deposited anywhere in the country is cleared in less than twenty-four hours.[26] Such speed is rarely found, even in advanced countries. Itaú claims that through a network of 104 data-processing centres set up all over the country[27],

90 per cent of all its transactions are electronically performed in real time (Itaú, *Annual Report*, 1988).

Achieving technological capabilities through such a process of innovation and diffusion fostered by user needs has also proved to be cost-effective (Erber 1988). It is true that in spite of its financial clout and extensive technological experience, the Brazilian IT industry geared to banking automation has not penetrated the external market sufficiently because the internal market has been large enough and the replacement market is already becoming important. It is certainly true that external markets were not considered strategic during the previous phases of automation but this does not mean that they are not competitive. For instance, Brazilian firms have successfully competed in international bids in Portugal against more established European firms. According to interviews, access to the external market is the next issue on the agenda of banking automation producers. Previous attempts showed that more sophisticated markets were difficult to penetrate for marketing reasons rather than technological ones. This is probably why some Brazilian firms are trying to overcome these problems by acquiring small hi-tech firms in Europe.[28]

Such problems notwithstanding, some results can be registered in terms of exports. In 1986, 1987 and 1988 exports of IT products of Brazilian national firms were respectively US$6 million, US$8 million and US$17 million (SEI 1989). These comparatively low figures suggest, however, that Brazilian products are gradually penetrating foreign markets. It is also worth emphasising that it is precisely firms (such as Itautec, Digilab, Sid and Elebra) and/or products related to the banking automation sector that are gaining access to foreign markets (SEI 1989:35). For instance, Itautec started to export in 1987 and has made some inroads into the European market. It has sold systems for banking and supermarket automation in Portugal. Itaucom, its micro-electronics subsidiary, started to export components in 1988 (SEI 1989). Sid has recently created a joint venture (Management and Technology for Banks) with three Hungarian firms in order to explore the Hungarian banking automation market. It is also automating the Moscow branch of the USSR's Industrial Bank as a first step to fully automating a 400-branch bank (Zaparolli 1989).[29] Some Latin American countries, such as Argentina, have also imported banking automation goods which form the bulk of Brazilian IT exports to countries such as India, China and Poland (Banhos 1989).

Achieving a degree of competitiveness in other segments of the Brazilian market (as in commerce and services automation) and in foreign markets where there is both more openness to international competition and more similarity to Brazilian markets (such as Portugal and other Latin American countries) would appear to be closely linked to the accumulation of intangibles during the innovation-diffusion process and the acquisition of externalities which are clearly location-specific.

Contrasting experiences

It is worth making some international comparisons here. The only country, other than Brazil, where user–producer relations played an important role during the process of banking automation is Japan. In their study of the introduction of IT to Japanese banks, Baba and Takai (1990) claimed that successful back-office automation (accompanied by organisational restructuring) was achieved through substantial collaboration with external organisations, such as equipment suppliers. Japanese computer vendors provided their banks with bank-specific systems engineers who remained inside the banks for a considerable period. Banks also collaborated with affiliated software houses to develop specific software and to maintain the systems. The overall result of such user–producer relationships is the development of an 'overall structure emerging from well-devised collaborations with external organizations allowing for the introduction of highly specific systems' (Baba and Takai 1990:7).

In contrast, other experiences such as in the United States and the United Kingdom, point to a very different situation. In the US, banks have been investing heavily in computer systems. The IT share of total capital stock in American banks has risen from 1.7 per cent in 1970 to 40.4 per cent in 1988 (Roach 1989). Recent estimates claim that US banks are investing approximately US$10 billion a year in such systems (*The Economist* 1989). Even if a comprehensive evaluation of American banks' IT strategy is not available, some evidence points to potential problems. Two series of papers on technology and the US services industry (Guile and Quinn 1988a and 1988b) conclude that such huge investments by American financial institutions are not paying off and productivity performance in these sectors has been very inadequate. It is suggested that the technology's disappointing productivity payback in the American services sector has been caused by a rather slow and deficient progress in programming and software applications (Roach 1988:136). A recent survey on international banking published by *The Economist* cites some cases where the systems bought could not satisfy the banks' requirements.

Citicorp, for instance, which spends US$1.5 billion a year on computer systems – more than any other bank – acquired the Trade and Analysis and Processing system from Morgan Stanley. After struggling for some time to make the system work Citicorp dropped it and wrote off almost US$50 million (*The Economist* 1989:33). This and other examples led Baba and Takai (1990), when comparing Japanese and US experiences, to conclude that banking automation in Japan is more efficient. The malfunction currently observed in US banks arises because 'the computer systems they had bought could not do the job they were required to do' (*The Economist* 1989:33–4).

Smith and Wield (1988) produced similar evidence for the UK. They point out that the users, British clearing banks, and equipment suppliers, have completely different views on the direction technological innovation should take. In some cases, suppliers 'attempted to get the banks to use more complicated systems involving more complex interactive systems such as those which are in place in the United States where customer behaviour and expectations differ' (Smith and Wield 1988:266). For example, Midland Bank which tried to introduce Hogan Systems' corporate banking accounting system for its retail banking eventually concluded that the system was not designed for a bank with 2,000 branches dealing with a multitude of small transactions (*The Economist* 1989). This is probably why there has been a sense of disappointment and surprise among senior bankers in the UK over new technology (Smith and Wield 1987). As Arthur (1985) pointed out, 'some systems show a lack of fit with branch banking and that problems then result from organizational change around the kit. IT as an "organizational technology" has proved inappropriate and if bankers are to blame then it is for accepting suppliers' productivity claims and design values.'

In fact, as Barras (1986:165) said, based on a survey of applications of computers in some service sectors: 'the suppliers are in a powerful position to influence the rate of adoption and application of the technology in user industries through their development and marketing strategies ... [and users are] ... at some stages implementing "leading edge" applications which quite often prove costly and inefficient'.

These international experiences help to underscore Brazil's achievements in banking automation. Even critics of Brazilian IT policies acknowledge that banking automation systems developed in Brazil are less expensive than similar foreign alternatives and that equipment such as banking terminals, ATMs and cash dispensers are the technological equivalent of those available on the international market (Frischtak 1989). Since they consider other segments of the Brazilian computer industry to be cost-inefficient, their policy prescription is that Brazil should specialise in the design of banking automation systems and source their hardware components internationally (Frischtak 1989). Such a proposal is contradictory to one of the main ideas developed here: namely that the capability to design and implement banking automation systems in Brazil has been structurally linked to the capability to technically understand, design and produce hardware and software for banking automation. The comparative efficiency achieved in banking automation in Brazil was part and parcel of a broader process of technological accumulation where the most important factor was a close relationship between the user and the producer of equipment. By trying to evaluate price competitiveness of individual equipment with different technical specifications analyses (such as Cline 1987 and Frischtak 1986, 1989) they fail to take into account that in the absence of user–producer contacts, the user will face significant after-sales costs which may render the cheaper equipment a burden.

A counterfactual

One question that remains is whether, in the absence of policies which fostered national IT firms, foreign firms would have provided the same sort of products tailored to local requirements. There are two possibilities in principle: foreign direct investment and licensing, by national firms, of foreign technology which could be adapted to local conditions later on.

There is little evidence that new investment by multinational subsidiaries would have occurred in hi-tech sectors. Investment location decisions in the new technological paradigm are strongly influenced by producer–user (both sophisticated, technically speaking) relations. This gives a premium to the proximity of a high-income (and sufficiently large) market which is more likely to pay off the costly investments.[30]

As a consequence, a new pattern of international investment occurred in the 1980s. This can be characterised as a small decline, early in the decade, followed by a severe reduction in the flow of foreign direct investment towards developing countries (with the single exception of China up to early 1989) and the reconcentration of the flow of foreign direct investment within the OECD area (Chesnais 1988). In fact, the share of developing countries in total foreign direct-investment flows from developed market economics has fallen consistently throughout the last decade from a peak in 1975 of almost 42 per cent to less than 17 per cent in 1986 suggesting a new trend in patterns of FDI in the Third World (Lütkenhorst 1988).

The same reduction has been observed in Brazil since 1982. New net foreign investment dropped throughout the period from almost US$2 billion in 1982 and was even negative in 1986 (Banco Central do Brasil, *Annual Report*, various issues).[31] It is plausible to assume that such a pattern was connected to changes in the economics of the location of manufacturing activities related to new technologies (Kaplinsky 1988).

In the absence of foreign direct investment, national firms could have tried to license technology. In fact, licensing has been one of the main channels used by Brazilian firms to gain access to foreign technology.[32]

The few examples of licensing attempts for hi-tech to Brazilian firms show that before the establishment of Brazilian policies for IT which constituted internal market protection, foreign firms were very reluctant to enter such agreements.[33] Only after policies were maintained for some time, did licensing of foreign technology in some areas prove to be feasible. As Tigre (1983) showed in his detailed study of the Brazilian industry, it was precisely because of the informatics policy of market protection that foreign firms began to enter into technology licence agreements with local firms.

But in general, foreign firms tend not to favour licence agreements of hi-tech to Third World countries both because the internal market potential is too low and also because such sensitive technologies have to be closely protected (OECD 1988 and Mytelka 1987). Tigre (1983:146) found that 'none of the firms which licensed Brazilian computer firms has technology agreements in other developing countries'. Moreover, the specificities of developments in hi-tech require some form of investment in fulfilling local needs which multinational firms try to avoid unless required to by competition.

In order to underline the argument put forward in this section it is worth pointing out that banking automation proceeded much faster in Brazil than in any other Latin American country where local policies were not implemented and multinational subsidiaries were able to sell their products in a relatively unconstrained fashion.

4 CONCLUSIONS

It is recognised that technology policies in countries such as Brazil have tended to focus much more on the generation of technology than on its diffusion. In trying to demonstrate that both processes are simultaneously determined in the new techno-economic paradigm this chapter strongly argues for policies which consider both aspects.

The example of banking automation in Brazil provides evidence that success was achieved precisely because, even in the absence of explicit policies which took into account user needs, other implicit policies permitted the banking system in Brazil to pursue a strategy which incorporated such linkages. Moreover, such policies provided opportunities for highly skilled technical people to work in R&D activities instead of becoming 'specialised' salespeople for multinational subsidiaries (Tigre 1983; Piragibe *et al.* 1985).

In a sector such as IT, where a move to a more mature stage has already happened internationally, there is nevertheless room for policies to foster producer–user relationships. The recent trend in the development of the computer industry towards smaller systems[34] means in fact a move in the direction of open systems, with barriers to entry being lowered and increased competition occurring as a result. Such open systems may allow space for countries like Brazil to explore technology paths where specific knowledge plays a significant role.

This chapter has argued that as the innovation process of new technologies is not independent but directly linked to the diffusion process its outcome is specific to the surrounding environment. In participating in the innovation/diffusion process since its early stages, a Third World economy can generate externalities which are unlikely to emerge without government intervention (Schmitz and Hewitt, Chapter 2 of this book).

It is widely recognised that technological and technical change within firms are cumulative processes (Pavitt 1988). Once this is perceived, technological trajectories cease to be random and what firms can hope to do in the future is heavily constrained by what they have been capable of doing in the past (Dosi 1988). This cumulative and self-generating aspect of technology is what makes it develop in particular directions (Rosenberg 1976:110–11). What is not sufficiently recognised is that for this process to come about in environments experiencing high degrees of specificity, users' knowledge may play an important role in the diffusion-innovation process.

Knowledge of user requirements has increased in importance as new information technologies and related organisational innovations reshape the basic parameters of designing, producing and marketing industrial goods and services. At the same time, constraints on the demand side rather than in the economics of production have become the main barrier to productivity enhancement (Ernst and O'Connor 1989). In other words diffusion of new technologies is being hampered by failing to fulfil users' requirements. Put another way, without sophisticated users who can conceptualise a need and translate it into a requirement for products and services, there will be scarcely sufficient pressure to improve producers' technological capabilities.

To be an efficient user of such technologies a country must possess a degree of familiarity with the design and production of hardware, components and the software which determine the performance of various information-based systems. The country must also know about the interface between IT and other technologies, in particular mechanical technologies. Learning-by-doing and learning-by-using are to a large extent complementary activities. The interactions among hardware, software and system suppliers are an important element of the learning process. Without a supplier sector, not only is one agent missing, but the synergies arising from the user–producer interaction are lost as well.

The example of the banking automation process in Brazil shows how – even in the context of a developing country – new firms in the new techno-economic paradigm can be in a position to pursue a very specific technological path when the surrounding environment does not act as a counterpoise. It has also shown how a series of intangible assets, crucial to performance and competitiveness, have been accumulated. These intangibles which are obviously difficult to quantify are also location-specific since they are related to local technological expertise and also to marketing or market feedback (Teubal 1984).

Finally, two points are worth making. First, success does not mean that everything achieved is desirable. It has already been mentioned that windfall profits obtained by Brazilian banks since the mid-1960s provided the financial means to invest heavily in IT. Even if a detailed account of such processes is beyond the scope of this chapter, it has to be recognised that policies which favoured concentration in the banking system and which allowed banks to make huge profits have had detrimental repercussions in other areas of the economy. And, the increased power of the banks may lead to problems in the manageability of the Brazilian economy in the future.

It must also be said that the tradition of Brazilian commercial banks not to engage in productive activities was broken when they made heavy investments in IT. After that, banks significantly changed these attitudes and are now playing a more positive role in channelling financial resources for production.

Second, the financial weakness of the Brazilian IT firms associated with growing R&D and production expenditures is a problem frequently raised in literature about the Brazilian experience (Erber 1985; Piragibe 1985). Entry of large financial conglomerates in the area may provide the required financial strength, synergy and scope economies since they are already involved in various segments of the electronic complex. There is also a possibility of some type of spillover to other economic sectors since one of the main macro-economic problems of the Brazilian economy is the pattern of financing, particularly investment financing. In this sense, the experience in banking automation can be a turning-point.

NOTES

1 This chapter has benefited from criticisms and suggestions from a number of people. Hubert Schmitz and Helena Lastres have contributed extensively through constant stimulus and detailed comments on earlier drafts. Carlota Perez, Chris Freeman, Dieter Ernst and Norman Clark have made very useful contributions. The same applies to the members of the Brazil Hi-Tech group at IDS, and participants of the Workshop 'Hi-tech for industrial development', IDS, University of Sussex, June 1990, whom I thank collectively. Obviously I am solely responsible.

2 It seems that even in countries such as the UK, large international firms supplying commodity-type products apparently perform very poorly in terms of reliability and after-sales servicing. A report ('What to buy for business') comparing the performance of nearly 200 PCs placed large multinationals such as IBM and Olivetti at the lower end of the scale for reliability and service and smaller firms (such as Dell, Tandon and Apricot) manufacturing desk-top PCs and Japanese firms (such as Toshiba and Sharp) manufacturing lap-top PCs at the top end. (*Financial Times*, 4 July 1989:31).

3 In Pavitt's (1984) terminology.

4 Bradesco acquired an IBM 1401 in 1961; in 1962 Banco Nacional bought a Burroughs B205; Itaú also acquired an IBM 1401 in 1965 and in 1968 Bamerindus started to use a UNIVAC 1005 (Dantas 1988).

5 As a comparison, in 1984 the number of branches of all British clearing banks was 13,000 (Heffernan 1984:39).

6 This is an example of how accumulated knowledge and experience can often act as a barrier to change (Georghiou *et al.* 1986) and how subsidiaries of multinational firms have difficulties in dealing with some problems of distant markets. In fact, even in advanced countries multi-national firms behave in the same way when markets are not considered big enough. For instance, Glaser (1988) showed how in the early 1970s Citicorp failed to raise the interest of major computer suppliers in developing the technology necessary to deliver on-line interactive transaction services.

7 Customer Information Control System, a software which monitors and controls communications between machines.
8 Itaú has broadened its range of activities since the late 1970s. Nowadays it controls firms in sectors such as chemicals, paper and pulp, wood, etc. (Itaú SA, *Annual Report*, 1988).
9 Lack of investment by EMBRATEL during the 1980s due to the government budget crisis meant an undercapacity of the telecommunications sector in the last few years. That is why Bradesco decided to launch its own satellite in 1989 and other banks are considering ways of following a similar path (*Exame Informática*, Vol. 4, No. 4, 5 April 1989:7 and *Info* 83, December 1989:28)
10 Personal information given by some systems analysts who were involved in the design of software for the automation process in one of the leading private banks during 1985.
11 This was peculiar to the situation before 1985 when, under the authoritarian regime, private firms in Brazil had a free hand in dealing with such issues. Even after 1985, as the legislation regarding jobs remained the same, there is at least one documented case when, as the Cruzado Plan of 1986 led to a sharp profit squeeze for the financial sector, at least 40,000 jobs were lost in banks (personal communication from representatives of the Brazilian Confederation of Workers in the Banking Sector).
12 Even if this is done under a very paternalistic approach which involves much more consultation than participation of the workers to be affected.
13 In Britain, for instance, an old legal requirement obliging individual entries and signatures to be verified before clearance sets definite limits and constraints to the process of bank automation (Child *et al.* 1983). As such legislation does not exist in Brazil, banks totally guarantee cheques to a certain limit and pay them even if the signature does not correspond to the holder's. They claim that it is less expensive to pay for cheques with adulterated signatures than to go through the slow process of checking every individual cheque.
14 By the mid-1980s, banks were annually processing approximately 9 million income tax statements, 11 million other tax transactions, 250 million social security payments (paid monthly), 850 million utility bills (also paid monthly) and so on (Frischtak 1989).
15 In 1979 banking automation revenues represented more than 60 per cent of all Cobra's sales (Botelho 1989).
16 In the words of A. Screiber, who was at the time consultant to one of the bank's data-processing centres and who later created Digirede, banking automation in the US and Europe was made through centralised systems. Abroad 'nobody wanted even to listen to distributed processing' (quoted in Dantas 1989:19–20).
17 One of the well known cases is of Digirede, a firm created with the financial support of Banorte to initially produce a minicomputer especially designed to operate at banks' branches with two 8–bit microprocessors Z-80, 320 Kbytes of memory capacity and a unity of flexible disc with a diameter of 20 centimetres (Dantas 1989).
18 A concentrator is a system consisting of a dedicated central processing unit which centralises and controls various teller and other terminals, printers and other peripherals at branch or regional level, maximising the

use of such equipment and selecting data to be stored at branch level and data to be sent to the main data-processing centre. This proved to be crucial for the decentralisation of banking automation in Brazil, particularly because costs of transporting data in a quasi-continent-size country such as Brazil are relatively expensive.

19 In 1984 sales of computers to the banking sector were US$5.54 billion world-wide. Approximately 45 per cent of this amount (US$2.49 billion) related to sales of mainframes with the remaining US$3.05 billion for other categories. The non-US market for the same categories (i.e. excluding mainframes) was estimated to be in the range of US$1.2–1.5 billion (Market Intelligence Research Company, cited in *Financial Times* 1989).

20 Between 1986 and 1988 the number of terminals installed increased fourfold and the number of automated banking points more than doubled (Botelho 1989).

21 Interviews with producers and users in May 1990 in Brazil.

22 Spin-offs have happened both in hardware and software. In 1985, for instance, a group of technicians from Sid (where Bradesco has a minority shareholding) created their own firm, Procomp, which quickly acquired a solid reputation in producing technologically advanced and low-cost cash terminals and dispensers and is now supplying such products to medium-sized banks. A lot of software houses have also been set up catering for some market niches (*Exame Informática*, Vol. 4, No. 4, 5 April 1989).

23 CPM is in a very healthy financial position, having achieved the highest rate of real growth – 249.2 per cent – of the 50 biggest Brazilian IT firms in 1988 (*Exame Informática*, Vol. 4, No. 7, 1989).

24 With an ATM the cost of supplying money in the UK is reduced from 30 pence to 5 pence, per operation (Child *et al.* 1983).

25 Cobra, Digilab, Digirede, Edisa, Itautec, Procomp, Racimec, Sid and Zanthus.

26 If cheques are deposited in agencies of the same bank (either Itau or Bradesco) they are cleared immediately since internal operations are 'on line' and terminals at the cashier's desk can check the signature and balance (*Exame Informática*, Vol. 4, No. 4, 5 April 1989). In fact recent evidence suggests that other banks in Brazil can provide the same service (interviews conducted in May 1990 with senior bank managers).

27 Comprising thirty-four mainframes and medium-sized computers and 8,500 teleprocessing lines.

28 A recent example is the acquisition by Sistema, the joint partner of Itau and Bradesco in Rima, of the West German IT firm Robotron in order to have better access to the European market and to keep its technology updated (information provided during interviews in May 1990 in Brazil and *Exame Informática*, Vol. 5, No. 5, May 1990).

29 It is interesting that the reasons claimed by the Soviets as to why they have chosen Brazilian firms to automate their banks are: the continent size of both countries; the existence in both countries of national, multi-branch banks and not (as in the US) of regional banks with a small number of branches (Zaparoli 1989).

30 For instance, Lütkenhorst (1988) demonstrated the importance of the absence of a sufficiently elaborate local engineering network which rendered local subcontracting and the provision of key industrial services

extremely difficult in Sri Lanka, in explaining the decision by a UK firm producing non-electrical machinery items to relocate back to the UK a production unit which was established in that country in the early 1980s.

31 There have been some attempts to interpret such patterns and which hold Brazilian policies towards FDI as responsible for it. They argue that Brazilian legislation is very restrictive and biased against foreign capital, a fact that would hinder MNCs from investing. This assumption is in fact totally contradicted by more serious in-depth analysis. Robinson (1976) has compared national policies towards foreign capital, investment and technology for fifteen Third World countries and concluded that Brazil presented the lowest level of restrictions. Apart from control towards technology inflows and investments in areas related to national security, entry of foreign capital is free with only small differences of treatment remaining for some minor subsidies. More recent analysis has confirmed that 'there are no sound arguments to support the view that the decline or stagnation of foreign investment [in Latin America] is caused by the inadequacy of national policies' (White 1988).

32 In Brazil where a relatively free entry of foreign capital exists even large firms totally owned by nationals tend to follow a strategy towards technical change that is more a reaction to multinationals' behaviour than a purely independent one. Elsewhere I have shown that one of the most important reasons of Brazilian private firms to rely more on licensing than on internal R&D investment is that there is always a possibility of a MNC subsidiary starting up internal production – for the same product line – and capturing markets already tested by local firms (Fung and Cassiolato 1976).

33 Dantas (1988) shows in detail how firms such as IBM, Data General and Digital refuse to license technology to Brazilian firms.

34 Coupled with the advent of UNIX and the growth of networks.

BIBLIOGRAPHY

ABICOMP (1988) *Catálogo da Industria Brasileira de Informática*, Rio de Janeiro: ABICOMP.

Adler, E. (1986) 'Ideological guerrillas and the quest for technological autonomy: development of a domestic computer industry in Brazil', *International Organisation*, Vol. 40, No. 3.

Arthur, C. (1985) 'Bankers begin to doubt technology', *Computer Weekly*, No. 961, 2 May.

Baba, Y. and Takai, S. (1990) 'Information technology introduction in the big banks the case of Japan', in C. Freeman, and L. Soete (eds) *New Explorations in the Economics of Technical Change*, London: Pinter Publishers.

Banco Central do Brasil, *Annual Report*, various issues.

Banhos, S. (1989) 'Corporate strategy in the computer industry in Brazil', M.Sc. dissertation, Brighton, Science Policy Research Unit, University of Sussex.

Barras, R. (1986) 'Towards a theory of innovation in services', *Research Policy*, Vol. 15, 161–73.

Beckler, D. (1982) 'The international dimension of innovations', in H. Miller and R. Piekarz (eds) *Technology, International Economics and Public Policy*, Boulder: Westview Press.

Botelho, A. (1989) 'Financial markets, State policy and the development of the Brazilian banking automation industry', paper presented at the OECD Workshop *Technological Change and the Electronics Sector – Perspectives and Policy Options for Newly Industrializing Countries*, Paris: OECD.

Chesnais, F. (1988) 'Multinational enterprises and the international diffusion of technology', in G. Dosi *et al.* (eds) *Technical Change and Economic Theory*, London: Pinter Publishers.

Child, J., Loveridge, R., Harvey, J. and Spencer, A. (1983) 'Microelectronics and the quality of employment in services', paper presented at the annual meeting of the British Association for the Advancement of Science, Brighton.

Clark, N. (1987) 'Similarities and differences between scientific and technological paradigms', *Futures*, February.

Clark, N. and Juma, C. (1987) *Long Run Economics, an Evolutionary Approach to Economic Growth*, London: Pinter Publishers.

Cline, W. (1987) *Informatics and Development: Trade and Industrial Policy in Argentina, Brazil and Mexico*, Washington, DC: Economics International.

Dados e Idéias, various issues.

Dantas, M. (1989) *O Crime de Prometeu: Como o Brasil Obteve a Tecnologia da Informática*, Rio de Janeiro: Abicomp.

Dantas, V. (1988) *Guerrilha Tecnológica: a Verdadeira História da Política Nacional de Informática*, Rio de Janeiro: LTC Editora.

Dosi, G. (1982) 'Technological paradigms and technological trajectories', *Research Policy*, Vol. 11 No. 3.

—— (1988) 'The nature of the innovative process', in G. Dosi *et al.* (eds) *Technical Change and Economic Theory*, London: Pinter Publishers.

Erber, F. (1985) 'The development of the "electronics complex" and government policies in Brazil', *World Development*, Vol. 13, No. 3.

—— (1988) 'A transformação dos regimes de regulação: desenvolvimento tecnológico e intervenção do estado nos países industrializados e no Brasil', thesis presented for the chair of Full Professor at the Economics and Administration College, Federal University of Rio de Janeiro, Rio de Janeiro.

Ernst, D. (1986) 'Global competition and the worldwide restructuring of the electronics industry – implications for Brazil', report to the Ministry of Science and Technology, Brasilia, Brazil.

Ernst, D. and O'Connor, D. (1989) *New Technologies, Global Competition and Latecomer Industrialisation – an Agenda for the 1990s*, Paris: OECD.

Evans, P. and Tigre, P. (1989) 'Paths to participation in hi-tech industry; a comparative analysis of computers in Brazil and Korea', *Asian Perspectives*, Vol. 13, No. 1.

Exame Informática, various issues.

Financial Times (1988) *Information Technology in Finance – a Survey*, 10 November.

Freeman, C. (1982) *The Economics of Industrial Innovation*, London: Pinter Publishers.

—— (1986) 'The challenge of new technologies', paper presented to the

OECD Symposium *Opportunities and Risks for the World Economy: the Challenge of Increasing Complexity*, Paris: OECD.
Frischtak, C. (1986) *The Informatics Sector in Brazil: Policies, Institutions and the Performance of the Computer Industry*, Industrial Strategy and Policy Division, Washington, DC: The World Bank.
—— (1989) 'Specialization, technical change and competitiveness in the Brazilian electronics industry', paper presented at the OECD Workshop *Technological Change and the Electronics Sector – Perspectives and Policy Options for Newly Industrializing Countries*, Paris: OECD.
Fung, S. and Cassiolato, J. (1976) *The International Transfer of Technology to Brazil Through License Agreements: Characteristics of Government Control System and Commercial Transactions*, Cambridge, Mass.: Center for Policy Alternatives, Massachussetts Institute of Technology.
Georghiou, L. *et al.* (1986) *Post-Innovation Performance: Technological Development and Competition*, London: The Macmillan Press.
Glaser, P. (1988) 'Using technology for competitive advantage: the ATM experience at Citicorp', in R. Guile and J. Quinn (eds) *Managing Innovation: Cases from the Service Industries*, Washington, DC: National Academy Press.
Guile, R. and Quinn, J. (eds) (1988a) *Managing Innovation: Cases from the Service Industries*, Washington, DC: National Academy Press.
—— (eds) (1988b) *Technology in Services: Policies for Growth, Trade and Employment*, Washington, DC: National Academy Press.
Hagerdoorn, J. (1989) *The Dynamic Analysis of Innovation and Diffusion – a Study in Process Control*, London: Pinter Publishers.
Heffernan, S. (1984) *New Technology and Competition in British Banking*, London: Centre for the Study of Financial Institutions.
Herrera, A. (1986) 'The new technology and the developing countries: problems and options', in R.M. MacLeod (ed.) *Technology and the Human Prospect: Essays in Honour of Christopher Freeman*, London: Frances Pinter.
Hobday, M. (1990) *Telecommunications in Developing Countries – the Challenge from Brazil*, London: Routledge.
Info 83, various issues.
Itaú SA (1988) *Annual Report*, São Paulo.
Itautec SA (1988) *Annual Report*, São Paulo.
Kaplinsky, R. (1988) 'Industrial restructuring in LDCs: the role of information technology', paper prepared for Conference of Technology Policy in the Americas, Stanford University, Stanford.
Langer, E. (1989) 'Generations of scientists and engineers: origins of the computer industry in Brazil', *Latin America Research Review*, Vol. 24, No. 2.
Lundvall, B.-Å. (1983) *Technology, Competitiveness and Small Countries*, Smaskrift No. 37, Aalborg: Aalborg University Press.
—— (1985) *Product Innovation and User–Producer Interaction*, Industrial Development Research Series No. 31, Aalborg: Aalborg University Press.
—— (1986) 'Technological revolutions and the international division of labour', paper presented at the Workshop on Innovation and Regional Development, University of Architecture, Venice.
Lütkenhorst, W. (1988) 'Challenges from new trends in foreign direct investment', *Intereconomics*, Vol. 23, No. 5.

Marques, N. (1987) 'Introdução da automação bancária no Brasil', *Estudos Apec – a Economia Brasileira e Suas Perspectivas*, Vol. 26.

Metcalfe, J. (1986) 'Technological innovation and the competitive process', in P. Hall (ed.) *Technology, Innovation and Economic Growth*, Southampton: Camelot Press.

Mytelka, L. (1987) 'Knowledge-intensive production and the changing internationalization of multinational firms', in J.A. Caporaso (ed.) *A Changing International Division of Labor*, Boulder: Lynne Rienner.

Nelson, R. and Winter, S. (1977) 'In search of a useful theory of innovation', *Research Policy*, Vol. 6, No. 1.

OECD (1988) *Economic and Wider Impacts of Biotechnology*, Paris: OECD.

Pavitt, K. (1984) 'Sectoral patterns of technical change: towards a taxonomy and a theory', *Research Policy*, Vol. 13.

—— (1988), 'International patterns of technological accumulation', in N. Hood and J. Vahlne (eds) *Strategies in Global Competition*, Beckenham: Croom Helm.

Perez, C. (1983) 'Structural change and the assimilation of new technologies in the economic and social systems', *Futures*, Vol. 15, No. 5.

—— (1986) 'Las nuevas tecnologias: una vision de conjunto', in C. Ominami (ed.) *La Tercera Revolucion Industrial: Impactos Internacionales del Actual Viraje Tecnologico*, Buenos Aires: Grupo Editor Latinoamericano.

—— (1988) 'New technologies and development', in C. Freeman and B.-Å. Lundvall (eds) *Small Countries Facing the Technological Revolution*, London: Pinter Publishers.

Peritore, N. (1988) 'Hi-tech import substitution: Brazil's computer industry and the possibilities for social transformation', *Alternatives*, Vol. 13, No. 1.

Petit, P. (1984) *Information Technology and Economic Perspectives – Automation of Services: the Case of the Banking Sector*, DSTI/ICCP/84.19, Paris: OECD.

Piragibe, C. (1985) *Indústria da Informática: Desenvolvimento Brasileiro e Mundial*, Rio de Janeiro: Campus.

Piragibe, C., Tigre, P. and Pena, V. (1985) *Recursos Humanos na Indústria Brasileira de Processamento de Dados*, Rio de Janeiro: Relatório de Pesquisa, Instituto de Economia Industrial, Universidade Federal do Rio de Janeiro.

Revell, J. (1983) *Banking and Electronic Fund Transfers*, Paris: OECD.

Roach, S. (1988) 'Technology and the services sector: America's hidden competitive challenge', in R. Guile and J. Quinn (eds) *Technology in Services: Policies for Growth, Trade and Employment*, Washington, DC: National Academy Press.

—— (1989) 'Pitfalls on the "new" assembly line: can services learn from manufacturing?', paper presented to the International Seminar on Science, Technology and Economic Growth, Paris: OECD.

Robinson, R. (1976) *National Control of Foreign Business Entry: a Survey of Fifteen Countries*, New York: Praeger Publishers.

Rosenberg, N. (1976) *Perspectives on Technology*, Cambridge: Cambridge University Press.

Santos, T. (1988) 'O desempenho do setor bancário após as mudanças na política econômica', *Estudos Apec – a Economia Brasileira e Suas Perspectivas*, Vol. 27.

SEI (1988) *Panorama do Setor de Informática*, Série Estatística, Vol. 1, No. 1.

—— (1989) *Panorama do Setor de Informática*, Série Estatística, Vol. 2, No. 2.

Smith, S. and Wield, D. (1987) 'New technology and bank work: banking in IT as an organizational technology', in R. Finnegan *et al.* (eds) *Information Technology: Social Issues*, London: Hodder and Stoughton.

—— (1988) 'Banking and the new technologies: cooperation, competition and the clearers', in L. Harris *et al.* (eds) *New Perspectives on the Financial System*, London: Croom Helm.

Teubal, M. (1984) 'The accumulation of intangibles by hi-technology firms', in D. Lamberton *et al.* (eds) *The Problem with Technology*, London: Pinter Publishers.

The Economist (1989) *Metamorphosis: a Survey of International Banking*, 25 March.

Tigre, P. (1983) *Technology and Competition in the Brazilian Computer Industry*, London: Pinter Publishers.

US Department of Commerce (1987) *A Competitive Assessment of U.S. Microcomputers*, Washington, DC: US Government Printing Office.

Utterback, J. and Kim, L. (1985) 'Invasion of a stable business by radical innovation', in P. Kleindorfer (ed.) *The Management of Productivity and Technology in Manufacturing*, New York: Plenum Press.

von Hippel, E. (1976) 'The dominant role of the users in the scientific instruments innovation process', *Research Policy*, Vol. 5, No. 3.

Walsh, V. (1988) 'Technology and the competitiveness of small countries: a review', in C. Freeman and B.-Å. Lundvall (eds) *Small Countries Facing the Technological Revolution*, London: Pinter Publishers.

White, E. (1988), 'The question of foreign investments and the economic crisis in Latin America' in R. Feinberg and R. Ffrench-Davis (eds) *Development and External Debt in Latin America: Bases for a New Consensus*, Notre Dame: University of Notre Dame Press.

Zaparoli, D. (1989) 'Como faz bem o ouro de Moscou', *Exame Informática*, 29 November.

4 Software strategies for developing countries

Lessons from the international and Brazilian experience

Fátima Gaio

1 INTRODUCTION

The diffusion of Information Technology (IT) has been paralleled by an increasing awareness of the key role played by computer software in this process. In the current literature on developing countries (LDCs), computer software is often regarded as an optimal point of entry into the IT 'complex'. The usual argument given is that software is a labour-intensive activity with relatively low rates of technical change, and that those countries more endowed with a skilled labour force could therefore exploit their comparative advantage of relatively cheap labour. However, little systematic evidence has been gathered to support this argument. Furthermore, this type of rationale can be regarded as oversimplistic and static. Neither the dynamics of the interdependences between hard and software at the technological level nor the evolution of the international market framework is taken into consideration.

The aim of this chapter is to critically examine some of these issues, by analysing the opportunities and constraints faced by Brazil in the promotion of economic and technological capacity in software.[1] For more than a decade, Brazil has been implementing explicit policy measures to foster IT within its economy under a broad 'import substitution' strategy. In addition, the country has a relatively well-developed industrial basis, a relatively large and well-educated workforce and a large internal market. Recognition of the crucial role of software has led the Brazilian Government to confront head-on the complex issue of devising policy measures to promote local software activities.

The Brazilian experience is discussed in the context of technological and institutional strategies pursued by industrialised countries which have maintained leadership in the supply of software for international markets. The objective is to show the complexities faced by developing countries in devising and implementing Government policies which seek to promote long-term market opportunities and dynamic technological capabilities in software.

The chapter is divided as follows: Section 2 contrasts the arguments put forward in relation to the potential opportunities for LDCs in software with the debate about the broad impacts of IT on these countries' prospects for growth. The purpose of Section 3 is to briefly conceptualise software in technological and economic terms. Section 4 analyses the international markets for software and computer services in order to examine whether in fact software has been a point of entry into the IT complex for LDCs and, particularly, for newly industrialising countries (NICs). Section 5 focuses on the policy strategy adopted in Brazil to promote software activities, examining why it has not chosen this area as a point of entry into the IT complex. This case study shows that a late entry into software at early stages of broad IT diffusion and an 'imitation' of what can be called the US model are likely to bear little fruit.

2 THE OPPORTUNITIES IN SOFTWARE FOR DEVELOPING COUNTRIES

The purpose of this section is to analyse the strategies currently recommended for developing countries' entry into software activities. This is done by contrasting arguments in support of these strategies with the emphasis on the impacts of IT on growth prospects of developing countries.

Sharing the theoretical underpinnings of comparative-advantage arguments, two sets of literature appear to outline opposite outcomes. On the one hand, the current debate about software opportunities and the resulting policy recommendations suggest that LDCs can create positive externalities in the area. On the other hand, the mainstream of existing literature on the broad impacts of IT upon the prospects for growth of LDCs claims that IT tends to widen the gap between developing and industrially advanced countries (IACs) at industrial, technological and social levels.

Software has been assigned a key role in reaping the potential benefits created by IT in LDCs (Ernst 1981; Hoffman 1982; O'Connor 1985). The basic argument given is that software could be an optimal point of entry into the IT 'complex' due to the following characteristics: the labour-intensive nature of its development process, relatively low capital requirements and the slow pace of technical change.

The broad strategies suggested for LDCs in the area have been twofold. First, to exploit the comparative advantage of relatively cheap skilled labour for generating export opportunities. Second, to exploit the niches created by a 'natural' market reserve of diversified local market conditions for the production of applications software which, in addition, would be more appropriate to local requirements (UNIDO Secretariat 1983; Kopetz 1984; Narasimhan 1984).

In some market segments such as cross-industry packages (e.g. applications and systems software for horizontal markets such as spread-sheets, word processors, data base management systems and operating systems), trade is increasingly becoming internationalised, led by a small number of large firms (Ernst 1981; O'Connor 1985). Furthermore, a few standardised products and economies of scale prevail as significant barriers for potential new entrants. Therefore, in such segments, the opportunities for local firms of LDCs would tend to be limited to minor adaptations, distribution and marketing activities. Nonetheless, reaping the comparative advantage of cheap skilled labour and generating export opportunities can be feasible through the establishment of joint ventures with leading firms of industrialised countries (O'Connor 1985; Katz 1987; Schware 1987).

The above debate, which is primarily concerned with opportunities for growth in software, states that LDCs could generate positive externalities in the area, reaping the benefits of its labour-intensive nature. In contrast, with regard to the broad impacts of IT on growth opportunities, the main current argument put forward is that the diffusion of IT would be detrimental to LDCs. Even the recent achievements of NICs could be seriously jeopardised by their inability to absorb the quantitative and qualitative leap that IT entails. The assumption is that the technological and industrial dynamism being undertaken by these countries over the past few decades has been based first on the development of technological capacity through incremental technical change and, second on labour-intensive production systems in mature industries through their comparative advantage with cheap unskilled and semi-skilled labour.

The theoretical basis of both these ways of thinking are found in the product cycle theory[2] which claims that LDCs suffer from a number of structural shortcomings. Among them are the lack of synergic elements, such as know-how and scientific, engineering and managerial skills, which have the effect of inhibiting industrial dynamism in the early phases. According to this theory, the factor endowments of LDCs (i.e. cheap unskilled and semi-skilled labour) become a competitive edge *vis-à-vis* industrialised countries only in the mature phase. At this stage, the system exhausts its technological dynamism, a few standardised products predominate and competition is focused on achieving economies of scale and minimising production costs.

These structural shortcomings would tend to be enhanced by the diffusion of IT, undermining even the scope for NICs sustaining economic growth,[3] since the major features of IT do not appear to be appropriate to the 'natural' factor endowments of LDCs. The science-based nature of IT, associated with an exceptional technological dynamism, a labour-saving bias and pervasive impacts on product and process innovations, would tend to generate negative externalities upon LDCs, such as a loss of competitiveness in international markets, a deepening of dependency on foreign technology and the generation of unemployment with structural characteristics.

Perez (1987) elaborates on this, remarking that a development strategy centred upon the exploitation of the 'natural' factor endowments of LDCs involves a vicious circle which is likely to perpetuate the disadvantageous conditions of these nations. Countries with a sizeable contingent of university-trained personnel can reap the benefits of a strategic entry into early phases of the innovation–diffusion process when capital and managerial requirements tend to be low. Scientific and technical skills are crucial inputs but information is more 'freely' available.[4] As the system evolves towards growth, capital requirements and know-how increase in complexity as well as the level of private appropriability, which are turned into significant entry barriers. In the mature phase, the system exhausts its dynamism and the know-how tends to be commercialised as a commodity. Thereby, an analysis based on static comparative advantages does not appear to outline the appropriate strategy for promoting sustained growth.

Neo-Schumpeterians regard the present period as a transition into the full deployment of the IT paradigm, strategically important for those countries which have already acquired a well-developed industrial infrastructure, a scientific and technical educational system and a significant absorptive capacity (Soete 1983, 1985; Perez 1985, 1987).

According to these authors, the widespread restructuring process required by IT is delayed, in particular, in those leading industrialised economies heavily committed to the previous paradigm while, for NICs, which aim at catch-up growth, this period of delay may create a conducive environment for designing and implementing appropriate strategies for growth. However, such a process requires an explicit and co-ordinated strategy which involves active Government intervention, appropriate choices of institutions and timing, considerable and sustained investment and learning efforts. Within this approach, the promotion of the IT 'complex' in itself is assigned a leading role and a selective approach is advocated.

The latter debate emphasises the importance of devising appropriate strategies for promoting indigenous activities. It is worth pointing out that such a task is of particular complexity in software, owing to its idiosyncratic, pervasive and highly differentiated nature. These aspects reinforce the convenience of a selective approach and the need to carefully address the costs and benefits involved in a policy strategy.

In sum, the strategies recommended for promoting software activities in LDCs reviewed in this section were questioned at a number of levels. It appears artificial and misleading to treat software in isolation as a labour-intensive and less demanding activity, disregarding the complex nature and essential role of software in the diffusion of IT. Following Perez's and Soete's line of argument, the theoretical basis of such strategies can be seen as a static analysis of comparative advantages, neglecting to take into account important factors such as the evolution of software in technological and economic terms, the dynamic interdependences that IT encompasses, or the appropriate institutional framework for promoting local activities in the area. Moreover, there is little empirical evidence in support of these recommended strategies. The following sections of this chapter seek to contribute to the understanding of these issues.

3 THE NATURE OF SOFTWARE AND ITS ROLE IN THE DIFFUSION OF INFORMATION TECHNOLOGY

The purpose of this section is to briefly review some fundamental characteristics of software at technological and economic levels.

Software is an increasingly intrinsic and pervasive element of IT-based systems. Its significance is defined at the technological and economic levels, which has important bearings upon the future prospects for the growth of LDCs. These two levels have been greatly influenced by the nature of IT evolution, which involves an exceptionally dynamic process coupled with synergic interdependences. At the industrial level, computers and telecommunications can be regarded as its system-building blocks (Genthon 1985; Freeman and

Soete 1986), while, at the technological level, software and microelectronics are the strategic elements to be analysed (Gaio 1990). The debate on technological and economic opportunities in software is of particular relevance to NICs which have been explicitly committed to promoting IT-based activities within a policy strategy of technological 'self-reliance'.

As the process of IT diffusion progresses, software also becomes increasingly pervasive as an essential but complex component. Its complexity stems from various levels. Its very nature is idiosyncratic, logical and non-material. The production of software is a design-intensive activity of intangible products which are independent of their carrier media, and information is the basic material. Its core function is to endow IT-based systems with structural 'plasticity'. The value of software is defined by the knowledge it incorporates into products and production systems in a coded form, with objective criteria such as: to be effective in carrying out computational operations which produce reliable results and to be correct in its capacity to represent the requirements of users. The idiosyncrasies generated by the non-material nature of software are further emphasised by the infant stage of both its highly multi-disciplinary scientific basis and software as an orchestrated economic activity.

In the early 1950s, with the advent of programs stored in digital computers, software could already be characterised as an independent technology; however, the process of becoming an independent economic activity was much slower. Well into the 1960s, software was mainly produced 'in-house' by large computer-user organisations and hardware manufacturers, and this trade was regarded as a subsidiary component of the computer industry. Therefore, from its outset, software could not be understood in isolation but only in its evolutionary interplay with user requirements and hardware as complementary assets. The 1970s can be seen as the turning-point for software. In this decade, three major tendencies are worth emphasising: the so-called 'software crisis'; a growing availability of hardware with increasing performance at decreasing costs; and the growth of an economic sector specialised in the production and distribution of software.

The idea of a 'software crisis' illustrated a growing awareness of the importance of software in its own right. Some commentators forecast a growing gap between demand and supply of software products and services, a scarcity of skilled labour, and escalating costs in development and in the maintenance of customised software. The demand/supply mismatch in software was paralleled by an acceleration of the electronics technological trajectory with the advent of semiconductors, microprocessors and the increasing level of integration of electronic components.[5] The dynamic nature of technological change in electronic components and computer hardware brought into focus the fact that software could hinder the diffusion of IT. In the late 1970s, the advent of microcomputers, as a result of a tendency to miniaturisation of micro-electronic components, also represented an important factor for accelerating the diffusion of IT, with a potential to achieve mass markets.

In IACs and, particularly, in the USA these trends provided strong inducements for the externalisation of demand for software and the growth of a producer activity of the service sector, with firms mainly specialised in the production and distribution of software products or packages. Market growth for production services interplay with complementary technologies and the relations of interdependence and integration of industry and services, are phases of activities which jointly contribute to the production of goods and services (Momigliano and Siniscalco 1982).

After the 1970s, in IACs, the concept of software as an independent economic activity appeared to be consolidated. However, the software industry has remained less than clearly defined as shown by the OECD (1989:15):

> recent modifications in the relationship between the structure of production of this industry and its market indicates the difficulties which arise with respect to its definition. In fact its structure is changing as a consequence of technological evolution and the categories of suppliers are expanding.

The existing literature on the software industry as an independent economic activity can be roughly divided into two quite opposing streams. In the first stream, qualitative analyses tend to emphasise the shades of complexity which surround the industry as a fragmented sum of mature and emerging market segments. As remarked by Gaio (1990:157):

> the software sector ... is characterised by a high degree of diversity, which stems from product differentiation combined with dissimilar user-producer patterns of interaction, constituting, in fact, a series of loosely related markets, among which both firms and skills may only have a limited degree of transferability.

In the other stream, quantitative interpretations draw on conventional and simplified structural analyses which do not reflect the level of segmentation and differentiation of the sector. In addition, this type of analysis tends to concentrate on the relatively more mature segments of this still emerging industry, usually set within a cluster of activities called computer services.[6] Although, at present, it is virtually impossible to define the precise boundaries between market segments, these relatively mature segments appear to be related to commercial data processing (DP), and general-purpose computer systems which correspond to some of the basic elements of early stages of the broad diffusion of IT.

In spite of its limitations, the latter type of analysis will be presented in the next section, since it is instrumental to examine the level of participation of LDCs in the international market and, in particular, to set the Brazilian case into perspective. However, it is important to keep in mind the limitations of data available at the international level and, even more so, in relation to Brazil,[7] in addition to the little knowledge about the nature and level of opportunities for local firms in national markets.

In short, taking into account that software production is an idiosyncratic design-intensive activity, with the crucial role of concentrating knowledge upon IT-based systems, it seems oversimplistic to treat software as a labour-intensive activity in which to apply static comparative advantages. The long-term range of opportunities in the area can only be grasped through an understanding of its technical and economic evolution, which has been influenced by dynamic complementarities with the hardware industry and the requirements of users' software attempts to model.

4 AN INTERNATIONAL PERSPECTIVE ON SOFTWARE AND COMPUTER SERVICES MARKETS

In current literature, software has often been regarded as an area of significant economic opportunities. This section presents a quantitative analysis of the international market for software and computer services with the purpose of examining the following questions: what has been the level and nature of opportunities generated for local firms across different national markets?; have LDCs reaped the potential benefits in software suggested by the current theoretical debate?

In the 1970s, software gained in importance. In the 1980s, in the context of industrialised countries, this economic activity appeared to be consolidated. However, as a consequence of a lack of thorough knowledge about the software industry, there has been a tendency to associate its structure with the model generated by the US, in general regarded as a benchmark. In fact the US model has not even been reproduced by any major industrialised economy (Gaio 1990). To examine the validity of this hypothesis, it is essential to characterise the US model and to present some indicators of the differentiated structure and performance of other national markets.

Currently, the US is the leading world market in computer services and is the major software exporter. Both the software industry and hardware producers have played a significant role in this performance, led by a small number of large firms. As shown in Table 4.1, in 1985 the US was responsible for about one half of the total world market of computer services, with revenues amounting to US$30.56 billion. Software accounted for 54 per cent (US$ 16.54 billion) of this

Table 4.1 A comparison of computer services markets' distribution of revenues, 1985 (US$ billion)

	US	*Japan*	*France*	*UK*	*Brazil*
Total US$ bn	30.56	7.07	3.40	2.67	1.79
% world rev.	55.9	12.9	6.2	4.9	3.3
Packages US$ bn	12.60	0.27	0.86	1.14	n.a.
% total SW	76	9	40	61	n.a.
Cust. & Cons. US$ bn	3.94	2.59	1.30	0.73	n.a.
% Total SW	24	91	e0	39	n.a.
Total SW US$ bn	16.54	2.86	2.16	1.87	0.53
% Mark. seg.	54	41	63	70	29
Train. & FM US$ bn	3.53	0.65	0.14	0.16	n.a.
% Mark. seg.	12	9	4	6	n.a.
Processing US$ bn	10.48	3.56	1.10	0.64	0.62
% Mark. seg.	34	50	32	24	35
Other US$ bn	n.a.	n.a.	n.a.	n.a.	0.64
% Mark. seg.	n.a.	n.a.	n.a.	n.a.	36

Key: Cust. & Cons. – customised or bespoke software and consultancy; Train & FM – training and facilities management; SW – software. For Brazil: processing includes facilities management; others include – first, the following services: training, consultancy, systems integration, maintenance, and other technical services, and second, computer services provided by local hardware firms in an aggregated form.

Note: A billion is used in this work as a thousand million. Data on total world market for computer services were estimated by IDC-C&L (1987) at US$54.7 billion. Data on software imports were estimated by SEI (1986) at US$552 million.

Source: Total world market, US, Japan, France and the UK elaborated from IDC-C&L (1987); Brazil elaborated from SEI (1986 and 1987), reproduced from Gaio (1990: Table 7.1)

amount, and 76 per cent of this was packages. The US internal market appears to be showing signs of maturation. In turn, revenues through international trade gain in importance. In this respect, software represents a major growth area and packages have the leading position.

In relation to the US independent software industry, although its structure remains fragmented, the number of firms is decreasing;[8] a tendency to concentration of revenues on a small number of firms can already be observed.[9] Such a tendency is linked to a significant level of acquisitions of firms and to the remarkable growth rates shown by some cross-industry packages (Katz 1987). The latter appears to be particularly pertinent to traded packages for microcomputers.[10] The independent industry has shown a strong presence across different market segments. However, its share has been higher in packages for microcomputers and 'applications solutions'.

US-based hardware producers have also shown a high level of concentration. In 1985, hardware producers were responsible for about one half of the total sales of packages (US$6.7 billion), with five of them accounting for around 90 per cent of this sum, and a stronger presence in the 'systems and utilities' segment (Grindley 1987). These firms have played a leading role in the positive international performance of the US, not only in terms of their direct participation in software trade but also in broadly influencing the dynamics and 'rules' of the software market.

The increasing importance of international trade for the US is emphasised by the differentiated growth rates of revenues accrued in internal *vis-à-vis* international markets. In the period 1984–1990, IDC (1986) estimates for the US computer services sector as a whole an average annual growth rate (AAGR) of about 18 per cent, the internal market having an AAGR of 16 per cent (US$23 to 55 billion), while revenues from international markets would grow at 28 per cent per year (US$4 to 19 billion). Packages stand out as the main source of external revenue, their share evolving from 62 per cent to 82 per cent (US$3 to 16 billion), with an AAGR of 34 per cent. 'Systems and utilities' and 'applications tools' hold the leading position, each segment with an evolving share of about 30 to 40 per cent (US$1 to 7 billion) of total international activity (US$4 to 16 billion), while the 'applications solutions' share is expected to increase at a much slower rate (US$1 to 2 billion).

While these figures are based on estimates, it is probably safe to conclude that the large-scale diffusion of cross-industry packages has contributed to the US maintaining its leadership as the major software exporter in international markets. The US can be characterised as the most conducive environment for generating standardised software products. The generation of such an environment has been influenced by a large internal market, closeness to sources of technical change in complementary technologies (e.g. computer systems, micro-electronics, telecommunications) and tight labour markets. Both the US-based independent software sector and hardware producers have played a significant role across market segments led by a small number of large firms.

Existing literature on the computer services markets of industrially advanced countries has often stressed software and packages in particular, as a source of structural change in computer services, and an area of significant economic opportunities. However, it is questionable whether the US phenomenon evidenced above has even been replicated by any other major industrialised market. This is exemplified by the differentiated structure of computer services observed across the markets of IACs and the lack of knowledge about the precise participation of local firms outside the US.

The distribution of revenues across market segments in computer services, the relative importance of software in national markets and its proportional allocation between packages and customised software have all varied markedly across the markets of IACs, and cannot be related only to the relative size of these markets (Table 4.1). In 1985, in Japan, the second largest market (US$7.07 billion), software accounted for 41 per cent (US$2.86 billion) of its total computer services revenues, this share being lower than that of other national markets of IACs, and the vast majority was customised (91 per cent). In contrast, it was in the UK, the third largest European market (US$2.67 billion), that software accounted for the largest share, with 70 per cent (US$1.87 billion) of total national revenues, 61 per cent of which were packages.

A contrasting pattern emerges when comparing the markets of LDCs with those of IACs. Although the markets of LDCs are expanding at a faster rate than those of the IACs, the former have remained with a small share of world revenues in both computer services and software. In IACs, on average, software can be characterised as a source of structural change in their national computer services markets, whereas in LDCs this pattern is not observable.

For instance, IDC-C&L (1987) estimated that, in 1985, worldwide revenues for computer services amounted to about US$54.7 billion. The US, Japan, and W. Europe were responsible for the vast majority (96.5 per cent), whereas other countries accounted for a mere 3.5 per cent share, with US$1.9 billion. According to SEI's estimates (1986, 1987), Brazil alone would be responsible for 94 per cent of the latter share, with US$1.79 billion, 29 per cent of which was for software (see Table 4.1). OECD (1989) presents a more optimistic picture of the participation of LDCs in the world computer services markets, with estimates for the period 1984–87 (see Tables 4.2 and 4.3). The participation of LDCs is increasing over time, with growth rates significantly higher than IACs. None the less, the latter has remained responsible for the majority share of the world computer services market.

With respect to IACs, in 1984 the nine largest national markets[11] accounted for about 90 per cent of the world market for both the computer services industry (US$43,435 million) and software (US$23,696 million). These shares decreased to around 86 per cent in 1987, with revenues amounting to US$66,876 million in computer services and US$41,185 million in software. In this period, revenue accrued from software was doubled, with an AAGR of 20 per cent, which was considerably higher than other services, with an AAGR of 9 per cent.

In relation to NICs, the estimates of the US Department of Commerce shown by the OECD (1989) were that, in 1984, Brazil, Mexico, S. Korea, Singapore, Taiwan and India together represented only approximately 2 per cent of the world market for both the computer services industry (US$1,030 million) and software (US$534 million). By 1987 these shares grew to about 6.5 per cent (US$5,068 million) in computer services and 5.4 per cent (US$2,576 million) in software. In a period of three years, revenues accrued in these national markets were quintupled, with software showing an AAGR of 69 per cent, slightly lower than other services, which were about 71 per cent. Within this group of countries, Brazil was responsible for the overwhelming proportion of revenues (e.g. in 1987, nearly 85 per cent of total software revenues).

Although software has shown growth rates higher than other services in IACs, the level and nature of opportunities for local firms generated by such a growth pattern remains unclear. While the significant international performance of the US in software and, in particular, packages is evidenced, the level of imports in other national markets has rarely been examined. As observed by the OECD (1989:55): 'In practice, the majority of data available at the country level represent worldwide revenues, thus including imports re-distributed by country-based firms, direct exports and revenues from foreign subsidiaries of nationally-owned firms . . .'

Table 4.2 The world market of software and computer services (hardware manufacturers included), 1984–7, in US$ million

Domestic markets	1984 Software	1984 Services	1984 Industry	1987 Software	1987 Services	1987 Industry	AAGR 1984–7 Software	AAGR 1984–7 Services	AAGR 1984–7 Industry
1 United States[1]	14,300.0	12,860.0	27,160.0	23,301.0	15,914.0	39,215.0	17.67	7.36	13.02
2 Japan[2]	2,368.2	2,575.3	4,943.5	4,522.5	4,168.6	8,691.1	24.07	17.41	20.69
3 France[1]	1,778.5	1,143.1	2,921.6	3,246.1	1,465.1	4,711.2	22.20	8.63	17.27
4 Germany[1]	1,487.6	838.9	2,326.5	2,824.5	1,039.9	3,864.4	23.83	7.42	18.43
5 United Kingdom[1]	1,469.9	745.4	2,215.3	2,856.4	1,004.3	3,860.7	24.79	10.45	20.34
6 Italy[1]	890.1	507.5	1,397.6	1,822.9	676.0	2,498.9	26.99	10.03	21.37
7 Canada[1]	487.6	522.4	1,010.0	824.6	684.9	1,509.5	19.14	9.45	14.34
8 Netherlands[1]	461.7	343.8	805.5	861.9	431.4	1,293.3	23.13	7.86	17.10
9 Australia[1]	452.1	203.2	655.3	925.3	306.1	1,231.4	26.96	14.63	23.40
10 Brazil[2]	363.5	337.7	701.2	2,186.2	2,031.4	4,217.6	81.86	81.87	81.86
11 Sweden[1]	280.8	305.7	586.5	555.5	405.6	961.1	25.53	9.88	17.90
12 Switzerland[1]	276.0	251.0	527.0	523.5	317.3	840.8	23.79	8.13	16.85
13 Spain[1]	242.3	131.1	373.4	548.0	184.8	732.8	31.26	12.12	25.20
14 Belgium[1]	241.7	216.0	457.7	452.2	278.1	730.3	23.22	8.79	16.85
15 South Africa[2]	208.0	32.0	240.0	364.0	54.0	418.0	20.51	19.06	20.32
16 P.R. of China[2]	175.0	n.a.	n.a.	968.0	n.a.	n.a.	76.85	n.a.	n.a.
17 Denmark[1]	166.8	262.8	429.6	323.2	335.7	658.9	24.67	8.50	15.32
18 Israel[2]	166.0	68.0	234.0	249.0	94.0	343.0	14.47	11.40	13.60
19 Norway[1]	155.8	238.0	393.8	315.8	320.5	636.3	26.56	10.43	17.34
20 Austria[1]	152.0	96.1	248.1	284.1	126.5	410.6	23.18	9.60	18.29
21 Finland[1]	150.5	175.3	325.8	290.9	233.2	524.1	24.57	9.98	17.17
22 Mexico[2]	59.0	6.0	65.0	117.0	13.0	130.0	25.64	29.40	25.99
23 Ireland[1]	54.3	27.2	81.5	n.a.	n.a.	n.a.	n.a.	n.a.	n.a.
24 Korea[2]	40.0	10.0	50.0	107.0	40.0	147.0	38.82	58.74	43.28
25 Singapore[2]	27.0	21.0	48.0	71.0	58.0	129.0	38.03	40.30	39.03
26 Taiwan[2]	26.0	29.0	55.0	57.0	51.0	108.0	29.91	20.70	25.22

27 Hong Kong[2]	25.0	n.a.	n.a.	61.0	n.a.	n.a.	34.63	n.a.	n.a.
28 Saudi Arabia[2]	25.0	n.a.	n.a.	49.0	n.a.	n.a.	25.15	n.a.	n.a.
29 Portugal[1]	22.0	17.1	39.0	n.a.	n.a.	n.a.	n.a.	n.a.	n.a.
30 Malaysia[2]	20.0	n.a.	n.a.	67.0	n.a.	n.a.	49.63	n.a.	n.a.
31 India[2]	18.3	92.8	111.1	37.7	298.9	336.6	27.22	47.68	44.69
32 Turkey[1]	6.9	7.4	14.3	n.a.	n.a.	n.a.	n.a.	n.a.	n.a.
	1984			*1987*			*AAGR*		
Total	*US$ mn.*	*%*		*US$ mn.*	*%*		*1984–7*		
(25 countries)									
Industry	48,281.5	100		78,199.6	100		17.44		
Software	26,269.4	54.4		47,667.3	61.0		21.97		
Services	22,012.1	45.6		30,532.3	39.0		11.52		

Notes: [1] These country data have been included by OECD. They represent value-added marketed domestic revenues. This information is based: for Western Europe on IDC Europa EUROCAST, June 1986 and on the 1985 Report for ECSA by IDC Europa for Ireland, Portugal and Turkey; for Canada and Australia on information provided respectively by IDC Canada and IDC Australia during 1987; for the United States on *Computing Services Industry 1986–1996. A Decade of Opportunity* prepared on the behalf of the UK Department of Trade and Industry by Coopers and Lybrand and IDC and on additional information provided by IDC Europa. The AAGR 1984–87 for data included by OECD has been calculated on the basis of projections made by the original source which has been deflated and re-inflated with the country-related inflation rates in Annex 6. As the country reports used by the US Department of Commerce were drafted in 1984, to obtain 1987 values the projections of 1985 figures have been done after conversion using 1984 exchange rates in order to avoid distortions introduced by the simple exchange rate. IDC Australia figures only reflect already current values.

[2] The US Department of Commerce (USDOC) assumes its data to be value-added marketed (i.e. excluding captive sales and hardware sales related to turn-key systems provision) and expressed with an approach which is similar to apparent consumption (i.e. domestic production less exports plus imports). As domestic revenues include imports by country-based firms, it has been assumed that they represent a reasonable 'proxy' of the domestic markets (see Annex 4). Figures provided by the US Department of Commerce have been taken from various reports and USDOC has provided estimates to fill in the missing information. The average annual growth rates (AAGR) to 1987, have been estimated by the individual authors of the above-mentioned reports for their respective countries. The values for Brazil, India and Japan have been recalculated adopting the corresponding IMF exchange rates (average of the period). The exchange rates used in consultancy reports were known only for these countries. It must be pointed out that the adoption of different exchange rate – unaveraged – can result in rather high deviations for certain countries.

Source: OECD based on US Department of Commerce, IDC Australia, IDC Canada and IDC Europa, elaborated from OECD (1989: Table 2, 24–5)

The precise level of imports in national markets remains unknown; only scattered estimates are available. In Western European countries, around 40 per cent of the software traded by the independent software industry alone would possibly come from imports.[12] In the UK market, according to Grindley's estimates (1987), about 70 per cent of total traded packages are imported. In Japan, the estimates were that in 1984 around 85 per cent of packages were imported.[13]

Correa (1989) claims that the existing analyses of the markets of both IACs and LDCs indicate a significant level of dependence on imported software, especially in systems and utilities and applications tools, whereas local firms tend to enjoy a 'natural' market reserve in applications solutions.

For NICs, the data examined suggest that the supposed comparative advantage 'enjoyed' in low-waged economies has had no significant impact in these markets. Correa (1989:13) remarks that:

> While, it is argued, newcomers face high barriers for entering into the production of hardware, with low capital investment and mobilisation of local qualified personnel it is relatively easy to exploit the growth potential of the software sector. Paradoxically, up to now a few NICs have evidenced their ability to successfully enter into some segments of hardware production (e.g. microcomputers and peripherals), while the efforts made to establish software capabilities have not had, at least up to now, significant results.

This is illustrated by the case of Taiwan. In 1987 total revenues with traded software amounted to US$170 million. Software exports accounted for 6.5 per cent (US$11 million) of this sum, which represented a mere 0.3 per cent of revenues accrued through exports by the computer industry (US$3,240 million) and around 1 per cent of the exports of personal computers. In addition, in 1986 around 75 per cent of the local market requirements of software were met by foreign firms (Chaponnière and Fouquin 1989).

In sum, at the international level, a dualism is emerging in the supply of commercial data processing software. The top league is composed of a small number of established firms which operate at international level, while a large number of small firms tend to be limited to specialised niches of local markets.

The main driving force behind this trend has been the wide diffusion of packages, in particular the cross-industry type, which is increasingly regarded as an internationalised trade led by products of US origin. In this area, where economies of scale play a significant role, the US has kept its competitive edge. The figures examined above denote that, although packages tend to prevail across national markets, no other major IAC has reproduced the US model. Gaio (1990) suggested that the Japanese and European industries should be viewed as alternative models, which will not forcefully evolve towards the US de-verticalization and 'commodification' of software production and trade. Furthermore, while the markets of LDCs are expanding at an accelerated rate, the opportunities for indigenous firms may have been limited.

Analyses of the importance of packages should be threefold. Packages have been the main factor responsible for the current leadership of the US in international markets. They represent a significant product and process innovation, and an optimising solution for the 1970s' software crisis and the early stage of the broad IT diffusion, characterised by the increasing availability of microelectronics-based systems for traditional applications (e.g. commercial data processing). On the other hand, the 'commodification' of software also led to a growing gap between users and producers of software innovations (OECD 1984; Valdez 1988).

Whereas commercial data processing software is approaching a mature stage, a new stage of IT diffusion is currently evolving and is possibly on the verge of its take-off stage, with a significant increase in demand for software in terms of volume and heterogeneity, and the emergence of largely unknown market segments.

For this forthcoming stage of diffusion of IT, vertically-oriented systems, user-producer interactions and economies of scope are predicted to gain in importance. In addition, it can be hypothesised that the diffusion of radical process innovations such as software engineering will have a significance for this period which is comparable to that of packages for the 1970s–1980s.[14] Taking into consideration the little evidence to indicate that radical process innovations in software are currently slowly diffusing in IACs, a new kind of 'crisis' may be under way. Indeed, such crises are the result of the technological desynchronisation of two major trajectories – software and micro-electronics – and their solution lies in the evolutionary interplay between these two trajectories (Gaio 1990). The US model of de-verticalisation and commodification of software production and trade may not provide the most conducive institutional framework for this new form of crisis. It is therefore necessary to interpret the leading performance of the US in software within a specific economic context and technological momentum.[15]

To sum up, an analysis of the international markets for computer services and software has shown that, the sector is already showing signs of maturation, where cross-industry packages and economies of scale tend to prevail. It appears that indigenous firms outside the US tend to be limited to specialised market segments of their national markets, such as applications solutions, customised software and other services. In addition, neither LDCs nor NICs have shown evidence of reaping significant benefits from their comparative advantage of cheap skilled labour. It is, therefore, questionable whether the assumptions of the product-cycle theory are pertinent to a design-intensive activity such as software.

In the next section, the case of Brazil is discussed to further exemplify the range of opportunities and constraints faced by NICs in promoting local software activities. As estimated by the US Department of Commerce (OECD 1989), Brazil has been responsible for the vast majority of revenues of the software and services markets of NICs, with outstanding growth rates. In 1987 the Brazilian computer services market was the fourth largest market (behind the US, Japan and France). It is worth remembering that fast market growth is different from growth in local production. In addition, although these figures can be questioned, they are quite illustrative of how the US, as the major software exporter, perceived the importance of the Brazilian market. Moreover, since the early 1970s Brazil has been explicitly engaged in promoting an indigenous IT complex at the economic and technological levels. After the mid-1980s Brazil faced significant pressures for liberalisation led by the US and particularly targeted at its software and services market (for details see Chapter 9).

5 THE BRAZILIAN STRATEGY FOR SOFTWARE AND COMPUTER SERVICES

Since the early 1970s, Brazil has been explicitly committed to promoting an indigenous IT 'complex'. Government intervention has attempted to combine the economic expansion of locally-owned enterprises with the upgrading of technological capacity, within an import-substitution framework. Although the stated objective has been to promote the IT 'complex' in an integrated manner, the strategies chosen have been gradual in nature and lacked inter-sectoral co-ordination.

IT evolution implies an exceptional dynamism and synergic interdependences at both industrial and technological levels. However, possibly the only feasible strategy for a developing country is to pursue targets in a selective manner. In the period of analysis, telecommunications and the computer industries have been selected to be at the centre of policy measures. The existing literature provides sufficient evidence of the positive achievements in these areas,[16] whereas micro-electronics and software lagged behind.[17]

This section is primarily concerned with examining the impacts of the Brazilian informatics policy on software activities. It is divided into three parts. The first presents a brief historical overview of the policy strategy endeavoured and the measures implemented. The second provides a critical assessment of the policy achievements, and the third discusses the Brazilian software and services market.

A chronological overview of the policy for software

This section presents a brief historical account of the period from the mid-1970s to the late 1980s, reviewing the policy guidelines formulated by Brazil for software and the instruments actually implemented.

In the period from the mid-1970s to the late 1980s, the aim of the Brazilian informatics policy has been to promote hardware and software in a co-ordinated fashion. In the early stage of the informatics policy there was already an awareness of the importance of devising a specific policy strategy for promoting local software activities, and the complexity of such a task.

Two independent studies commissioned by the Government essentially reached the same conclusions in terms of policy guidelines (CAPRE (Commission for the Co-ordination of Electronic Data-Processing Activities) 1979 and SCSS (Special Commission for Software and Services) 1981). In accordance with the general principles of the informatics policy, and based upon 'infant industry' arguments, both studies stressed the importance of protecting and promoting the emergence of a local software industry, with emphasis on the production and trade of packages. This recommendation was justified by the significant growth rates shown by the US and European software industries, especially in the segment of packages.

To reap these potential economic benefits, there was a need to foster the indigenous software industry, which was at an incipient level. According to the SCSS estimates (1981), the majority of local production was customised software of the administrative type, concentrated on 'in-house' developments undertaken by large IT user firms. These studies also held that the convenience of such a strategy was reinforced by the advent of minicomputers, in combination with a trend towards distributed processing, which tend to accelerate the diffusion of IT into medium and small firms and the externalisation of demand for software products and services. Moreover, the expansion of local supply of software products in this market segment was regarded as an important inducement for the diffusion of locally manufactured minicomputers.

Characterising software as the incorporation of knowledge into IT-based systems, the SCSS (1981) remarked that 'free market' rules in this area would not only have negative impacts on the balance of payments but also hinder employment generation effects and relevant experiences for local qualified personnel. Furthermore, it was thought to jeopardise the country's 'autonomous' technological strategy. Nevertheless, CAPRE (1979) also emphasised the complexity of establishing an effective import control system, asserting that the low level of private appropriability in software is intrinsic to its intellectual and intangible nature, and facilitates illegal reproduction and transportation (e.g. 'piracy').

The policy instruments recommended by these studies were: Government procurement; the establishment of special credit lines for funding software production; the control of imports combined with the guarantee of intellectual property rights; the promotion of technical education and the collaboration between university and industry. In the end, the policy measures implemented diverged considerably from the recommendations described above.

By the mid-1970s, CAPRE and INPI (National Institute for Industrial Property) considered that the adoption of a legal mechanism (e.g. patents, copyrights) for protecting intellectual property rights (IPRs) in software was not convenient for Brazil. The main arguments put forward were that the adoption of a legal mechanism would have a number of negative impacts such as accelerating the remittance of foreign exchange and hindering the emergence of a local software industry. In addition, it was claimed that there was no international consensus over the matter. This position of not complying with any legal form of protection for intellectual property rights was held until 1988. It was set by the Normative Act No. 15/1975, according to which software imports were only permitted as technology transfer agreements, requiring the approval of INPI.

In 1982, a Registry of Computer Programs was established under the co-ordination of SEI (Special Secretariat for Informatics) (Normative Act No. 22). Its aim was to control the supply of imported software for the internal market. This measure was combined with Government procurement to induce demand for local software products (Normative Act No. 23).

In 1983, SEI conditioned the approval of manufacturing projects for general-purpose microcomputers to the adoption of operating systems developed locally (Normative Act No. 27).

In 1985, FINEP (Funding Agency for Studies and Projects) and BNDES (National Bank for Economic and Social Development) set up special credit lines for funding software projects to circumvent the shortcomings faced by software houses with the lack of capital assets as a guarantee for access to funding.

In October 1984, the so-called 'Informatics Law' (Law No. 7,232) was approved by the National Congress and in 1985 ratified by Decree No. 92,187. In a simplified manner, this law can be interpreted as a consolidation of the IT policy strategy being followed. Although it included the legal guarantee of an eight-year market reserve for local IT products and services, no detailed treatment was assigned to software, which was to be regulated by a special law.

Finally, in December 1987 the so-called 'Software Law' (Law No. 7,646) was approved by the National Congress and in May 1988 ratified by Decree No. 96,036. This law conditioned the trade of software in the internal market to its registration at SEI. The registration of programs of foreign origin is only granted for a period of three years, which is liable to renewal, but requires the systematic proof of no similar local product. In conformity with the international trend, copyrights were guaranteed for twenty-five years, as the legal form of protecting IPRs.

To sum up, there was early recognition in Brazil of the importance of formulating co-ordinated policies for hardware and software. For the latter, emphasis was given to the promotion of a local software industry, package production and trade. The following part provides a critical analysis of the policy strategy reviewed.

A critical assessment of the policy achievements

As shown, in the period from the mid-1970s to the late 1980s, the objective stated by the Brazilian informatics policy was to formulate complementary policies for hardware and software. The broad strategy recommended for software was to promote the emergence of a local software industry, with particular emphasis on the production and trade of packages, perceived as an area of significant economic opportunities. However, the measures implemented in this period have had limited impact upon software, while hardware can be characterised as one of the major growth areas within the Brazilian IT complex. This section attempts to investigate the reasons for this imbalanced outcome.

In this period, the participation of the local computer industry in the domestic market was consolidated. For an in-depth analysis of the accomplishments of this sector, see Chapter 2. In 1986, operating in the small processor segments (e.g. mini- and microcomputers), these firms were responsible for about 50 per cent of the total revenues of the hardware industry. In addition, the local microcomputer industry has been the most dynamic segment of this industry (SEI 1987). By contrast, in the same year SEI (1986) estimated that local demand for software was largely met by imports, particularly in the mainframe and microcomputer segments.

A common effect observed in an import-substitution strategy is that it modifies the composition of imports by inducing demand for complementary inputs. The case of software illustrates this effect, since hardware and software function in a complementary manner. The policy strategy that has matched the expansion of demand for computers in the domestic market with supply inducements to local production also fostered the externalisation of demand for software products and services, without generating a comparable conducive market for the adoption of local software products. In other words, the informatics policy tended to focus on the promotion of the supply of local products, while not dealing with the diffusion process. Such a strategy encompasses a fundamental misunderstanding of the pervasive role of software in the process of diffusion of IT.

As described in the previous section, in the period analysed the three basic policy instruments implemented for software have been: a registry of computer programs combined with Government procurement for local products; the control of imports combined with a refusal to comply with any legal form of protection for IPRs; and the conditioning of approval of manufacturing projects for microcomputers to the adoption of locally developed operating systems. In 1986 SEI remarked that these instruments have had limited impacts on the promotion of local software activities, as evidenced by the great proportion of imported software traded in the domestic market.

The control of software imports set up in 1975 was limited to transactions involving remittance of payments. In practice, this instrument has only hindered the initiatives of local firms attempting to import through legal channels, whereas it has little influence on intra-firm operations of local subsidiaries of MNCs with their parent companies. According to SEI (1986), the main source of trade of imported software has actually been the local subsidiaries of MNCs, as illustrated by the Registry of Computer Programs. In 1984, of the 8,561 programs registered, 80.3 per cent were imported products traded by MNCs. In 1986 this share dropped to 74.5 per cent of the total amount (11,729 programs), while products imported through technology transfer agreements and traded by local firms remained with the negligible shares of 0.7 per cent and 0.6 per cent, respectively. In these years, locally developed products accounted for 19 per cent and 24.9 per cent, respectively. In 1986, the vast majority of local products registered were applications solutions (71.7 per cent), followed by support software (applications tools) with 16.5 per cent and basic software (systems and utilities) with 11.8 per cent.

However, this limited scheme of import controls combined with the lack of recognition of any legal form of protection for intellectual property rights has created a certain level of externalities, particularly visible in the segment of packages for general-purpose microcomputers. This was especially evidenced by the local developments of operating systems. Despite being one of the most internationalised market segments, this area was selectively targeted by the Brazilian government for promotion.

In Brazil, the broad pattern adopted for microcomputers has been to follow international standards of key micro-electronics components and operating systems, available to be purchased or licensed on a third-party basis. As noted by O'Connor (1985), in terms of hardware, the so-called 'open-architecture' strategy provided a conducive environment for the entry of NICs into the computer industry, allowing indigenous developments of an imitative type without reliance upon licence agreements. This contrasts with the argument put forward by Hoffman (1982) that the science-based nature of IT would tend to undermine the scope for 'unpackaging' technological systems.

The Brazilian case has shown that this imitative strategy is also feasible in software but did not quite take off. SEI (1986) considered that government intervention through the Normative Act No. 27 had been particularly effective in the 16-bit and 32-bit microcomputer families exemplified by the local development of 'MS-DOS-likes' and 'UNIX-likes'. This was mainly undertaken by local computer producers but also created opportunities for a small number of independent software houses.[18] However, the rapid growth of the local computer industry has not generated a broad conducive environment either for local software products or for the promotion of a local software sector.

Farias (1986) interpreted this differentiated outcome as not conflicting with the broad principles of the Brazilian informatics policy, since the local supply of software would not have been capable of meeting the increasing demand generated by the local mini- and microcomputer industry. In turn, the effective control of imports of software products could have jeopardised the growth of the local computer industry. Farias' view reflects an implicit position of the Brazilian IT policy makers, which is further exemplified by the maintenance of undefined market rules for the sector until 1988.

Notwithstanding the predominance of imported packages in the domestic market, the absence of clear market rules and the Brazilian refusal to adopt any legal form of protection for intellectual property rights have generated increasingly strong internal and external political pressures polarised around protectionism and total liberalisation of the software market. For a thorough discussion of the political pressures being faced by Brazil in this period, see Chapter 9. According to Gaio and Segre (1989) these pressures were intensified after 1984, with the approval of the 'Informatics Law', which did not include detailed legislation on software. Finally, the 'Software Law' ratified in 1988 enshrined the prevailing practices in the Brazilian market, providing a legal framework and broadening the scope of liberalisation.

In sum, the small internal progress observed in the promotion of local software activities has been the result of a number of intertwined factors. The policy instruments adopted have not been strong enough to cope with the complexities involved in fostering the sector, which are inherent in the idiosyncratic and highly differentiated nature of software. These problems have been enhanced by the political pressures faced by Brazil towards the liberalisation of its internal market. The following section investigates the extent and nature of opportunities in software which have been generated for indigenous firms.

The Brazilian software and services market

This section provides a more detailed analysis of the Brazilian computer services sector, and the role played by software. The empirical evidence points to a number of important aspects. First, the case of Brazil does not support the conventional theoretical view that economies endowed with cheap skilled labour are liable to reap static comparative advantages in software. Second, in a market where imported packages accounted for the majority share of software trade, applications solutions have been an important area for indigenous firms, but not sufficient for sustaining their growth. Lastly, local computer services firms tend to diversify their service lines whereas the software industry in itself remained at an incipient stage, providing evidence of the inappropriateness of the US model.

The structure of this sector is illustrated by data shown in Tables 4.3 and 4.4. Table 4.3 presents an aggregated view of the Brazilian computer services sector for 1985–6. According to SEI (1986, 1987), local private computer services firms were responsible for an evolving share of about 25 to 32 per cent of the total market (from about US$1.8 billion to US$2.0 billion) and the relatively high growth rate of 42.6 per cent. By 1987, total revenue of the sector amounted to about US$2.3 billion,[19] and traded software accounted for around 52 per cent of this market (US$1.2 billion), 90.6 per cent of which came from imports.[20] Subsidiaries of MNCs stand out as the main channel for those imports, being responsible for 71.5 per cent of the total software traded in this year, with an estimated AAGR of 30 per cent (SEI 1986).

The picture shown in Table 4.3 is further exemplified by a survey carried out by ASSESPRO (Brazilian Association of Processing Services Enterprises) in 1988, with members only (see Table 4.4). This association is basically comprised of local producers of the private sector. In 1987, total revenue accruing to software and

Table 4.3 Computer services in Brazil: distribution of revenue by type of firm, 1985–6, in US$ million

Type of firm	*1985*	*1986*	*GR (%)*
Local comp. serv.	930.61	1255.66	34.9
% total	52.0	62.4	
Public	484.23	618.53	27.7
% total	27.1	30.7	
Private	446.68	637.13	42.6
% total	24.9	31.7	
Local hardw. man.	459.45	234.72	−48.9
% total	25.7	11.7	
MNCs	400.0	520.0	30.0
% total	22.3	25.9	
Total revenues	1790.06	2010.38	12.3

Key: GR – growth rate for the period 1985–6; comp. serv. – independent computer services firms; hardw. man. – hardware manufacturers; MNCs – subsidiaries of multinational hardware manufacturers, and software houses.

Notes: Revenues of local private computer service firms were added up to a share of 20 per cent of the total software revenue with imports (US$500 and 650 million), as estimated by SEI (1986).

Source: Reproduced from Gaio (1990:Table 7.9), MNCs elaborated from (SEI 1986); local firms elaborated from SEI (1986 and 1987)

Table 4.4 Computer services in Brazil: the independent local private sector – distribution of revenue by market segment, 1985–7, in US$ million

Main Activity	*1985*	*1986*	*1987*	*AAGR %*
Total revenues	253.9	354.1	450.3	33.2
Software	22.6	27.8	35.0	24.4
% total	8.9	7.9	7.8	
Processing	193.6	268.9	320.2	28.6
% total	76.3	75.9	71.1	
Training	2.0	4.3	8.0	100.0
% total	0.8	1.2	1.8	
Consultancy	4.6	9.3	16.5	89.4
% total	1.8	2.6	3.7	
Trade hw./sw. & maintenance	13.3	22.9	20.8	25.1
% total	5.2	6.5	4.6	
Others	17.8	20.9	49.8	67.3
% total	7.0	5.9	11.0	

Key: AAGR – average annual growth rate; trade hw./sw. – trade of computers, peripherals and software; others – include system integration, and other technical services.

Note: Data are for ASSESPRO members only, 254 firms declared their total revenue by main type of activity. There is no account of diversification of service lines.

Source: Reproduced from Gaio (1990:Table 7.11), elaborated from ASSESPRO (1988)

services amounted to US$450.3 million. According to ASSESPRO's estimates, revenues with software amounted to 28 per cent of this total, while software houses alone were responsible for a mere 8 per cent share (US$35 million). As remarked by ASSESPRO, this differentiated outcome is due to a significant level of diversification of services among associated firms. The main declared types of complementary activity were: 67 per cent of firms with 'applications solutions', 58 per cent with consultancy, and 45 per cent with data-processing services.

Notwithstanding the negligible share of the local software sector, according to ASSESPRO's survey the industry is expanding in a fragmented mode, with the majority of firms being small. In the period 1985–7, their main source of revenue was 'applications solutions' with an evolving share of 73 per cent to 79 per cent (US$16 to 28 million) of total revenue of the sector and a relatively high AAGR of 29 per cent.[21]

To further characterise the nature of opportunities and constraints in software faced by local private computer services firms, it is useful to recall some of the results shown by a survey conducted by Gaio (1990) in late 1985 to early 1986. This research included not only local producers but also representatives of imported software. Some of its findings are comparable to those of ASSESPRO's survey. For instance, according to the former, firms tend to diversify their service lines considerably. In 1985, their main sources of complementary activity were software production and consultancy respectively. In this year, traded software accounted for 32 per cent of the total revenue accruing to these firms. Imports were responsible for 46 per cent of this share, the vast majority (90 per cent) concentrated on software representatives, whereas local production was found dispersed among a number of different firms.[22]

Among Gaio's findings, two features are worth emphasising. First, the prevailing mode of delivery was packaged software (90 per cent of total revenue), whereas customisation was mainly undertaken by system integrators. Second, a distribution infrastructure appears to be developing, primarily linked with imported packages. In the existing literature, distribution channels have often been stressed as a key component for promoting the software industry.

However, in the Brazilian market, the predominance of imports has jeopardised the scope for local production at a number of levels. Not only the existing distribution channels but users also have shown on the one hand a marked preference for finished products of well-known international brands and, on the other hand, a reluctance to interact with 'unsettled' products of local producers.

User–producer interactions are of particular relevance for the development and evolution of applications-oriented software. The local producers interviewed have also remarked that the production of applications packages involves systematic and extensive efforts, with activities such as redesign, customisation, technical support and training. In the absence of user–producer relations, the outlay for such activities would have to be absorbed by producers or software houses. As a consequence these activities become too risky and potentially unprofitable.

In fact the local software houses interviewed were facing a series of inter-related shortcomings. In particular, they assessed package production as a highly risky enterprise and an erratic source of revenue. Identifying viable world niches is not an easy alternative; it requires a great deal of knowledge, sometimes even work experience in user firms.

The majority of firms classified their mode of production as a 'knowledge-intensive' craft. The entrepreneurs showed an awareness of the limitations involved in this process and regarded an increase in the capital-intensity of their firms as a potential competitive edge for exploiting both the internal and external markets. The basic argument put forward was that to decrease production costs was important but not essential, since their market niches tend not to be price-sensitive, while to increase the flexibility and reliability of products were the core aspects.

In spite of the awareness shown by these firms, they adopted a defensive and short-term business strategy owing to the fierce competition with imported packages combined with an internal market in which close user-producer relations are hard to establish. The small size of firms is combined with low capacity for investing in both product and process technologies.

The Brazilian policy strategy for software and its outcome can be contrasted, therefore, with the general theoretical recommendations about appropriate strategies for LDCs. The empirical evidence of Brazil shows that the early stage of mass market diffusion of IT is largely met by general-purpose packages (e.g. cross-industry applications and systems alike), mainly from external sources. The Brazilian case has also provided evidence that late market entry into these segments is likely to bear little fruit.

The Brazilian case should be situated in the context of a market which is already showing signs of maturation, with oligopolistic tendencies, where the scope for firms outside the US has been largely limited to specialised niches of national markets. As opposed to the theoretical arguments that LDCs hold comparative advantages in mature industries, Brazil has not given evidence of 'enjoying' the supposed benefits provided by their cheap skilled labour. In fact, the only activity which has been reallocated to Brazil was distribution, which generated very limited externalities to other types of activities in software.

The recommended strategy for LDCs concentrating on applications solutions has proved to be feasible. Indeed applications software has been an important source of revenue not only for software houses but also for computer services firms as a whole. However, this strategy has proved not to be sufficient for promoting the sustained growth of a dynamic independent sector in these early stages of IT diffusion when cross-industry packages tend to prevail. Moreover, it can be affirmed that the notion of a local independent software sector composed of firms specialised in software trade has only been pertinent to imports, while for local production it remained at an 'infant' stage.

To sum up, an analysis of the Brazilian computer services and software market provided evidence of the limited results achieved by the policy strategy pursued in software, which focused on the promotion of a local software industry, package production and trade. Such a strategy has shown the inappropriateness of the US model as a benchmark to be followed, and reflected a misunderstanding of the fundamental role played by software in the diffusion process of IT, which is influenced by dynamic complementarities with the hardware industry and the requirements of users.

6 CONCLUSIONS

This chapter discusses possible strategies for the entry of developing countries into the software market. The focus of this discussion is the prevailing view that software provides an optimal point of entry for developing countries into the IT complex. This view is disputed in this chapter which argues that it is based on a fundamental misunderstanding of the nature of software and on the use of inappropriate benchmarks.

The analysis of the international market for commercial data processing software has pointed to a number of controversial issues. Software has often been suggested as an area of significant economic opportunities, not only for IACs but also for LDCs. There is evidence that the US has maintained international leadership in software trade as a consequence of its competitive edge in cross-industry packages. As a result, there has been a tendency to view the US model as a benchmark. However, this pattern does not seem to have been replicated by any other major industrialised economy, nor has the specific nature and level of local activities in these countries been clearly examined. The fact that the market is already showing signs of maturity, with oligopolistic tendencies, means that the issues of scope for competing, and dependence on imports are not only pertinent to LDCs but also to IACs.

Brazil is an interesting case to examine, as a major NIC which has been explicitly committed to promoting an indigenous IT complex at both economic and technological levels. In contrast with the prevailing theoretical view, Brazil has not chosen software as a point of entry into the IT complex. Instead, the Brazilian experience reveals the complexities of promoting local software activities in early stages of IT diffusion. When cross-industry packages tend to prevail, the scope for local firms being in a position to compete has been limited.

The perceived strategy of concentrating on applications segments has proved to be feasible but not sufficient for promoting a dynamic local software industry. By contrast, Brazil has not shown any significant benefits from static comparative advantages of low-waged skilled labour. Thereby, taking into account the empirical evidence from Brazil, the assumptions of the product cycle theory do not seem to be appropriate to a design-intensive activity such as software production. The distribution of imported packages was the only phase reallocated to the domestic market led by local subsidiaries of MNCs. In fact, the predominance of imported products in the internal market had negative implications on the extent of learning undertaken by local firms. Both the existing distribution channels and users have shown a marked preference for products of well-known international brands, and resistance to interaction with local producers.

In Brazil, the policy strategy devised for software has been to foster the local software industry, with emphasis on packate production and trade. This strategy followed a common pattern observed in the informatics policy of focusing on the production of hardware. The diffusion of IT was neglected and policies for applications-oriented software were neither designed nor implemented. Case studies presented in this book on banking automation and petrochemicals (see Chapters 3 and 5) suggest that such a type of software is crucial for the achievement of structural competitiveness. They also show that, in the absence of close user-producer relations, the development of applications oriented software may be too expensive and risky for producers.

In accordance with Perez's and Soete's line of argument for NICs such as Brazil, windows of opportunity may be generated by an early entry into emerging verticalised markets where the country has already accumulated an industrial and technological capacity. However, to effectively exploit these potential opportunities requires a strategy which orchestrates a number of complex aspects such as: significant and sustained investments and learning efforts; the appropriate choice of markets, timing and institutions capable of acting as synergic forces to promote both the generation and diffusion of innovations in IT-based systems. Within this institutional framework, it is important to emphasise the convenience of fostering a 'three-pronged' collaboration between users, producers and the scientific community and that software activities should not be dealt in isolation but in terms of interdependence and complementarity with the requirements of users and the hardware industry.

NOTES

1 This chapter draws heavily upon Gaio (1990).

2 For analyses of the product cycle theory see, for instance, Hirsch (1965) and Vernon (1966) for arguments about its implications on the location of production facilities and international trade, and Utterback and Abernathy (1975) for analysis of evolutionary interdependencies between product and process innovations.

3 For some advocates of this line of argument see, for example, Rada (1980, 1982), UNCTAD (1984), Erber (1984) and Kaplinsky (1982, 1985 and 1987).

4 The convenience of a strategic entry into early phases has been evidenced by Hobday (1986) who characterised Brazil as one of the first countries on the international scene to be committed to a fully digital telecommunications system, and by Park (1987) in the case of South Korea's entry into the development and production of dynamic random access memory devices, competing with the US and Japan. Both cases cannot possibly be explained on the basis of 'natural' factor endowments. Quite the opposite, they have been the result of explicitly orchestrated national strategies.

5 An in-depth conceptualisation of the notions of paradigms and trajectories is found in Dosi (1982). For a detailed analysis of the electronics evolution see also Dosi (1984).

6 The computer services sector is composed of a wide variety of services, and software is usually considered as part of this cluster of activities. In the current literature there is no consensus over market segmentation. However, the classification of IDC-C&L (1987) presented here illustrates the range of services comprised by the sector, which is divided into three main categories. The first is processing services which include all the services supplied to a range of customers and priced on the basis of the rate of use of resources (e.g. computer and communications facilities). The second category is professional services supplied on a customised or contractual basis (e.g. data-processing consultancy, training, facilities management and customised software). The latter type may be regarded as a separate category. The third is software products or packages produced for sale to a variety of users.

7 Among the shortcomings encountered in quantitative analyses, the following are worth emphasising. Data are unreliable, often being estimates, whereas comprehensive statistics about national markets are not yet available; and there is no agreement over types of software or market segmentation.

8 According to Fertig (1985), in 1984 there were around 3,000 firms, while in 1985 this number was reduced to 1,300.

9 In 1984, the top fifteen independent suppliers accounted for 34 per cent of the total sales of the industry (Future Views 1986, No. 15).

10 The level of concentration observed in microcomputer software is not only in relation to a small number of firms with a large share of total revenues, but also in terms of specific products. For instance, in 1986, of the 27,000 programs for microcomputers available in the US market, only four were responsible for more than 50 per cent of the total retail sales of this segment (*Computer Systems News* 1986, Nov.).

11 Countries were ranked by the size of their software market in 1984.
12 Estimated by INPUT Europe in OECD (1989:59).
13 Figures elaborated by IDC Japan in OECD (1989:59).
14 For details about the importance of software engineering for the technical and economic feasibility of customised or small-scale software production, which is of particular pertinence to vertically-oriented applications, see Gaio (1990).
15 The US model is characterised by a high level of de-verticalisation and commodification of software production, which has been the most appropriate institutional framework for the early stage of mass market diffusion of IT, but also led to two concomitant trends: a growing gap between users and producers (Valdez 1988 and OECD 1984) and what OECD (1984:14) coins as a 'duality in the computer world, with a "leading edge" community becoming more and more marginalised within an increasingly sluggish industrial environment . . .' These two tendencies may be turned into barriers for the generation and diffusion of vertically-oriented software products and services which require stronger user-producer interactions coupled with the diffusion of radical process innovations in software. For a detailed analysis and evidence to support this approach see Chapters 3 and 7 of Gaio (1990).
16 For more details on the Brazilian policy for telecommunications see Hobday (1986) and for the computer industry see, for instance: Tigre (1985, 1987), Piragibe (1985) and Frischtak (1986).
17 Erber (1985), Tigre (1987) and Meyer-Stamer (1990) provide critiques of the lack of inter-sectoral co-ordination in the Brazilian IT policy and its negative impacts on the promotion of the electronic components sector.
18 Gaio (1990:Chapter 6) presents a detailed analysis of the development efforts undertaken by Brazilian computer firms in operating systems. The author contrasts the scope for learning provided by licensing agreements with 'non-institutionalised' mechanisms of technology transfer and examines the prospects of the latter maintaining its validity as a significant source for learning.
19 Figures estimated for 1987 were calculated on the basis of 1986 data presented in Table 4.3, applying 12.3 per cent as a steady growth rate.
20 The traded software market figure was the result of adding up SEI's (1986) estimates of imports with ASSESPRO's estimates of local production.
21 In this period, 'applications tools' were responsible for a share of about 13 per cent (US$3 to 5 million) and an AAGR of 28 per cent, whereas 'systems and utilities' held a decreasing share of 9 per cent to 8 per cent, with an AAGR of −7 per cent (ASSESPRO 1988).
22 In fact, data-processing bureaux stand out as the main local producer of software, with about 41 per cent of total revenue of this segment, followed by software houses, with 25 per cent, and system integrators, with 23 per cent (Gaio 1990).

BIBLIOGRAPHY

ASSESPRO (1988) *PERSSONA Survey*, working paper, Rio de Janeiro.
CAPRE (1979) *Balanço das Atividades da CAPRE em Relação a Software até 15.04.1979*, working paper, Rio de Janeiro.

Chaponnière, J.R. and Fouquin, M. (1989) 'Technological change and the electronics sector – perspectives and policy options for Taiwan', working paper, Paris: OECD.

The Computer Marketing Newsletter (1986) London, October.

Computer Systems News (1986) November.

Correa, C.M. (1989) 'The legal protection of software – implications for latecomer strategies in newly industrialising economies (NIEs) and middle-income Countries (MICs)', working paper, Paris: OECD.

Dosi, G. (1982) 'Technological paradigms and technological trajectories: a suggested interpretation of the determinants and directions of technical change', Research Policy, Amsterdam: North-Holland Publ.

—— (1984) *Technical Change and Industrial Transformation: The Theory and an Application to the Semiconductor Industry*, London: MacMillan.

Erber, F. (1984) 'Microeletrônica: reforma ou revolução', *Revista Brasileira de Tecnologia*, Vol. 15, No. 1, Brasilia.

—— (1985) 'The development of the "electronic complex" and government policies in Brazil', *World Development*, Vol. 13, No. 3, Elmsford, NY.: Pergamon Press.

Ernst, D. (1981) *The Software Market Conditioning Factors and Possible Future Trends. An Analysis undertaken from a Third World Perspective*, Vienna: UNIDO.

Farias, O.L.M. (1986) *A Situação do Mercado de 'Software' no Brasil*, working paper, Rio de Janeiro: INPI.

Fertig, R. (1985) *The Software Revolution*, North Holland: Elsevier Sciences.

Freeman, C. and Soete, L. (1986) *L'onda Informatica – nuove tecnologie e occupazione*, Milano: del Sole 24 Ore.

Frischtak, C. (1986) *The Informatics Sector in Brazil – Policies, Institutions and the Performance of the Computer Industry*, Washington, DC: Industrial Strategy & Policy Division, the World Bank.

Future Views (1986), Vol. 5, No. 15, Dallas, Texas.

Gaio, F.J. (1990) *The Development of Computer Software Technological Capabilities in Developing Countries – a Case Study of Brazil*, D.Phil. thesis, Brighton: Science Policy Research Unit, University of Sussex.

Gaio, F.J. and Segre, L.M. (1989) *Uma Análise da Política de Software no Brasil*, paper presented at the Congresso Iberoamericano sobre Desarrollo de la Informatica, Quito, CREI/CALAI, 13–15 Nov.

Genthon, C. (1985) *L'Industrie Electronique Mondiale – Innovation, Intervention Etatique*, Ph.D. thesis, Grenoble: Université des Sciences Sociales de Grenoble.

Grindley, P. (1987) 'The U.K. software industry: a description of the industry and evaluation of policy', working paper, London: London Business School.

Hirsch, S. (1965) 'The United States electronics industry in international trade', *National Institute Economic Review*, London, November.

Hobday, M. (1986) 'Digital telecommunications technology and the third world: the theory, the challenge, and the evidence from Brazil', D.Phil. thesis, Brighton: Science Policy Research Unit, University of Sussex.

Hoffman, K. (1982) 'Microelectronics and industry in the third world: policy issues and research priorities', working paper, Brighton: Science Policy Research Unit, University of Sussex.

IDC (1986) *Software and Services Industry – Review and Forecast*, Cambridge, Mass.: ADAPSO.

IDC-C&L (1987) 'Computer services industry 1986–1996 – a decade of opportunity', working paper, London, Department of Trade and Industry.

Kaplinsky, R. (1982) *Electronics, Comparative Advantage and Development*, UNIDO, London: Frances Pinter.

—— (1985) 'Electronics-based automation technologies and the onset of systemofacture: implications for third world industrialisation' *World Development*, Vol. 13, No. 3, Elmsford, NY.: Pergamon Press.

—— (1987) 'Technological revolution and the international division of labour in manufacturing: a place for the third world?', paper presented at the EADI Conference on New Technologies and the Third World, Brighton: Institute of Development Studies, University of Sussex, July.

Katz, R. (1987) *La Industria del Software en Estados Unidos: Structura y Comercializacion del Producto*, Buenos Aires: CALAI.

Kopetz, H. (1984) *Guidelines for Software Production in Developing Countries*, Vienna: UNIDO.

Meyer-Stamer, F. (1990) *From Import Substitution to International Competitiveness – Brazil's Informatics Industry at the Crossroads*, mimeo, Berlin: German Development Institute.

Momigliano, F. and Siniscalco, D. (1982) 'The growth of service employment: a reappraisal', *Banca Nazionale del Labore*, No. 142, September, Rome.

Narasimhan, R. (1984) *Guidelines for Software Development in Developing Countries*, IS.439, Vienna: UNIDO.

O'Connor, DC (1985) 'The computer industry in the third world: some policy options and constraints', *World Development*, Vol. 13, No. 3, Elmsford, NY.: Pergamon Press.

OECD (1984) 'Software: a new industry', working paper, Paris.

OECD (1989) 'The internationalisation of software and computer services', Paris, No. 17.

Park, Y. (1987) 'The national system of innovation in Korea, with an introduction to the semiconductor industry', M.Sc. dissertation, Science Policy Research Unit, University of Sussex.

Perez, C. (1985) 'Microelectronics, long waves and world structural change: new perspectives for developing countries', *World Development*, Vol. 13, No. 3, Special Issue on Microelectronics, Elmsford, NY.: Pergamon Press.

—— (1987) 'New technologies and development', paper presented at the Conference on New Waves of Technology, International Conditioning and Domestic Industrial Responses, Turin: University of Padova, May 28–9.

Piragibe, C. (1985) *Industria de Informática: Desenvolvimento Brasileiro e Mundial*, Rio de Janeiro: Editora Campus.

Rada, J. (1980) *The Impact of Microelectronics: A Tentative Appraisal of Information Technology*, Geneva: International Labour Office.

—— (1982) 'Technology and the north-south division of labour', in R. Kaplinsky (ed.), *Comparative Advantage in an Automating World*, IDS Bulletin, Vol. 13, No. 2, Brighton: Institute of Development Studies, University of Sussex, March.

Schware, R. (1987) 'Software industry development in the third world: policy guidelines, institutional options, and constraints', *World Development*, Vol. 15, No. 10/11, Elmsford, NY.: Pergamon Press.

SCSS (1981) *Relatório da Comissão Especial de Software e Serviços*, Brasilia: SEI.

SEI (1986) *Diagnóstico do Mercado de Software no Pais*, working paper, Brasilia.

—— (1987) *Panorama do Setor de Informática*, Vol. 7, No. 16, Special Edition, Brasilia.

Soete, L. (1983) 'Long cycles and international diffusion of technology', working paper, Brighton: Science Policy Research Unit, University of Sussex.

—— (1985) 'International diffusion of technology, industrial development and technological leapfrogging', *World Development*, Vol. 13, No. 3, Elmsford, NY.: Pergamon Press.

Tigre, P.B. (1983) *Technology and Competition in the Brazilian Computer Industry*, London: Frances Pinter.

—— (1985) *Indústria Mundial de Informática: Análise das Tendências Econômicas e Tecnológicas Recentes*, working paper, Rio de Janeiro, IEI/ Universidade Federal do Rio de Janeiro, Oct.

—— (1987) *Indústria Brasileira de Computadores – perspectivas até os anos 90*, Rio de Janeiro: Editora Campus.

UNCTAD (1984) *New and Emerging Technologies: Some Economic and Commercial Development Aspects*, TD/B/C.6/120, Geneva, 2 Aug.

UNIDO Secretariat (1983) *Problems of Software in Developing Countries*, Vienna: UNIDO.

Utterback, J. and Abernathy, W.J. (1975) 'A dynamic model of process and product innovation', *Omega*, Vol. 3, No. 6, Elmsford, NY.: Pergamon Press.

Valdez, M.E.P. (1988) 'A gift from Pandora's box: the software crisis', Ph.D. thesis, Edinburgh: University of Edinburgh.

Vernon, R. (1966) 'International investment and international trade in the product cycle', *Quarterly Journal of Economics*, May, Cambridge, Mass.

5 Why the market reserve is not enough

Lessons from the diffusion of industrial automation technology in Brazilian process industries[1]

Ruy de Quadros Carvalho

> *It is not possible to come to grips with the complexities of technology, its interrelations with other components of the social system, and its social and economic consequences, without a willingness to move from highly aggregated to highly disaggregated modes of thinking. One must move from the general to the specific, from technology to technologies. One must be even prepared to 'dirty one's hands' in acquiring a familiarity with the relevant details of the technology itself.*
>
> Nathan Rosenberg (1982)

1 INTRODUCTION

This chapter is concerned with the diffusion of electronic automation technology among Brazilian industrial users and its implications for industrial modernisation. It is argued that the new technologies have been adopted by Brazilian firms in an inadequate fashion, referred to here as *passive modernisation*.

Although the diffusion of industrial electronic equipment has been relatively quick in Brazilian process industries, they are far from reaping the potential technological and economic benefits. In the petrochemical industry, the focus of this case study, the adoption of digital control systems (based on Information Technology) has been mainly regarded as a solution to the problems resulting from the previous crop of control equipment available in Brazil, rather than as a technological strategy to improve firms' competitive position. The major determinants of this situation are the technological fragility of both users and suppliers, and the lack of institutional support in the process of diffusion.

The main policy conclusion is that, if upgrading of Brazilian industry is to be pursued, the Brazilian information technology policy should not be restricted to the development of the electronics industry, but should widen its scope to the actual use of its products in the economy. This requires the integration of the policy for the electronics industry into a broader industrial policy strategy.

The chapter starts by directing the reader's attention to the importance of the diffusion of the new technology in Brazilian process industries (Section 2). Section 3 presents the empirical investigation on the diffusion of digital control technology in the Brazilian petrochemical industry. The research was based on a survey of eighteen petrochemical firms, users and non-users of the new technology, located in the states of São Paulo and Bahia. The survey comprised extensive interviews with engineers and industrial managers, visits to plants, and informal conversation with workers.

It is shown that the diffusion of digital control techniques has been relatively fast given that this occurred during a period of uncertainty and unstable economic activity (Section 3). This diffusion process is not unproblematic, however. In the assessment by user firms, a great many of the difficulties they face in the adoption of digital control technology originate in problems related to the supply of the new technology. Major problems are the high prices and sometimes inadequate quality of Brazilian equipment, and the lack of technological capability of suppliers.

While the adoption of new technology is extensive, in most cases it amounts to little more than *passive modernisation*, that is equipment replacement motivated by obsolescence. The new techniques are used mainly to perform conventional control functions. Only a few firms were adequately exploiting the potential of the new technology: that is, improving efficiency through the use of more complex software for plant optimisation and using the digital basis for process simulation. This is referred to as *under-utilisation* of the new technology.

The major determinants of passive modernisation and under-utilisation are then examined. They stem mainly from the technological limitations of both Brazilian suppliers and users of the new technology. The effects of these technological limitations are aggravated by insufficient supply of engineering services and lack of institutional support to the process of diffusion.

The theoretical and policy implications of the case study are presented in Section 4. The broad conclusion is that countries in which import substitution was the basis of industrialisation face specific problems in order to adapt to the requirements of the new technologies. The technological limitations of both suppliers *and users* have been obstacles to the development of synergies in their interaction. Thus, it is concluded that effective assimilation of previously imported technologies is an important factor for the successful adoption of the new technologies.

Finally, it is argued that a market reserve for suppliers of digital equipment is not enough and that policy should be equally concerned with diffusion. Facilitating resources are required in order to enable firms to successfully adopt new technologies and improve their competitiveness. Such support should be provided in a context in which government seeks to create a competitive environment and integrates information technology policy with other industrial and macro-economic policies.

2 THE MARKET FOR INDUSTRIAL ELECTRONICS EQUIPMENT IN BRAZIL: PROCESS INDUSTRIES AS MAJOR USERS

This section shows that process industries[2] are the main users of micro-electronics-based industrial equipment in Brazil, a fact little explored by academic literature and the media.[3] Demand from process industries accounted for at least two-thirds of the Brazilian market for industrial electronic equipment during the 1980s and is likely to continue to be high in the 1990s.

The Brazilian market for electronic industrial automation equipment boomed in the 1980s, despite the economic crisis. The average annual rate of growth (imports excluded) between 1984 and 1988 was 25 per cent, the value of annual sales rising from US$86 million to US$255 million in this period (SEI 1989).

Growth in this market has been a story of business success for national firms. Between 1981 and 1984, SEI extended its market reserve policy in order to protect Brazilian producers of industrial electronic equipment. Imports of such equipment have been kept under strict control by SEI; they have been permitted only in the case of products which Brazilian firms cannot supply. Hence the bulk of the local market for industrial automation equipment has been supplied by national firms. These multiplied rapidly from ten firms in 1982 to over ninety in 1987. Subsidiaries of foreign firms have participated only in the segment of CAD/CAE systems, in which they supply large systems.

Throughout the 1980s there was great concern in Brazil with the economic and social implications of the diffusion of electronics-based automation in local industry. Research fuelled by this concern focused on the engineering industries such as machine tools, motor vehicles and electronics.[4] However, the actual diffusion of micro-electronics-based industrial equipment in Brazil suggests that the major users of the new technology are process industries rather than engineering industries. Electronic equipment for process control accounts for over 70 per cent of the total market for industrial electronics-based automation[5] equipment.

Even if we do not take Programmable Logic Controls (PLCs) into account (since they are utilised in both engineering and process industries), sales of process control electronic equipment (US$79.4 million) were more than twice those of equipment for automation of manufacture (US$37.7 million) in 1988. Sales of Digital Distributed Control Systems (DCSs) (equipment to 'computerise' process control in the chemical/petrochemical, steel, pulp and paper and food industries) alone were comparable in monetary terms to sales of CNCs, robots and CAD systems together. The smaller and cheaper Control and Supervision Systems, used mainly in the electricity and mining industries, sold as much as CNCs, for which a vast range of applications exists in machinery for discrete production.

In 1989, sales of DCS rose to US$100 million (*Química e Derivados*, Jan./Feb. 1990), making this equipment the single most sold industrial electronic equipment in the Brazilian market and revealing an acceleration of the diffusion of industrial electronic equipment in the process industries, particularly in the pulp and paper and in the chemical/petrochemical industries.

Apparently this trend seen in the Brazilian process industries is reproducing diffusion trends in developed countries (OECD 1985). Therefore, what seems to be different in Brazil is not the rapid diffusion of industrial electronic equipment in process industries, but the much slower response of engineering industries in adopting electronics-based industrial automation.[6]

The concern in this section is to emphasise that, *in Brazil, the diffusion of industrial electronic equipment in process industries has been considerably more dynamic than in the engineering industries*. This has occurred for the following reasons.

The first is that investment in computerised control systems need not be high. In process industries, the adoption of digital control does not entail major changes in *manufacturing* equipment, but rather the incorporation of new *control* equipment to plants which are already highly automated. Thus, investment in digital controls represents on average only 10 to 12 per cent of global investment in plant.[7] Given that efficiency gains affect the global performance of the plant, investment return is rapid.

Second, the bulk of the users of process control equipment operate in the energy and intermediate goods industrial sectors. Generally, they are highly capital-intensive firms, with the financial ability to undertake investment of the type under discussion here. It should be noted that state participation, and particularly Petrobrás participation, is very pronounced in the industrial sectors which are the major users of industrial electronics equipment for process control.[8]

Finally, there were major technical and economic difficulties in the local supply of the previous control systems, motivating an early replacement by digital systems. This was an unintended outcome of the policy of the former Brazilian CDI (Council for Industrial Development), aimed at the substitution of imports of conventional control equipment.[9]

Considering the influence of these forces, it is not surprising that Brazilian process industries forecast a high and steady demand for industrial electronic equipment in the 1990s (SEI 1988). The building of new plants and the modernisation of existing ones, in the pulp and paper, steel, chemical/petrochemical, and electricity sectors, will account for the bulk of the market. This is why DCSs are expected to continue to be the equipment with the highest sales (*Guia de Automação Industrial* 1990, No. 7).

Therefore, the micro-electronics-based modernisation of Brazilian process industries has already reached significant proportions, and is likely to become even more important in the near future. This process of technical change entails major implications for both user firms and suppliers of automation equipment and services. These will be addressed in the following section which is based on a study of the adoption of digital control technology in the Brazilian petrochemical industry.

3 THE PROCESS OF DIFFUSION OF DIGITAL CONTROL TECHNOLOGY IN THE BRAZILIAN PETROCHEMICAL INDUSTRY: QUANTITATIVE AND QUALITATIVE ASPECTS

The purpose of this section is to present the main findings of the case study and to analyse the social and economic forces behind them. In addition to a quantitative account of the diffusion of hardware, a qualitative assessment is also presented. This is a necessary task, since the new technology can be employed in different ways and for different purposes. The analysis of economic and institutional determinants follows at the end of this section.

The quantity of new technology incorporated by Brazilian firms: rapid diffusion

The diffusion of the new control techniques in the Brazilian petrochemical industry began in the early 1980s and has been rapid during a period of uncertainty and low economic activity. At the time of our research, 15 per cent of all petrochemical plants had already adopted DCSs in their plants. More importantly though, in the years to come, 80 per cent of the planned investment in control equipment, either in new plants or in the modernisation of existing plants, is to be realised using digital techniques. The diffusion of industrial process computers has been much slower, however. This reveals that, in spite of relatively rapid diffusion of digital control hardware, the current systems are unable to carry out the more complex tasks of integration and control.

According to data provided by ABIQUIM (the Brazilian Association of Chemical Industries), in 1987/88 approximately 10 per cent of the total number of loops installed in Brazilian chemical/petrochemical plants were digitally controlled, 8 per cent being controlled by DCSs. This is a modest share, yet it is impressive if we consider that in 1982 only one chemical/petrochemical firm in Brazil used this technique. This number had increased to eighteen firms by the time of this research. Evidence of the importance of electronic industrial equipment in the Brazilian petrochemical industry also comes from the use of Programmable Logic Controls (PLCs) in safety loops. ABIQUIM data show that, in 1987, 22 per cent of the total number of points of logic control in the chemical/petrochemical industry were controlled by PLCs.

Most users of DCSs in the Brazilian chemical industry are large firms producing petrochemicals. All but one are firms with more than 500 employees, based in capital-intensive, highly automated plants. Most of them operate continuous-flow production processes. Within the population of petrochemical firms defined for this research, fourteen companies have already adopted DCSs. Out of these, six are firms associated with the Petroquisa group (controlled by Petrobrás, the state-owned oil monopoly). Two are fertiliser producers controlled by Petrobrás, four are MNC subsidiaries, and only two are wholly controlled by Brazilian private groups. This is in line with the comments in this chapter on the high profile of state-controlled firms (and particularly that of Petrobrás and affiliates) in the shaping of demand for industrial electronic equipment in Brazil.[10]

Data on planned investment in process control equipment are perhaps more revealing of the trend for digital, integrated control systems to completely replace conventional techniques in the Brazilian chemical industry. For instance, investment intentions documented by ABIQUIM at the time of the research and confirmed by fieldwork showed that, for the period 1987–90, 97 per cent of all planned investment in room control equipment (approx. US$100 million) was to be realised in digital controls, being 86 per cent in complex, integrated control systems (DCSs). Investment in conventional control techniques was only to be significant (approximately US$25 m.) in field controls.

Table 5.1 shows the diffusion of digital process control technology within my sample of petrochemical firms. Firms A to H are users of DCSs. Firms F and G are recently built plants (1988), having introduced DCSs comprising all process loops from the start. Firm C

Table 5.1 Hardware technologies used in process controls[a] in selected Brazilian petrochemical plants, 1988

	% of process loops per type of technology			*% of safety controls*		*Number of process computers*[f]
Firms	*Analog pneumatic*[b]	*Analog electronic*[c]	*DCS*[d]	*Conventional relays*	*PLC*[e]	
A	–	84	16	54	46	3Mn
B	–	50	50	50	50	1Mn, 1Mc
C	38	–	62	15	85	–
D	2	48	50	100	–	1Mc
E	74	6	20	50	50	–
F	0.5	1.5	98	–	–	–
G	–	–	100	50	50	–
H	75	5	20	98	2	–
I	80	20	–	30	70	12Mc, 2Mn
J	35	64	–	87	13	4Mc
K	99	1	–	100	–	–
L	95	5	–	(nd)	(nd)	–
M	80	20	–	100	–	2Mc
N	95	5	–	90	10	–
O	50	50	–	100	–	–
P	100	–	–	100	–	–
Q	100	–	–	–	–	–
R	–	100	–	–	–	

Notes: a. It includes equipment used in room control functions; it does not comprise field instrumentation.
b. Analog, pneumatic instrumentation used in room control.
c. Analog, electro-electronic instrumentation used in room control.
d. Distributed digital control systems.
e. Programmable Logic Controls.
f. Mn stands for minicomputers; Mc stands for microcomputers.

Source: Author's survey of firms

first introduced a DCS under a project to expand the plant, based on the introduction of a new process for higher value-added products. Other users have introduced DCSs within projects for the modernisation of their control equipment.

So far most DCS users have partially adopted the new technique, that is, only a percentage of their process loops are digitally controlled. Though partially covering the plant, DCS is always a comprehensive system, in that it controls all loops (therefore, all industrial operations) in the equipment of the plant in which it is introduced. For instance, in the case of firm C, 62 per cent of its monitoring and control loops, which correspond to the major product stream of the plant, are controlled/monitored by one DCS. This firm expected to introduce a second DCS before 1991, in order to modernise control of its older area and to integrate 100 per cent of plant control into a digital control *system*. Therefore, partial adoption of digital control may only mean a step towards the full 'electronisation' of plant control.

This would be exactly the case with most partial users of DCSs in the sample, if their investment plans for the modernisation of process control until 1992 were confirmed (Table 5.2). Combined with Table

Table 5.2 Planned investment in process control equipment in selected Brazilian petrochemical plants, 1988–92 (US$ million)

	Value of investment per type of technology				*Type of investment*
*Firm**/ *year*	*Anal. pneu.*	*Anal. elect.*	*DCS*	*Process computer*	
A (88/89)			15		Complete replacement
B (88/91)			12	1.6	Complete replacement
C (88/91)			2.9		Complete replacement
E (88/90)			23		Complete replacement
I (88)		0.056		O.06	
J (88/92)			19.4		Modernisation
K	Future investment in digital control; no fixed date				
L	Future investment in digital control; no fixed date				
M			0.3		Cont. pilot plant
P					
Total		0.056	72.6	1.66	

* Firms D, N, O, Q, and R have not given information on this topic.

Source: Author's survey of firms

5.1, data from Table 5.2 reveal two important tendencies. First, most firms which were already users of the new technology (except for H) presented plans to completely replace conventional control with decentralised digital systems. This suggests that, once it has begun, the tendency is for a full-length, radical change of control technology.

Second, it seems that non-users will continue to lag behind for some time. With the exception of firm J, non-users have not yet made plans for the future introduction of digital technology. Firm J is a MNC subsidiary, which is drawing on its parent company's experience to define its strategy for a complete change of control instrumentation. Firms K to N, also part of the Petroquisa group, declared that they are aware they will have to change their control equipment in the years to come (their current controls are based on the most conventional and problematic pneumatic technology) and that they are very likely to shift to digital controls, though not necessarily integrated ones (DCSs). But no clear decisions have yet been made.

To conclude this section, it is worth stressing the considerable importance already attained by micro-electronics-based industrial automation in Brazilian petrochemical industries. Further evidence is given by the extensive use of PLCs in safety control (see Table 5.1).

However, the survey has also revealed that there has been very little diffusion of industrial process computers (Table 5.1) and of advanced control software. This was confirmed in the studies carried out by ABIQUIM (ABIQUIM 1986). The low level of diffusion of process computers reflects the fact that the new technology has been used mainly for conventional functions, as well as the limits of suppliers' capability in basic software. We will return to these points later in this chapter.

The next section examines what user firms have considered to be factors holding back diffusion. Although users and non-users have considered Brazilian digital control systems to be more reliable than conventional ones, they have pointed to a number of problems in their experience with the new technology, which have inhibited a faster diffusion. Among these, they have stressed those which have been often associated with the effects of Brazilian IT policy: price and quality of equipment, and the technological capability of suppliers.

How user firms assess Brazilian suppliers of industrial electronic equipment

Before examining the adoption of new technologies in petrochemical user firms, we should ask how these users assess the suppliers of their

Table 5.3 Rating* of obstacles to the adoption of DCS/computerised process control in selected Brazilian petrochemical firms

High cost of Brazilian equipment	126
Obstacles created by PNI	112
Firm's staff lacks confidence in suppliers' future technological capability	92
Benefits do not off-set costs	92
Inadequacy of human resources policy for informatics area	87
Shortage of the right skills for devising and implementing digital control project	86
Inadequacy of government industrial policy	67
Quality of Brazilian hardware	66
Shortage of services	63
Maintenance problems	59

Source: Survey of firms.

* Total number of respondents: 16.

technology. One of the most striking issues during fieldwork was the engineers' criticism of the Brazilian informatics policy. This is also revealed in the high rating given to the item 'obstacles created by PNI (Política Nacional de Informática)', in their assessment of the obstacles to a greater diffusion of digital control technology (Table 5.3).

The formulation is admittedly vague, but is presented in this form in order to attain the respondents' general feeling about the policy. Such a result is useful for a political assessment, at least. One would expect to hear staff of MNC subsidiaries blaming the policy. But most firms in the sample belong to the Petroquisa group, in which the influence of nationalist values inherited from Petrobrás by engineers is very strong.

Criticism is more concrete in the form of the other points listed in Table 5.3. In fact the table works in two directions, since the factors receiving lower scores could be interpreted as the ones in which Brazilian industrial electronic equipment performs best. In this section we will comment on those related to the supply of industrial electronics equipment.

High cost and low quality of Brazilian equipment

The first three factors in Table 5.3 have been singled out as the most problematic by all groups of firms in the survey: national users (eight firms), national non-users (seven firms), and MNC subsidiaries (three firms). The high cost of Brazilian digital control systems was at the top of the list of most firms. Prices of installed Brazilian DCSs varied between US$2 and US$5 million, depending on the size of equipment. Given the large range of sizes, as well as of alternative peripheral components, it is difficult to compare national prices with international ones. Some of the respondents mentioned that national equipment, including installation services, were two and a half to four times more expensive than installed imported equipment. A similar evaluation was made in an interview with a technical advisor to ABIQUIM.

In spite of their higher prices, a number of users and potential users would prefer equipment assembled in the country to imported equipment, owing to maintenance convenience. However, high prices have contributed to the delay of decisions to introduce digital control equipment. This is an aspect particularly singled out by non-users. Expensive equipment is an additional factor in causing potential users to delay decisions in this area until the obsolescence of conventional equipment, rather than pursuing a strategy based on technological upgrading. In this sense, the higher costs of Brazilian digital control systems have contributed to holding back diffusion.

As far as prices are determined by production costs, a number of cost determinants deserve comments. Producers of digital equipment complain about two factors: the high tariffs imposed by the Brazilian government on imported electronic components and the small volume of the Brazilian market for digital control systems, which does not make up for minimum scales of R&D.[11] To a great extent, these problems are less a result of market protection of an infant industry in general and more the way it has been managed.

Consider the issue of market size and minimum scales. Its effects would have been lessened if the policy of the licensing of producers had taken into account this aspect and, thus, had licensed fewer producers. Nevertheless, five producers of DCSs were licensed by SEI to produce DCS with licensed technology. A sixth producer entered the market with its own technology, and two more submitted requests to enter the market with licensed technology at the time of this research. Since these have been accepted also, the Brazilian market for DCS, US$100 million per annum, will be shared by eight producers, not taking into account three other producers who entered the lower segment of the market, with small-size DCSs based on microcomputers. This accounts for average annual sales of four medium-size DCSs per firm.[12]

Would not this proliferation of firms bring about 'healthy competition in an expanding market'? This argument reveals a lack of comprehension of the basic characteristics of the Brazilian market for DCS and other industrial electronic equipment. First, competition is limited from the first sale, because each Brazilian producer has based its product design on a distinct technology[13] and these technologies are not compatible among themselves, unless through costly investment in additional equipment and software. Once a user firm has installed the first DCS, it is bound to order from the same supplier for future equipment investment. This is why users commonly say that they are 'captive' to suppliers, a feature that has raised a fierce debate between Brazilian users and producers about the adoption of communication protocols.

Second, in spite of very good prospects for the next five years, high rates of growth in the internal market for DCS will decline at some point in the future. As is the case with other electronic goods used in industrial automation (and in contrast to data-processing equipment for general use), applications of DCS are limited to a relatively small number of industrial sectors. Once the process of replacement of conventional controls is concluded, the dynamism of the internal market will fade substantially. Thus, an outlet to external markets is a requirement for long-term survival in this segment of the electronics industry. As such, this feature should be the main one to be taken into account in a policy for this segment. The market for industrial electronic equipment is not the market for microcomputers; thus, policies which have worked in the latter may not succeed in the former.

The question of an export outlet touches upon the problem of tariffs on components and the coordination of trade and industrial policy. Producers of DCSs complain that high tariffs on the value of imported electronic components are a major cause for final products being expensive. This seems to reveal a lack of coordination within the government departments. Since SEI defines deadlines for the nationalisation of production of most components of digital controls, it would seem reasonable that tariffs on components should be lowered during the transition phase, in order not to overburden end users. In the medium and long term, however, tariff reduction could be a useful tool to motivate Brazilian producers to export.

A further critical point raised by user firms is the quality of equipment. Problems here are more localised within certain specific products and suppliers. In general, the quality of performance of Brazilian digital control equipment seems to be very uneven, varying between good and bad according to the specific product, as the assessment by ABIQUIM's Process Control Commission indicates (Table 5.4).

Table 5.4 Quality assessment of Brazilian digital control equipment by the ABIQUIM process control commission, 1986

Product	*Assessment*
Process minicomputers	Insufficient
Process superminicomputers	Insufficient
Process microcomputers	Insufficient
Single-loop digital controls	Insufficient
Programmable logic controls (PLC)	Excellent
Data-loggers	Insufficient
Sequential event controller	Good
Process chromatographers	Very bad
Digital distributed control systems	Good

Source: ABIQUIM (1986)

However, it is important to note that the most diffused equipment (PLCs and DCSs) are rated as very good and good, respectively, in ABIQUIM's assessment. In our survey of petrochemical firms, problems with the quality of DCSs were raised, but seemed localised to two particular suppliers. Apparently, the most problematic areas in Brazilian DCSs are not in the hardware but in basic software.

Basic software for process control is also the motive behind ABIQUIM's assessment (as insufficient) of industrial micro-, mini-, and superminicomputers. In this case, basic software was simply non-existent in the Brazilian market. As will be seen, limitations in the supply of basic software for process control is a major bottleneck in the development of applications of digital control technology in Brazil.[14]

The problem of technological capability of producers

Proliferation of firms, and the consequent non-economic R&D scales, is not compatible with the technological strength necessary to develop external markets. The issue here is technological updating, a crucial weapon in the competition for servicing external markets. As Table 5.3 shows, user firms are sceptical about the technological capability of Brazilian producers. Producers recognise this as a major challenge.

Technologies embodied in Brazilian digital control systems were, and remain, close to the technological frontier when the first licences were approved (1983). The problem, as producers and users have put it, is that the pace of innovation in this area is very fast. The technological trend is for the greater integration of DCSs with process computers, and with digital process analysers (in what is called 'hierarchical systems of digital control'). If Brazilian producers are to follow this path, they will need to be capable of adapting basic software.

The capability of adapting basic software is crucial for the quality of service producers can supply. Basic software is at the core of digital control systems technology, and is responsible for the basic functions of the system, such as enabling the configuration of the system to work, generating control screens, and processing and recording information. In contrast to most basic software for microcomputers, the basic software for DCSs are complex pieces, which 'require an enormous engineering effort to be digested', as a respondent put it. Nevertheless, 'digesting' licensed basic softwares is an indispensable step *if producers are to adapt digital control systems to users' needs*, and not merely transfer the licensed package.

Insofar as our respondents are well informed, basic software is one of the weakest points with Brazilian producers of digital control systems. This is a point which requires further investigation, as does the whole issue of industrial automation suppliers' capabilities. Yet user firms have given enough indication of the problems. A great number of engineers declared that they did not consider suppliers capable of adapting/updating basic software, although they attributed importance to this quality. Some of them expected suppliers to access new developments through new technology licences in the future. According to an informer's detailed comment, the only supplier which has undertaken some development in this area is the one whose system is based on a network of microcomputers. In three plants in our sample, problems which led to a stoppage of the system (and of the plant) were primarily attributed to the supplier's lack of knowledge of basic software.

The example of basic software is useful in illustrating the technological challenges national producers have to face. According to users and producers, developing capability in this area is something that requires enormous investment. In these circumstances, reaching a minimum R&D scale seems to be a necessary step in the creation of a competitive industry in this segment of the market.[15]

Some users expressed concern about the absence of an institutional mechanism to monitor/control suppliers' commitment to R&D activities. In the words of one respondent, 'suppliers are doing little in terms of technological improvement; today, the policy is one of protecting suppliers without requiring them to meet performance targets'. The real commitment of suppliers to R&D is an issue beyond the aim of this paper. However, this statement reveals a perception by users, that market reserve should be combined with the requirement of producers' commitment to achieve defined standards in terms of quality, costs and technological capability.

The limitations of the supply of digital control techniques is one of the factors which shapes the way in which the new technology is used in the Brazilian petrochemical industry. Brazilian petrochemical firms have not exploited the most important opportunities offered by digital control technology, under-utilising the potential of the new technology. The logic of such a strategy is analysed in the following section.

Why and how firms adopt the new technology: passive modernisation and under-utilisation

Given the limitations of the supplier side, what leads Brazilian petrochemical firms to adopt digital control systems? What are their strategies towards the new technology? This section shows that, with a few exceptions, firms investigated have pursued a *passive strategy of modernisation*, in which the adoption of the new technology is primarily determined by the need to replace obsolete or faulty existing control systems. Within such an approach, the introduction of new technology is regarded as equipment replacement and digital control systems are used mainly to perform conventional control strategies. Passive modernisation has resulted in *under-utilisation* of the full potential of the new technology.

The point of departure of management's logic behind passive modernisation is that the adoption of digital control systems is primarily motivated by the complete obsolescence of existing control equipment and the greater problems presented by alternative technologies. This is clear from the importance attributed to the 'need to change obsolete controls' by engineers involved in the decision-making process concerning the adoption of digital control techniques, when asked to rate motives for adoption and to comment on them. Five firms out of nine pointed to the obsolescence of their previous control equipment as the single most important factor. In some of these firms, the condition of pneumatic controls had deteriorated to the point of jeopardising plant performance. The importance of the obsolescence factor as a driving force was definitely confirmed amongst non-users. Firms K to O declared that the timing and the nature of a future decision about changing control equipment was subordinated to the evolution of the (already detected) deterioration of their conventional systems.

Firms A and D were the only ones in which the decision to introduce digital control systems in existing plants was primarily motivated by the specific economic benefits and technological opportunities expected from new technology. None of them were under pressure to replace faulty instruments; with the change of control systems they primarily aimed at a 'reduction of raw material costs' (firms A and D) and at 'increased control over the production process'. At firm A, cost reduction was expected from the introduction of advanced softwares for plant optimisation, a feature which has been part of its automation plans from the beginning.

In the passive modernisation approach, once the decision about replacement (or to equip a new plant, as in the case of firms F and G) has been made, the decisive factors in favour of digital technology *vis-à-vis* conventional technologies have been *price*, *reliability*, and the *maintenance convenience* of control equipment. For reasons mentioned in Section 2, the local production of conventional process control instruments in Brazil has left a great number of problems for users in the petrochemical industry.

Many respondents stressed the problems of conventional controls supplied by Brazilian firms. Brazilian instrumentation industry, they said, has not been able to internalise the production of a crucial number of components, particularly those depending on fine mechanical technology. It follows that the supply of conventional control instruments to the chemical industry remained dependent on Brazilian products with a high import content or on imports of instruments *per se*. In each case, costs have been high owing to tariffs imposed on imports. Moreover, dependence on imports very often has resulted in delays of much needed replacements, due to the usual bureaucratic barriers to importation in Brazil. Finally, a number of respondents have also criticised the quality of Brazilian conventional control equipment. All these replacement problems have contributed to the acceleration of obsolescence of all of the conventional control systems.

Considering this background, users of DCSs have perceived digital technology as a better deal, in economic and technical terms, than the conventional alternatives. As an engineer from firm H put it: 'We had to change the control equipment anyhow, because of obsolescence; we have gone for digital controls because they are cheaper and better than analog instrumentation.' This is not entirely accurate; according to ABIQUIM data, in the case of control systems comprising more than 100 loops (which is the case of most petrochemical plants), digital control systems available on the Brazilian market are slightly more expensive than conventional controls. Other respondents in user firms confirmed that adopting conventional equipment would have cost slightly less.

Given the small cost differential, the decisive factor leading to the option for digital systems has been the technical performance of the equipment. Here, engineers stressed the expected easier maintenance and better performance of Brazilian digital systems *vis-à-vis* their conventional counterparts. Firms expected digital systems would reach a higher level of national content, a strategic aspect as far as maintenance is concerned, given the problems already mentioned with imports of parts. Effective maintenance is a crucial issue in highly capital-intensive, automated plants, such as the ones in the petrochemical industry. They cannot afford unexpected downtime.

Finally, firms were also aware of the potential intangible benefits of the new technology (a factor stressed by engineers), such as increased control of the production process, improvement in product quality and increased plant safety. For such a small price difference, and given the maintenance advantages, it was worth paying to see whether the intangible became tangible.

However, in the choice between DCS alternatives, only firms A and B have decided on the basis of the technological potential of equipment (for instance, compatibility of hardware to run optimisation software available for a specific process). Price and the need to sustain the market for an associated supplier have been the most decisive factors. Two of the respondents clearly declared that they have decided in favour of a certain supplier on the grounds that such a supplier had offered the cheapest deal.

What we can conclude from such analysis is that the adoption of digital *equipment* has been very much influenced by the import substitution policy context of the country. The former CDI policy for import substitution of conventional control equipment and SEI's market reserve policy have worked together to create a leap from pneumatic control technology directly to digital control technology, with a relatively rapid diffusion of the latter. CDI policy has worked in a negative way, since it contributed to the acceleration of the obsolescence of conventional equipment; besides, it was not able to reduce the costs of conventional equipment or to overcome the problems that undermined the confidence of users.

In this context, the policy implemented by SEI in this sector – namely, the inclusion of digital control systems in the market reserve for national firms and restrictions on imports – has had the effect of organising this problematic market, offering an alternative to the difficulties resulting from the previous policy for control instruments.

With a promising market for digital control systems, a number of private local groups were attracted to enter it in the early 1980s. Access to technology was not problematic, since international producers were, and still are, interested in some form of access to the Brazilian market. Market reserve itself had contributed to the Brazilian private sector's early accumulation of some technological capability in IT, an important starting point in the process of nationalisation. These factors explain users' relatively higher confidence in Brazilian digital control systems *vis-à-vis* their conventional competitors.

In short, if the Brazilian IT policy has problems, it cannot be criticised for a lack of supply of technologically updated industrial electronic equipment for process control. On the contrary, it has been a major force behind the organisation of such a supply. From the user's point of view, the *internalisation of equipment production* has been a very positive outcome of this policy. If this were its main objective in the area of industrial automation, the policy can be considered successful.

However, if the internal supply of digital control equipment has been a major contribution in avoiding downtime in petrochemical production, it has not (and will not) *by itself* lead user firms to a real technological upgrading, one which will allow them to reap the best technological opportunities offered by the new systems technology (Aliperti 1987).

In fact, the basic limitation of passive modernisation is that firms adopting this strategy have been using the new control systems to carry out basically the same control functions as those of conventional controls. In this sense, they are *under-utilising* the new technology since they have not explored its more sophisticated functions, those which can lead to the upgrading of plants in terms of integration, efficiency, product quality and innovation.

What is referred to as under-utilisation of digital control technology in this chapter is precisely the non-exploitation of the most innovative possibilities opened up by the application of IT to industrial process control. Firms under-utilise the new technology when they employ it only to perform ordinary control functions, when they do not take advantage of the new techniques to perform more advanced tasks such as plant optimisation, process simulation, and plant integration.

As Table 5.5 (Column I) shows, only three of the sample of

Table 5.5 Use of digital control technology, automation strategy and capability in process technology in a sample of Brazilian petrochemical firms, 1988

Firms	*Current use of digital control technology*	*Responsible for automation*	*Automation plan*	*Technological capability in process design*
A	Ordinary control + data acquisition + imported optimisation softwares tendency register	Own automation division	6 years, 3 stages	Full assimilation and major innovation in 2 processes; partial assimilation in the others
B	Ordinary control + data acquisition + tendency register	Own department for computerisation engineering	3 years	5 new processes developed and patented; partial assimilation of the others
C	Ordinary control + register	Instrumentation division in engineering department in charge	No	All processes, partial assimilation, minor innovations
D	Ordinary control	Ad hoc informal organisation; maintenance in charge	No	Partial assimilation of 1 (one) process; no assimilation of the other
E	Ordinary control + tendency register	Ad hoc, informal group, various departments	3-stage strategy	Little assimilation of major process; partial assimilation and minor innovations on others
F	Ordinary control + register	Mostly in hands of supplier (associate)	No	Process engineers on training–expectation – they will master design in future
G	Ordinary control	No particular area in charge	No	Partial assimilation; few innovations
H	Ordinary control	Current initiatives come from maintenance department	No	No assimilation of major process/partial assimilation of other
I	Ordinary control + data acquisition + advanced control + digital direct control + optimisation of plants	Process control group at plant level; inter-department committee at national level	5 years	All processes proprietary

petrochemical firms (A, B and I) were using digital control technology to perform tasks beyond ordinary control functions. Firm A (associated with the Petroquisa Group) and firm B had already reached the stage of using the system for data acquisition (a data bank with information on process performance). At the time of research, firm A was implementing an optimisation software.[16]

Firm I, a subsidiary of a multinational chemical group, is a unique case in the sample (and possibly in Brazil) in that it has introduced computerised control without changing conventional control equipment, introducing an interface between the latter and a computer network. Firm I's reports suggest that it is one of the most advanced firms in Brazil in terms of the adaptation and use of advanced software for process control, drawing on solutions developed in its foreign plants.

All the other user firms in the sample were employing digital control technology to perform basically the same control tasks they used to perform (or would perform, in the case of new plants) under conventional control equipment (Table 5.5, first column).

The performance of ordinary control was said to have improved with the new control systems in most, but not all, firms. Engineers confirmed this by showing that digital controls are more accurate than analog ones, enabling greater precision in control operations. They also praised the automatic tendency register which comes with most DCSs.

But the most emphasised quality of the new systems was their potential to considerably increase management's control over the labour process, particularly through the automation of supervisory tasks and through the register of operators' interventions.[17] It would seem concern with control could only assume such importance in engineers' assessment of the new technology in the absence of more sophisticated and profitable uses of the new techniques.

To sum up the argument, we can draw on the words of one of the experts interviewed: so far the use of digital control systems in the Brazilian petrochemical industry is reminiscent of 'using cannons to kill birds'.

The under-utilisation of digital control technology was found to be in line with firms' strategies towards automation. Table 5.5, second and third columns, show that firms A, B and I again were the only ones to present a strategy for automation with clear long-term objectives, an automation plan and the setting of specific organisational arrangements to deal with this issue.

In this respect, firm A reported a very ambitious and well

documented automation plan, which aimed at global automation and integration of the firm, with large utilisation of optimisation techniques, over six years.[18]

Firm B, which was still concluding the implementation stage of the first DCS at the time of research, also presented a broad strategy towards the new control technology. It had planned to reach global automation and integration of the plant within three years. In parallel with the Automation Division initiative of firm A, this company was putting a special division in charge of development/adaptation of optimisation and simulation software.

In all the remaining user firms, the issue of planning for the adoption of digital control technology received less attention or resources than in firms A, B and I. In most cases, the matter was the responsibility of ad hoc informal groups, with the participation of different areas of engineering and maintenance departments. These groups were concerned, rather, with the purchase of equipment and its implementation, with little or no attention given to software applications.

Among non-user firms, considerations about future adoption of digital technology were the exclusive concern of maintenance departments. The very fact that maintenance departments had such a high profile in the process of decision-making regarding the adoption of digital technology (in contrast to the more important role of process engineering departments in the case of firms A, B, and I), is further evidence of passive modernisation. The adoption of digital control systems in these firms was regarded merely as the replacement of control equipment. Within this approach, in which adoption is subordinated to the evolution of the deterioration of existing equipment, there is no use for a comprehensive plan for factory automation and integration.

However, this lack of planning is not a neutral factor. Experts have insisted that hasty, non-planned decisions as to what equipment to adopt and how to utilise it may lead to future problems, either hindering the future increase of automation and integration or requiring the scrapping of previous investment (Vibranovski 1987; Aliperti 1987). In this sense, the simple adoption of a DCS does not lead, by itself, to the technological upgrading of a firm, just as the diffusion of hardware does not say much about the process of modernisation of an industry.

Explaining passive modernisation: some suggestions on the technological limitations of users and suppliers

In the previous sections, we have seen that although diffusion of hardware has been relatively fast, in most firms it has occurred within a strategy of passive modernisation. The adoption of new technology is rather a response to pressures to replace old equipment and so the potential of the new technology has not been adequately exploited. The aim in this section is to present an interpretation of the main economic and institutional forces which have brought this about.

Literature on the diffusion of micro-electronics-based equipment in Brazilian engineering industries has emphasised the recent deterioration of the macro-economic environment and the relative costs of capital (high) and labour (very low) in Brazil as the major economic factors holding back faster diffusion (Laplane 1988; Peliano *et al.* 1988; Prado 1989). However, in the shaping of the pattern of diffusion in the petrochemical industry, these factors do not seem to have had as important a role.

As discussed in Section 2, the volume of capital required by investment in digital control systems is relatively low in the context of capital-intensive, process industries. Moreover, the share of direct labour costs in the global operational costs of the petrochemical industry is much lower than the average in the engineering industries. Thus, relative factor prices have not been as important an influence on the pace and quality of diffusion in this industry as they have in the case of engineering industries.

The increasing uncertainty faced by the Brazilian economy has certainly contributed to holding back investment in new technologies. It could be argued that passive modernisation has been a product of economic crisis, in the sense that firms have assumed a conservative attitude towards investment and, therefore, have only considered the adoption of digital controls under strong pressure of obsolescence. However, this argument could not explain why the actual users of digital control systems did not invest a bit more in order to benefit from higher-level applications, which could have yielded *faster return* over total investment.

Thus, not dismissing the relevance of the recent economic crisis in Brazil, it could be suggested that the major determinants of the automation strategy prevailing in the firms investigated are rather related to structural features of Brazilian industry. It would appear, *passive modernisation of industry and under-utilisation of the new technology stem from the technological limitations of both Brazilian suppliers and users of new technology*. Moreover, the effects of these technological limitations are aggravated by the insufficient supply of relevant engineering services and by the lack of adequate institutional support to the process of diffusion.

Let us examine each of these aspects. We have already seen some of the major technological shortcomings of Brazilian suppliers of digital control equipment. They have restricted the possibilities for users to adopt automation strategies aiming at the optimisation and higher level of integration of plants. Suppliers' lack of capability in basic software has limited the chances of tailoring the configuration of DCSs to users' needs, resulting in the adoption of inadequate or over-elaborate systems design (Vibranovski 1987).

At the time of this research, none of the suppliers of digital control systems were able to offer engineering services aiming at plant integration. Only one of the suppliers was offering some support for user firms to draft plans and strategies for automation. Therefore, the very nature of supply has reinforced the idea that the new technology is embodied in a piece of equipment.

However, if the technological fragility of suppliers has been an important variable in explaining passive modernisation, the evidence of this research suggests that users' own technological limitations are no less important. Here we are referring to users' own ability to identify the opportunities of new technologies, to make an appropriate choice of equipment and software and, most important, to make the best possible use of them. This idea corresponds to what Edquist and Jacobsson (1988:185–9) put forward as the 'information and knowledge factor' in the process of diffusion, referring to users' level of information about the new technology and to their capability to develop/adapt and use adequate applications. What is suggested here is that passive modernisation and poor utilisation of digital control systems are also a product of the inability of users to take a different route.

The limitations of users' knowledge in respect to the development, adaptation, and use of new technology applications are rooted in factors which are both internal and external to firms. In the case under investigation, the major internal factor has been users' limited capability in the core technology of the industry, that is, technology for process design. In turn, this limitation is reflected in users' overall technological strategy.

The *development* of software for plant optimisation and process simulation in the chemical industries requires a great deal of knowledge of the specific process to which the software applies, since this is based on the modelling of process dynamism. This is why software available in the international market has usually been developed and licensed by firms which are also the proprietors (or their associates) of the respective processes. Even the adaptation and use of licensed software requires a considerable knowledge of the process.

The situation of firms A, B, and I is a good illustration of this interaction between the knowledge of process technology and the capability in computerised advanced control techniques. Since its process is a very diffused one, firm A found it more convenient to license an optimisation software from a foreign specialist company (which is associated with the proprietor of the process formerly licensed by firm A), rather than developing one in-house. Nevertheless, the advanced stage of assimilation of process technology by firm A (see Table 5.5, fourth column) was crucial in order to enable this firm to participate in the adaptation of the imported software within a training program provided by the foreign supplier. The training program was meant to lead firm A to become autonomous in maintaining and further improving the licensed software.

In the case of firm B, at least part of the optimisation software this firm intended to use would have to be developed in-house. This could not have even been thought of, were not firm B one of the two most advanced Brazilian petrochemical firms in terms of process design (this firm has developed a number of proprietary processes). Firm I is a subsidiary of a large international group which utilises its own process technologies.

However, firms A and B, in their effective assimilation of imported process technologies and in the importance they attach to technological innovation in their competitive strategy, are exceptions in the context of the Brazilian petrochemical industry. Other research has already shown that the process of technological learning in this industry has been generally limited (partial assimilation). Particularly in respect of the core technology of the industry, most firms have not gone beyond the stage of minor improvements, that is, very few firms have assimilated technology to the point of becoming capable to develop new processes or products (Teixeira 1985, 1987; Bastos 1989).[19]

This situation is a product of historical circumstances and of the objectives of the policy which have oriented the development of this industrial sector in Brazil. The bulk of Brazilian petrochemical plants were built during the 1970s, in the last (and problematic) phase of the import substitution strategy. As such, the main objective of the implementation of a local petrochemical industry was to quickly build up productive capacity to supply the internal market. The development of local technological capability in process design and basic engineering did not receive priority in the initial phases of implementation. Teixeira (1985) has shown that it was not before the third (and last) wave of investment that Brazilian negotiators and firms started to seek technological transfer agreements which included some form of transfer of knowledge of process design.

Soon after this was completed, the Brazilian petrochemical industry faced the downturn of the early 1980s. In the initial years, massive government subsidies enabled output excess to be diverted to the international market. Later, under pressure of subsidy-cuts, firms accelerated the search for increased efficiency via minor improvements, which led to savings of energy and raw materials and to an impressive increase of effective capacity (Carvalho *et al.* 1989; Teixeira 1987). According to Teixeira (1987), the potential for minor innovation has now been mostly exhausted and the time will have come for more ambitious investments in R&D. However, owing to the managerial, technological and financial limitations of Brazilian petrochemical firms, only a few have managed to place technological innovation at the centre of their corporate strategy.

In this respect, the sample of firms examined here is representative of the situation in the industry. As Table 5.5 shows, with the exception of firms A and B, national petrochemical firms have only partially assimilated the process technologies on which their plants are based. But what is most relevant for us in Table 5.5 is *the correspondence between firms' level of assimilation of process technology and the level of sophistication in adopting digital control technology*. Firms with greater accumulated capability in process design (A and B) showed more advanced use of and more ambitious plans for automation, whereas firms at a lower level of capability in process technology tended to be the ones using new technology only for ordinary control.

This tendency seems to have a dual significance. On the one hand, it reflects the fact that the most advanced use of digital control technology requires that user firms have certain skills in process design. Thus, firms which have not progressed much in the assimilation of process technology have not been *capable* of searching for, adapting, and using advanced techniques based on digital controls (e.g. plant optimisation).

On the other hand, this tendency also suggests that firms with more aggressive strategies in respect of core technology show greater willingness to adopt more complex applications of the new control technology. Regarding this point, it is important to add that firms with partial assimilation of process technology are not bound to use the new control techniques exclusively for ordinary control. Certain applications can be devised precisely as tools to help users to develop a deeper knowledge of the process (for instance, the development of a data bank with information on process behaviour). If these opportunities have been seldom explored by users of DCS in the Brazilian petrochemical industry, this could be due to managerial rather than technological limitations.

Users' limitations have not just been a product of their intrinsic weakness. They have also stemmed from the poor development of basic externalities required by the process of diffusion. Two aspects stand out in this respect: the precarious diffusion of information about the new technology and the shortage of human resources with adequate skills.

Lack of information about the new technology was singled out by user firm engineers as a serious obstacle to adoption on a larger scale. One respondent in a user firm admitted that nobody knew how to specify the equipment needed when the decision was taken to introduce digital controls. He added that even seeing the equipment in operation in another firm was not an easy procedure to organise. Users have complained that suppliers have not provided further information about the technical and economical possibilities of the new techniques.[20]

In turn, suppliers of equipment have blamed users, arguing that they lack 'information technology culture' (SEI 1989). This would lead to the choice of the wrong equipment. As one supplier put it, 'today users purchase equipment knowing less than 20 per cent of what they should know. Thus they are not able to properly assess equipment and end by choosing the cheapest'.[21] Suppliers tend to see this as a problem of the development of human resources.

The fact is that users do lack information about the technical and economic aspects of the new technology. The flow of information to users, either via the training of engineers or via the dissemination of technical information is precarious, depending mostly on the initiative of each firm or industrial association. There is no institutional support to the dissemination of information for users. The Brazilian IT policy has taken the diffusion of information about IT for granted. The importance of this diffusion was stressed by Edquist and Jacobsson (1988), particularly for local firms in developing countries, which have much less access to technological information than multinational companies.

Shortage and/or inadequacy of human resources are also real

constraints. Most user firms have recently created posts for electronics engineers in maintenance departments. A smaller number of jobs were created in engineering departments for electronics engineers and process control engineers who would be in charge of the planning and implemention of digital control systems. Firms have not found the right people for at least 30 per cent of these jobs.

In the case of engineers, shortage seems to be a consequence of inadequacy of education too. The area of instrumentation and process control engineering is interdisciplinary and has been traditionally a weak one in Brazil. Implementing/maintaining digital systems requires electronics engineers who master knowledge of industrial instrumentation and control, in general, and of digital controls, in particular. This is a knowledge engineers acquire only superficially at university. Firms which are starting with the new technology are not able to train engineers in-house. So, they look for experienced engineers, who are very rare in the labour market.

Finally, user firms cannot get much help from engineering consultancy firms. As far as sophisticated applications are concerned, Brazilian engineering firms seldom master the knowledge of the process dynamism models necessary to develop such applications (owing to the same limitations suffered by user firms themselves) (ABIQUIM 1986). However, according to a number of respondents, even for less demanding support tasks, such as the elaboration of the detailed project of implementation of a DCS, external consultancy firms have not shown familiarity with the new technology.

To conclude, the lack of technological strategy and the technical limitations of user firms combine with the precarious state of externalities to shape a pattern of diffusion in which most users do not (and, to a certain extent, cannot) exploit the best opportunities of the new technology.

A further question to ask is to what extent the features of diffusion found in the petrochemical industry can be generalised to other process industries. It is hard to answer this question, owing to the lack of comparable research covering other industries. However, there is some evidence that such a generalisation is valid at least in the case of other Brazilian process industries based on complex processes (e.g. steel and electricity). According to the report of SEI's Commission No. 29 for Industrial Automation (SEI 1988), the industries mentioned have stated problems with the use of digital controls which are similar to those presented above, particularly as to the difficulties in developing/purchasing more sophisticated software applications.

4 CONCLUSIONS

One of the broadest conclusions this case study seems to suggest is that countries such as Brazil, in which the behaviour of economic agents and the practice of industrial policy is rooted in decades of experience with import-substitution industrialisation, face particular problems in adapting to the requirements of restructuring their economies on the basis of exploiting new technologies.

The diffusion of digital controls in the context of the Brazilian economy has occurred under typical import-substitution conditions. The central concern of Brazilian information technology (IT) policy in this area has been with *the production of industrial electronic equipment* by national firms. Suppliers' effective assimilation of technology, particularly in the area of software, has been a distant secondary goal in spite of all words spoken to the contrary. The quality of use of the new technology – marked by under-utilisation within a strategy of passive modernisation – shows that, with a few exceptions, fast diffusion of equipment has neither substantially altered user firms' corporate strategies, nor led to substantial technological upgrading.

The limitations of the newly industrialised countries (NICs), previously oriented to import-substitution industrialisation, in dealing with the new technologies have already been commented on by Unger (1988). He stresses that the interaction of user and producer in industrial innovation, an aspect emphasised elsewhere in this book, is rather weak in most of these countries, owing particularly to the lack of a sufficiently strong capital goods sector.

The point this research could add to the debate is that the limitations these countries face in dealing with information technology relate not only to the suppliers' weakness, but also to the users'. Our case study shows that the technological fragility of suppliers *and users* of the new technology hinders the development of synergies in the user–supplier interaction.

The question of users' technological limitations takes us to the discussion of the importance of accumulated technological capabilities for the diffusion of IT applications. This research shows that, as far as *users* of IT industrial applications in the NICs are concerned, a firm's previous accumulation of technological capability is a major advantage in the exploitation of the opportunities of the new technology. The development of applications of IT in industries other than electronics requires the generation of knowledge which draws heavily on previously accumulated experience with the core technology of the industry.

The degree to which this is the case seems to depend on the complexity of the core technology. In industries based on simpler processes/products (e.g. machining of non-complex parts, the food industry, etc.), the development of IT applications seems to be usually less problematic and even tends to be standardised, whilst in the more complex sectors (chemicals, steel, cars), applications are more customised and their development requires a great deal of formal knowledge relative to the specific process/product. What this research has shown is that the most advanced applications in complex industries require *effective assimilation* of the core technology, rather than just a 'superficial' knowledge.[22]

The technological fragility of firms is an economic factor which should be understood in relation to the environment created by import-substitution policies. Our case study shows how the feeble technological strategies of both suppliers and users originated in an environment created by policies concerned primarily with import substitution, but with little concern for competitiveness. As Perez put it,

> Firms were not designed to evolve. The majority were meant to operate mature technologies, supposed to be already optimised . . . Firms were not expected to reach competitiveness on their own. The import substitution policies and instruments got managers accustomed to having profitability determined by exogenous factors, such as tariff protection, subsidies, and other forms of government help, rather than by the firm's own capability to increase productivity or quality.
>
> (Perez 1988)

How the protected environment created by import-substitution policies has affected the capability of firms to innovate is an issue which deserves priority in the current research agenda. However, as far as policy is concerned, the available evidence is enough to suggest that sectoral policies designed to support (with fiscal and credit incentives, for instance) the development of technological activities in firms can be undermined by the absence of competitive pressures upon them.

This point touches upon an urgent and challenging policy issue faced by countries such as Brazil, which are seeking to move from import substitution to competitive restructuring: namely the need to coordinate industrial and technological policies with macro-economic policies, and particularly with trade policy. Temporary protection and planned liberalisation, as in the case of Japan (Nagaoka 1989), or the binding of protection to export performance, as in South Korea, have proved to be, in the historical contexts of these countries, useful instruments to create a competitive environment for local firms. However, industrial and technological progress in these countries seems to owe a lot to the fine co-ordination, at sectoral level, of trade policies with policies for the promotion of industry and technology (Nagaoka 1989; Luedde-Neurath 1988).

Therefore, an active industrial policy seems to be necessary, if the objective is to steer competitive restructuring in such a way as to make the most of the economic and technological potential of local firms. As Perez has argued, the simple elimination of state intervention in order to let the best firms survive, 'risks drowning many potential winners before giving them a chance to learn how to swim' (Perez 1988:25). Removing the obstacles and allowing competition is important, but firms require learning time and support, in the form of 'facilitating resources' (Perez 1988), in order to become innovative and competitive. These resources are adequate financing and the development of externalities (for instance, provision of human resources and of an adequate infra-structure of technological services) necessary to enable firms to successfully generate/adopt new technologies and new organisational techniques. The role of industrial policy is to provide these resources or to create conditions in which they can be provided by the private sector.

In providing facilitating resources for the diffusion of information technology, special attention should be given to user firms outside the electronics sector, particularly in the case of developing countries. At least this is what the findings of this study suggest. Brazilian user firms are still struggling to master the technologies they have licensed in the past. Their knowledge is imperfect, their engineering areas are fragile. They do not have the appropriate conditions to spontaneously develop the right internal environment, the 'IT culture', to incorporate new technology correctly. Suppliers too are swamped by their own difficulties in mastering licensed technology and are unable to provide help for users in respect of their deficiencies. Such a situation requires specific state intervention aimed at 'bridging' the gap between users and producers.

One basic finding of this research is that in order to be 'good users' of industrial applications of information technology, firms have to master the core technologies on which production is based. The implication of this for technology policy in the NICs is that concern with new technologies, which certainly have a highly strategic role in the process of restructuring, should not (and need not) increase at the cost of a lack of policies for and concern with the effective assimilation of 'mature' technologies. The Korean experience seems to offer a very useful lesson in this respect, since they have developed very aggressive policies and institutions for the electronics sector without abandoning their efforts to complete the 'course of adoption' of imported technologies in other industries (Enos and Park 1988).

To conclude, the task of promoting the diffusion of IT outside the electronics sector is complex and may involve a number of programs and institutions. An information technology policy concerned with this issue should not limit itself to the development of the electronics industry, as the Brazilian informatics policy has done so far. Certainly promoting the electronics industry should be at the core of any industrial policy concerned with the development of indigenous capability for the generation and use of IT-related innovations. However, in the face of the need to adapt the rest of the economy to the new technology, the policy should widen its scope and integrate the objective of development of the electronics industry with other objectives of industrial policy, otherwise there is a risk that IT policy may be regarded by the potential beneficiaries outside the electronics sector as a policy to attend to the interests of a particular industrial sector. IT policy is too important an issue to be reduced to such a politically weak position.

NOTES

1 I wish to thank CNPQ/PADCT for funding research on which this chapter is based. Help from the following institutions and individuals was decisive in fieldwork: Flávio Rabello, Luciana de Almeida, Maria Conceição Costa and Leda Gitahy, from DPCT/UNICAMP; Francisco Biato, from IPEA; Edvaldo Ribeiro from ABIQUIM; and the managers and engineers of the petrochemical firms. Helpful suggestions on a previous draft were made by Martin Bell, José Cassiolato, Tom Hewitt, Adilson de Oliveira and Hubert Schmitz. However responsibility for the views expressed here is exclusively mine.

2 Process industries are defined here as those based on production processes aimed at the transformation of the physical/chemical structure of materials, such as oil refining, chemicals, steel, pulp and paper, electricity, glass, food and drink, etc. In most cases, production in such industries is considerably automated and organised in continuous flow systems.

3 For a fuller discussion of these issues, see Carvalho (1990).

4 See for example: Tauile (1984 and 1987), Silva (1988), Fleury (1988), Peliano *et al.* (1988), Laplane (1988), Prado (1989), Neder *et al.* (1988), Carvalho (1987), Schmitz and Carvalho (1989), Carvalho and Schmitz (1989). The emphasis on engineering industries in these studies is not

surprising, since electronics-based innovation affects these industries more radically. For thorough discussions of the economic and social implications of the diffusion of electronics-based innovations in industry see, for instance, Kaplinsky (1984) and Edquist and Jacobsson (1988).

5 SEI groups (electronic) industrial automation equipment produced in Brazil into two main categories: products for process control and products for manufacturing automation.

6 In fact, comparative data on the diffusion of NCMTs, robots and CAD/CAM systems suggests that Brazil is lagging behind not only *vis-à-vis* developed countries, but also with respect to some of the Asian NICs in which, as in Brazil, the diffusion of the new technology in industry only took off in the 1980s. For empirical evidence, see Carvalho (1990:5–6)

7 Information obtained in interviews with plant engineers and with the technical advisor to ABIQUIM.

8 The very fact that state-owned/state-controlled industries are major users of industrial electronic equipment in Brazil represents an important policy tool for IT and industrial modernisation policies, one which, in my view, has been little exploited.

9 This is a long story, which has not yet been properly investigated, but is briefly explained in Carvalho (1990), a previous extended version of this chapter.

10 For an account of the political development of Petroquisa, see Suarez (1986).

11 Interview with the President-Director of UNICONTROL, *Química e Derivados*, Aug. 1987.

12 The situation in the PLC market is similar. Presently there are twenty producers (for an annual US$40 million market); but another twenty have submitted requests to be licensed by SEI.

13 Yokogawa, Leeds, Fisher, Hitachi, and Asea-Brown Boveri are the proprietors of technology licensed early by current producers. Honeywell and Valmet account for the new licences recently approved by SEI. Comsip is the national supplier operating with in-house developed technology.

14 Since 1982, users' associations and state-owned user firms have demanded that SEI create a permanent assessment commission which would be in charge of testing and setting standards for Brazilian electronic equipment for process control. Such a group would include representatives of government agencies, a user firm association, technical societies, and state-owned firms (Siderbrás, Eletrobrás, Petrobrás). At the time of this research, this was still a proposal.

15 As the director of ABCPAI, the association of Brazilian suppliers of industrial automation equipment and services, puts it: 'We cannot expect that the licensee of technology will be able to carry out significant product improvement in a few years. In such a small market as ours, investment in research is too heavy a burden for firms to carry' (*Guia de Automação Industrial*, 1990, No. 7).

16 Firm A's engineers have estimated that the optimisation project would yield a 0.1 per cent saving in raw materials and a 1 per cent saving in energy, in a pessimistic forecast. Not taking into account other tangible economies, such as increase in effective capacity, engineers expected an

annual return of US$8 million, with a period of return no longer than two and a half years. Software represented 40 per cent of the global investment of the project.

17 This seems to reveal that Taylorist management conceptions still prevail among managerial/engineering staff in these plants, in spite of all the advances in automation. For further analysis of the labour implications of the diffusion of digital controls in the Brazilian petrochemical industry see Carvalho *et al.* (1989) and Carvalho (1989).

18 See Carvalho *et al.* (1989) for a detailed examination of Firm A.

19 For the conceptualisation of the process of adoption and assimilation of foreign technologies in developing countries, see Bell (1984), Fransman (1986), and Enos and Park (1988).

20 See 'Falta diálogo entre fabricante e usuário', in *Revista Petro & Química*, May 1987.

21 Interview with the President-Director of Ecil P&D, *INSTEC*, April 1987.

22 A parallel could be made here with the issue of equipment design limitations of Brazilian producers of capital goods for process industries, as proposed by Erber (1982). He showed that the lack of local capability in process design is usually associated with the lack of local capability in equipment design.

BIBLIOGRAPHY

ABIQUIM, Associação Brasileira da Indústria Química e de Produtos Derivados (1986) 'Palestras da Comissão de Controle de Processos da Abiquim no II Conai', *Documento Técnico* No. 02/86, ABIQUIM, São Paulo.

Aliperti, J. (1987) 'Modernização de sistemas de controle – o conceito global', *Instec* (June).

Bastos, V.D. (1989) 'A questão tecnológica nas joint-ventures petroquímicas Brasileiras', MA thesis, Coppe, Rio de Janeiro: Universidade Federal do Rio de Janeiro.

Bell, M. (1984) 'Learning and the accumulation of industrial technological capacity in developing countries', in M. Fransman and K. King (eds) *Technogical Capability in the Third World*, London: Macmillan.

Carvalho, R.Q. (1987) *Tecnologia e Trabalho Industrial*, Porto Alegre and São Paulo: L & PM.

—— (1989) 'Labour and information technology in newly industrialising countries: case studies in Brazilian industry', paper prepared for the Workshop 'Changes in the labour process in developed and developing countries', The Hague: Institute of Social Studies.

—— (1990) 'Limits of modernisation with low technological capability: a study of the diffusion of industrial automation technology in Brazilian process industries', *Textos para Discussão Interna – IPEA*, No. 181, Brasília: Instituto de Pesquisa Econômica Aplicada/IPEA, May.

Carvalho, R.Q. and Schmitz, H. (1989) 'Fordism is alive in Brazil', in *IDS Bulletin*, Vol. 20, No. 4, Brighton: Institute of Development Studies, University of Sussex.

Carvalho, R.Q., Rabelo, F.M., Almeida, L.T. and Costa, M.C. (1989)

'Microeletrônica, capacitação tecnológica, competitividade e trabalho na indústria petroquímica Brasileira', in *Anais do Seminário Padrões Tecnológicos e Políticas de Gestão na Industria Brasileira*, Universidade de São Paulo, São Paulo.

Edquist, C. and Jacobsson, S. (1988) *Flexible Automation: The Global Diffusion of New Technology in the Engineering Industry*, Oxford: Basil Blackwell.

Enos, J.L. and Park, W.H. (1988) *The Adoption and Diffusion of Imported Technology: The Case of Korea*, Beckenham: Croom Helm.

Erber, F. (1982) 'Technology issues in the capital goods sector', Geneva: UNCTAD.

Ernst, D. and O'Connor, D. (1989) *Technology and Global Competition – The Challenge for Newly Industrialising Economies*, Paris: OECD.

Fleury, A. (1988) 'The impacts of microelectronics on employment and income in the Brazilian metal-engineering industry', prepared for the International Labour Office, Geneva, February.

Fransman, M. (1986) *Technology and Economic Development*, Brighton: Wheatsheaf.

Kaplinsky, R. (1984) *Automation – the Technology and Society*, London: Longman.

Laplane, M.F. (1988) 'Competitive assessment of Brazilian industrial robots and computer numerical control industries', mimeo, Instituto de Economia, Universidade de Campinas.

Luedde-Neurath, R. (1988) 'State intervention and export-oriented development in South Korea', in Gordon White (ed.) *Developmental States in East Asia*, London: Macmillan.

Nagaoka, S. (1989) 'Overview of Japanese industrial technology development', *Industry and Energy Department Working Paper*, Industry Series Paper No. 6, the World Bank Industry and Energy Department, March.

Neder, R.T., Abramo, L.W., Bicalho de Souza, N.H., Diaz, A., Falabella, G., Silva, R.A. (1988) *Automação e Movimento Sindical no Brasil*, São Paulo: Hucitec.

OECD, Organisation for Economic Cooperation and Development (1985) 'Software: an emerging industry', *Information, Computer, Communications Policy*, No. 9, Paris: OECD.

Peliano, J.C., Carvalho, R.Q., Bicalho de Sousa, N., Cassiolato, M.M., Schmitz, H., Gitahy, L. and Neder, R. (1988) *Automação e Trabalho na Indústria Automobilística*, Brasilia: Editora Universidade de Brasilia.

Perez, C. (1988) 'The institutional implications of the present wave of technical change for developing countries', paper prepared for the World Bank Seminar on Technology and Long-Term Economic Growth Prospects, November.

Prado, A.J.C. (1989) *A Difusão de Automação Flexível na Indústria Brasileira de Autopeças*, MA thesis, Instituto de Economia, Campinas: Universidade de Campinas.

Rosenberg, N. (1982) *Inside the Black Box: Technology and Economics*, Cambridge: Cambridge University Press.

Schmitz, H. and Carvalho, R.Q. (1989) 'Automation and labour in the Brazilian car industry', *Journal of Development Studies*, Vol. 26, No. 1, October.

SEI, Secretaria Especial de Informática (1988) *Relatório da Comissão Especial no. 29 – Automação Industrial: Proposta de Plano Setorial de Automação Industrial*, Brasilia: SEI.

—— (1989) 'Panorama do setor de informática', *Séries Estatisticas*, Vol. 2, No. 1, SEI, Brasilia, August.

Silva, E.B. (1988) 'Labour and technology in the car industry: Ford strategies in Britain and Brazil', Ph.D. thesis, University of London.

Suarez, M.A. (1986) *Petroquímica e Tecnoburocracia – Capítulos do Desenvolvimento Capitalista no Brasil*, Sao Paulo: Hucitec.

Tauile, J.R. (1984) 'Microelectronics, automation and economic development: the case of numerically controlled machine-tools in Brazil', Ph.D. thesis, New York: New School for Social Research.

—— (1987) 'Microelectronics and the internationalisation of the Brazilian automobile industry', in S. Watenabe (ed.) *Microelectronics, Automation and Employment in the Automobile Industry*, Chichester, Sussex: John Wiley and Sons.

Teixeira, F.L.C. (1985) 'The political economy of technological learning in the Brazilian petrochemical industry', Ph.D. thesis, Brighton: Science Policy Research Unit, University of Sussex.

—— (1987) 'Dinâmica Empresarial e Tecnológica das Empresas do Complexo Petroquímico de Camaçari', *Anais do Encontro Anual da ANPEC*, Salvador: ANPEC.

Unger, K. (1988) 'Industrial structure, technical change and microeconomic behaviour in LDCs', in G. Dosi *et al.*, *Technical Change and Economic Theory*, London and New York: Pinter Publishers.

Vibranovski, S. (1987) 'Temos competência, faltam confiança e investimentos', *INSTEC*, February.

TECHNICAL MAGAZINES

Guia de Automação Industrial No. 7, February 1990.
INSTEC, February 1987.
INSTEC, April 1987.
INSTEC, June 1987.
Química e Derivados, January/February 1990.
Química e Derivados, August 1987.
Revista Petro Química, May 1987.

6 Automation and work organisation

Contrasts between Brazilian and British car factories[1]

Elizabeth Bortolaia Silva

1 INTRODUCTION

This chapter looks at the relationship between productivity, product quality and competitiveness in order to assess the extent to which international competitiveness depends on the immediate adoption of the latest automation technology. Two car factories have been compared, a Brazilian one (São Bernardo) and a British one (Dagenham). Both are owned by the Ford Motor Company (FMC).[2] Both plants compete in the international market but have not been automated to the same degree. The comparison sheds light on the relative importance of automation and work organisation in competitive strategies. Therefore this analysis contributes to the debate on how firms in Brazil have been affected by the policy of protecting national producers of electronics-based automation technology.

This chapter first outlines the questions posed by international competitive strategies in the car industry in the 1980s. It then looks at the strategies pursued by FMC in Brazil and Britain. A discussion of the changes introduced in the early 1980s in production processes in Ford's plants at Dagenham and São Bernardo follows. These sections provide a basis for exploring some ideas about technological 'imperatives' for improving productivity, product quality and competitiveness. The analysis of similarities and differences in the production processes and in productivity and quality raises the question of relevance of work organisation and, in particular, issues of flexibility in the deployment of labour. The chapter concludes by questioning the technological 'imperative' and drawing out some policy implications for strengthening competitiveness at company and country levels.

2 INTERNATIONAL STRATEGIES IN THE CAR INDUSTRY

New product design, new production techniques, increasing economies of scale, continuous capital concentration, and expanding markets have characterised the motor industry since the beginning of this century. A particular feature of the latest phase in the industry's history is that it has involved major technological and organisational change.[3] This was related to transformations in the pattern of competition. From the late 1960s, the long established positions of European and North American producers were undermined by higher Japanese rates of productivity. The so-called 'Japanese challenge' came simultaneously with increased oil prices and a growing demand for smaller and more economical cars. International competition in the industry intensified because the Japanese significantly reduced costs and were thus able to offer cheaper cars. New technologies, new systems of organising work and relations to suppliers (Just In Time production) enabled them to offer cars which were not only cheaper, but also more reliable, and to respond more quickly to changes in product markets.[4] As a result Western producers faced a severe crisis. Since then, technological advances and organisational innovation have come to be the core of the worldwide competitive process.[5]

This is the context in which the American multinational, Ford Motor Company, developed a strategy to face the competitive challenge of the 1980s which involved both the British and Brazilian subsidiaries.

3 FORD IN BRITAIN AND IN BRAZIL

The role of each country in the Ford Motor Company's production strategy in the 1980s and the resulting challenges are best understood in the light of a summary of the history of motor vehicle production in the two countries.

British entrepreneurs were involved in the development of the first motor cars. Ford established its Dagenham operations in 1931 to meet UK and European demands. Since then its UK production has been increasingly integrated with the European market, turning Dagenham into an important subsidiary within the company's international competitive strategy. In the late 1950s Dagenham achieved economic viability through economies of scale provided by the European market. However, in the 1980s competition amongst European producers challenged Ford's market share and intensified Dagenham's need to respond to pressures for increasing efficiency.[6]

Although Brazil has been involved in assembling motor vehicles since the beginning of this century, production of cars did not start until 1958. Ford launched its first car on the Brazilian market in 1967. It was produced in São Bernardo. Unlike the UK, domestic market protection policies insulated the Brazilian car market until 1990. Competition in Brazil was largely determined by the strategies of multinationals based in the country. Historically this resulted in a backward technological pattern in the Brazilian domestic industry compared with the prevailing international pattern.[7]

In order to increase its competitiveness, the Ford Motor Company and its competitors developed an international strategy of rationalising production through technological change, new products and design, and new approaches to industrial relations and labour utilisation.

Both the Brazilian and British subsidiaries were involved in this international production strategy, and have experienced major changes in their operations since the early 1980s. However, although the strategy of integrating production in the global market involved both subsidiaries (the Brazilian one for the first time) their roles in corporate plans were very different.

Ford's goal since the 1980s has been to produce high quality products at a cost low enough to be internationally competitive. Dagenham's role in the early 1980s was as the chosen location for the production of the Sierra, which involved major technological renewal, particularly in the bodyshop.

The Brazilian subsidiary was designated to produce the Escort in the early 1980s. Thus major plant rebuilding and technical changes were implemented at the São Bernardo site. On the face of it, the competitive challenge was the same but, in contrast to Dagenham, São Bernardo's production had hitherto played only a marginal part in competition in developed markets.[8]

A comparison between Dagenham and São Bernardo is instructive for three basic reasons. First, both sites experienced a process of technological change linked to the launch of a new model for the international market. (At Dagenham the model was a brand new one; at São Bernardo an existing model was introduced from scratch.) Second, the timing of the changes corresponded very closely. The Sierra was launched in October 1982, the Escort in July 1983. Third, there was a sharp contrast in the degrees of automation involved: for example, 120 robots were installed at

Dagenham's Body Assembly but only seven robots in the equivalent process in São Bernardo.

Dagenham was Ford's largest production site in Europe in terms of employment. In 1985 19,600 people were employed there and 11,845 (60.4 per cent) were hourly-paid. The hourly workers in the three plants analysed in detail here (body shop, paint shop and trim and final assembly), amounted to 64 per cent of the total workforce on site in December 1985. Fiesta and Sierra models were produced there during most of the 1980s.[9] Although the Dagenham plant was still located on the old site, first constructed in 1931, Ford managers considered it to be 'in line with the best world standards in terms of technology improvements'.[10]

The São Bernardo estate was the third manufacturing site for the Escort, following West Germany and England. In January 1986 over 12,000 people were employed there. Almost half of these were hourly workers directly involved in the car manufacturing areas with which this study is concerned.

A detailed study of car production in Dagenham and São Bernardo makes it possible to explore the extent to which high productivity and high quality depend on high levels of automation.[11] The following section examines these issues on the basis of a *process by process comparison*. This implies looking closely at the three main processes of car manufacturing, i.e., 1. Body Construction, 2. Paint, and 3. Trim and Final Assembly.

4 TECHNOLOGY, LABOUR AND PERFORMANCE AT DAGENHAM AND SÃO BERNARDO

The relationship between technology and labour is controversial. Some studies stress that technological innovations determine the way work is organised and the outcomes for production. Others analyse culture as a major stimulus or hindrance to technological change, and see forms of work as culturally determined regardless of the technology in use. Other views see an interdependence of technology and broad political, social and cultural issues in shaping particular forms of work.[12] Accordingly, there is no consensus concerning a necessary relation between high levels of automation and high productivity. Productivity is dependent upon a wide range of factors related to the overall context of production. Thus, factories can present similar productivity levels with very different levels of employment and automated processes of production.

A comparison of the same manufacturing process in different sites and countries, of the sort undertaken here, illuminates the complexity of relations between technology, labour and performance. The comparison between Dagenham and São Bernardo in this section concentrates on the characteristics of the manufacturing areas, car model mix and design, and levels of productivity. The processes discussed are those of Body Construction, Paint, and Trim and Final Assembly. The information concerning these production processes was collected in 1985/6. Any technological improvements or changes implemented at Dagenham or São Bernardo after September 1986 have not been considered.

A comparison of productivity levels is affected by differences in the division of labour in each site. The proportion of workers allocated to each of the three production areas was similar on each site but there were two basic distinctions. First, the proportion of 'direct' (production workers) and 'indirect' workers (in support activities such as cleaning, handling material, maintenance and quality control) varied, with a much higher proportion of 'indirect' workers at Dagenham. Second, Dagenham had a 'general services' category which was allocated to Body Construction or to Trim and Final Assembly as production demanded. The latter, in particular, significantly affected the measurement of productivity (See Tables 6.1 and 6.2). Output per worker in each area is the criterion adopted to measure productivity. The measurement of output in relation to working hours was not feasible because of differences in working time for particular

Table 6.1 Direct and indirect car workers by area: Dagenham[a]

	Body construction					
	Stamping & sub-assy[b]	*Body assy*	*Paint*	*Trim and final assy*[c]	*General Services*[d]	*Total*
Direct (1)	826	749	700	2,131	0	4,406
Indirect (2)	987	492	298	497	972	3,246
Total	1,813	1,241	998	2,628	972	7,652
%	23.7	16.2	13.0	34.5	12.7	100.0
Ratio (1)/(2)	0.8	1.5	2.3	4.3	n.a.	1.4

(a) Numbers refer to average hourly workforce actually working from January to September 1986.
(b) Operations defined as 'sub-assemblies' are included in Stamping.
(c) Includes Trim manufacture.
(d) Includes maintenance, production control, quality control and apprentices. Refers only to 'General Services' allocated to Body Assembly and Trim and Final Assembly areas. Leaves out 615 'general services' workers allocated to other areas not covered in this study.

Source: Ford-Britain

operations, different shift patterns and varying use of overtime. Given these constraints, varying ratios of output per worker reflect the way that the production process was organised and the differences in job structures that existed in Dagenham and São Bernardo.

Table 6.2 Direct and indirect car workers by area: São Bernardo[a]

	Body Construction				
	Stamping[b]	*Body-assy & sub-assy*	*Paint*	*Trim and final assy*[c]	*Total*
Direct (1)	792	1,220	621	1,874	4,507
Indirect (2)	325	433	233	480	1,471
Total	1,117	1,653	854	2,354	5,978
%	18.7	27.7	14.3	39.4	100.0
Ratio (1)/(2)	2.4	2.8	3.7	3.9	3.1

(a) Numbers refer to average hourly workforce actually working from November 1985 to June 1986.
(b) Operations defined as 'sub-assemblies' are included in Body-Assembly.
(c) Includes Trim manufacture.

Source: Ford-Brazil

The following sections consider each of the three major production areas in turn, comparing levels of automation and productivity in the two plants. Given the concerns of this book, the origin of the automation technology of the Brazilian plant should be mentioned. The robots were imported, whereas the programmable controllers (for automatic transfer) and computers (for integrating production and control) came mainly from domestic suppliers.

The body shop at Dagenham and São Bernardo

Car body construction involves *Stamping* and *Body Assembly* which, although related, constitute separate sets of activities usually located on different shopfloors. Sub-assembly of stamped parts is usually done either in Press shops or in Body Assembly shops. At Dagenham sub-assemblies were done in the Press shop. At São Bernardo they were done in Body Assembly.

These different job allocations made direct productivity comparison of these sub-areas difficult. Particularly in Stamping, comparisons were constrained by the variety of product (weight and kind of steel or the size and complexity of the stampings) as well as the international integration of Dagenham with Ford's Spanish subsidiary for the supply of stamped parts. Moreover, there were more stamped parts to be assembled in São Bernardo, because of the production of older models with more transport, fitting, fixing, and labour-intensive welding than in Dagenham.

The Body Assembly plant in São Bernardo was organised in three separate production areas up to the Respot Line (final welding). One was modern, with seven robots and automatic transfer. It was built especially for the Escort, launched in 1983, which, on average, accounted for half of daily production. The other two very traditional production areas had some of the operations renewed when the whole plant was rebuilt to provide space for the new assembly areas for the new model. Traditional processes were used for the other half of daily production, which consisted of a mix of the other three models (five nameplates) Corcel/Scala, Del Rey/Belina, and Pampa.[13] From the Respot Line onwards, Escorts and the three other models were all mixed until they were delivered to the dealers.

In Dagenham the Body Assembly shopfloor was entirely renewed for the Sierra launch in 1982. Some operations were done in the traditional way on some of the Fiesta models such as in the Respot Line (in a carousel). Except for this, all the jobs were carried out in new facilities, which included 120 robots of four different types, and a system of fully automatic transfer and overhead conveyors.

Measuring productivity on the basis of workers per car per day, São Bernardo appeared to be slightly less efficient. The 2,770 workers in São Bernardo's Body Construction made an average of 750 cars per day, resulting in an average of 3.7 workers per car. Productivity in Dagenham was more difficult to estimate because of the flexibly allocated 'general services' category. If one-third of these workers were located in Body Construction (the remaining two-thirds in Trim and Final Assembly), the average of 995 cars per day would be produced by 3,378 workers, resulting in a ratio of 3.4 workers per car. Only if two-thirds of 'general services' workers were located in the Body Construction area would the ratio of output at Dagenham be equal to that of São Bernardo (see assumptions (a), (b) and (c) in Table 6.3). Therefore, Dagenham had a maximum advantage of only about 10 per cent over São Bernardo in terms of output per worker. Any higher proportion of 'general services' workers would make Dagenham's Body Construction levels of employment/output more closely equivalent to São Bernardo.

Looking at employment allocation in the Body Construction subareas, it would be plausible to assume that the lower manning[14] level in Dagenham's Body Assembly was compensated by a higher manning level in its Press shop.

Table 6.3 Output per worker per day and other characteristics of the production process in Dagenham and São Bernardo

Areas	*Dagenham*			*São Bernardo*		
	Workers (1)	*Output (2)*	*(1)/(2)*	*Workers (1)*	*Output (2)*	*(1)/(2)*
1. Body Construction (Stamping and Body Assembly)	3,378[a]	995	3.4	2,770	750	3.7
	or 3,540[b]	995	3.6			
	or 3,702[c]	995	3.7			
2. Paint	998	995	1.0	854	750	1.1
3. Trim and Final Assembly (includes Trim Manufacture)	2,952[a]	995	2.9	2,354	750	3.1
	or 3,113[b]	995	3.1			
	or 3,276[c]	995	3.2			
Total	7,652	995	7.7	5,978	750	8.0

(a) Assuming that 1/3 of the 972 workers in the 'general services' category were employed in the area, in which case the proportion in the other area (either Body Construction or Trim and Final Assembly) would be higher than the ones hypothesised in the table. Excluding the fluctuating 'general services' workers, hourly employment in Body Construction was 3,054 and in Trim and Final Assembly it was 2,628.
(b) Assuming that 1/2 of the 972 workers in the 'general services' category were employed in the area, in which case ... (as above).
(c) Assuming that 2/3 of the 972 workers in the 'general services' ... (as above).

Source: Ford-Britain and Ford-Brazil (see Tables 6.1, 6.2 and Silva, 1988, op.cit., Table 5.4)

Other aspects of the production process also affect productivity. The design of the car models is crucial. The best example relates to welding which is the main job involved in Body Assembly. New design specifications have reduced the number of welds required by the use of larger stamped parts. Better welding material and the accuracy of robot jobs have also resulted in a reduction of the number of welds applied. In the Cortina, the model replaced by the Sierra, 3,508 spotwelds were manually applied. The Sierra required only 2,274 spotwelds. However, the Escort produced in São Bernardo required 3,738 spotwelds, even more than the old Cortina.[15] The greater welding requirement can be attributed not just to the Escort's older design, but also to the production facilities in São Bernardo. When welding is done manually more welding spots are necessary to achieve similar quality specifications.

Since 82.6 per cent of spotwelds were applied manually in the

Escort and only 22.1 per cent were applied manually in the Sierra, the amount that workers had to do to assemble a given number of car bodies was much greater in São Bernardo than in Dagenham.

Thus the higher level of automation in Dagenham's Body Assembly should favour considerably lower employment levels. But in fact this was not the case. The impact of technological modernisation on employment was quite different in the two sites.

At São Bernardo, although overall employment levels rose, the new techniques of Body Assembly production reduced employment in relation to output. Thus, at São Bernardo, the underbody, body side and body frame of the Escort, which, on average, amounted to half of daily production, required eighty-eight workers. For the production of the other half (involving three other models), 228 workers were required to carry out the equivalent jobs.[16] But at Dagenham, the higher levels of automation had a much less dramatic effect on the number of jobs. Even though at Dagenham 54 per cent of daily production was Fiestas for which not all the automated equipment was used, Fiesta production still used more automated equipment than the older models in São Bernardo.

In this light, it has to be asked why employment levels in the more automated Dagenham Body Assembly were so high, compared with employment levels in São Bernardo. For instance, even with 120 robots in Dagenham's Body Assembly compared with seven in São Bernardo the level of output per worker was very similar. The causes are discussed in later sections. First we will look in detail at the other two processes, Paint, and Trim and Final Assembly, and then explore the reasons for the differences in productivity.

The paint shops at Dagenham and São Bernardo

The paint equipment in use at Dagenham was acquired about two years before that at São Bernardo. Although they were both new, they were renewed in a period of great technological advance in this area and there was a gap of about two years between the renewal of Dagenham and São Bernardo. The equipment in use at São Bernardo may have been acquired after problems which Dagenham continued to face had already been solved through technological improvements. Paint processes were basically equal.

São Bernardo Paint shop's target productivity was slightly higher than Dagenham's. With 998 workers Dagenham's expected production was sixty cars per hour, and with 854 workers São Bernardo's expected production was fifty-two cars per hour – an almost identical figure in planned cars per worker per hour.[17] Also there was little difference in terms of actual productivity. In the months of my fieldwork, on average, the 998 workers in Dagenham were painting fifty cars per hour while the 854 workers in São Bernardo were painting 43.2. Thus actual productivity was almost identical.

The level of Right First Time Quality favoured São Bernardo where an average of 16 per cent of car bodies had to be repainted. In Dagenham the average was 19 per cent. Despite the small difference between the two, this implied raw material and working hours savings at São Bernardo, having a positive impact on overall efficiency.

The similarities between the Paint areas of both sites were closer than those between the other two manufacturing areas. Paint processes do not vary as much as other production processes, being one of the 'rigidities' in modern production. Management and workers cannot compensate for failures in the equipment or in the paint as raw material to the same extent as they can in other manufacturing processes. Coating applications of specific substances and chemical reactions are more independent of human intervention. This explains the closer similarity between paint processes and equipment in Dagenham and São Bernardo as compared to Body Construction and Trim and Final Assembly.

If certain productivity and quality levels are to be achieved, it may be impossible to carry on the Paint production process with very different production organisation in terms of equipment, employment levels and division of labour. But productivity and quality differences in Right First Time performance at Dagenham and São Bernardo were also linked to different training for Statistical Process Control. This was more developed at São Bernardo as part of a newly introduced Employee Involvement Programme.[18] Different patterns of labour relations accounted for these differences in the training and involvement of the workforce. Unions had a say in this matter in Britain, and had not co-operated with its introduction in Dagenham. Thus, workers' organisation and their relationship to management at each site further explains these differences.

Trim and final assembly at Dagenham and São Bernardo

In Trim and Final Assembly the operations were similar in each plant, but while few of these jobs were automated in Dagenham, the São Bernardo plant was designed according to modern manufacturing concepts.

In Dagenham, the installation was not flexible and only two car models could be trimmed and assembled at a time because the work was organised along two lines, each programmed for a particular car model. In São Bernardo the four car models (six nameplates) came down from the Paint plant, mixed up, for trimming and final assembly along one line. The work organisation there was more complex and made more extensive use of computers. Programmable Logic Controls produced in Brazil were extensively installed on the shop-floor. However, some operations at Dagenham, such as windscreen fitting, were automated, while similar operations at São Bernardo were not.

Productivity levels appeared to be very close. In this case the more modern facilities in São Bernardo were offset by the greater number of tasks required to produce four models compared with two in Dagenham. Let us look at how the production process was organised and compare the performance in each site in detail.

The variety of the model range produced the most significant variations in the way production was organised in this area. Some of the differences in the assembly process were closely related to the design of the car. For example, the side panel of the Escort was a single stamped sheet, whereas in the older models the side of the car was assembled from many different pieces. This was so because modern car designs simplify the assembly process, as illustrated by the new Escort's single unit for what were previously separate wings, bumpers, claws and supports.

The modern design of the new Trim and Final Assembly plant in São Bernardo allowed maximum flexibility for product renewal and further automation. For example, changing to the new Escort in 1986 required only a two-week lay-off in the Sub-Assembly plants. It was also relatively easy to introduce new models to São Bernardo.

Productivity levels in Trim and Final Assembly (Trim manufacturing included) were apparently lower in São Bernardo than in Dagenham. However, if some proportion – half or even two-thirds – of the workers in the 'general services' category was added to the indirect workers in Dagenham, the output of cars per worker would have been the same as in São Bernardo, or higher (see assumptions (a), (b) and (c) in Table 6.3).

Both the production process and performance at São Bernardo compared favourably, in terms of flexibility, though not necessarily in terms of productivity with the traditional assembly concept employed at Dagenham. The levels of output per worker were very similar in Britain and Brazil, although no final statement can be made with the data available.

To summarise, a comparison of the three basic production processes at Dagenham and São Bernardo indicates that:

1 In Body Construction productivity levels in the two sites were very close. The percentage of automated welding applied to the Sierra's body at Dagenham was over four times higher than the automated welding applied to the Escort body at São Bernardo. There were 120 robots installed in Dagenham's Body Assembly compared to seven at São Bernardo's.
2 In Paint the levels of productivity were almost identical. The Paint shop at São Bernardo was slightly more modern than the one in Dagenham. A small difference in the rate of Right First Time quality favoured São Bernardo.
3 In Trim and Final Assembly, productivity levels in the two sites were, again, almost the same. The São Bernardo plant was more modern than the one at Dagenham. Model mix complexity was higher at São Bernardo.

On the whole, there was a higher degree of automation at Dagenham than at São Bernardo. There was a greater product mix in São Bernardo with more older models. Even so São Bernardo and Dagenham had very similar levels of productivity. Comparing the two plants thus helps to assess the role of technology. The relative weight to be given to technology in achieving international competitiveness is explored further in the following section.

5 PRODUCT QUALITY IN SÃO BERNARDO AND DAGENHAM

Productivity and quality are key factors in comparing the performance of car plants which are automated to a different degree. The previous section focused on productivity and showed the difficulties of international comparisons. This section deals with performance in terms of product quality. To this end it is possible to draw upon Ford Motor Company's own international comparisons.

Ford had a worldwide system of quality assessment. It evolved over time. The system employed throughout the 1980s was called UPAS – Uniformity Product Assessment System which replaced the less sophisticated Uniformity Quality Audit in 1980 as a means of improving the sensitivity of the audit to customer demands. It was based on customer evaluation and was done daily in the plants. The measurement of quality is very complex.[19]

UPAS was the plant-based daily product quality assessment, which fed into a much broader system used by FMC to assess the operations of its factories around the world. This Audit system was carried out once a year. Until 1986 Ford issued the results of the worldwide Audit to its subsidiaries and the results were made known to many senior employees. However, the company stopped the internal publication of its worldwide comparisons because, according to one director interviewed, 'it became too contentious. Sites ended up competing with each other and aiming to achieve the others' level. Also, it was hard to push for improvements above the best level. Since 1986, the emphasis has been on continuous improvement, not any longer on improvements up to a certain level reached by someone else.'[20]

Ford's approach to the process of making high-quality vehicles was expressed by a Brazilian manufacturing area manager as 'quality is made by production, not inspection'. Quality control was done alongside all the manufacturing processes from the choice of raw material and the allocation of contracts to suppliers,[21] to the moment that the car was delivered to the dealers. At this stage all the quality standards were checked against their targets.

The Ford Motor Company Quality Audit in January 1986[22] examined fifty-nine sites of different car companies worldwide. The Escort produced in São Bernardo was ranked eleventh amongst the best quality cars in the world. The Sierra produced in Dagenham was ranked fourteenth. In addition, we have already seen that data on Right First Time Quality rates (available only for the paint process) showed that São Bernardo's performance was better.

It must be remembered that this high ranking was achieved even though a major section of the Brazilian factory (Body Assembly) was considerably less automated. The results obtained at the São Bernardo factory were described by management as basically the result of good management and labour discipline.

At Dagenham, relations between management and labour led to different results. During 1984 many meetings were held to discuss policies for improving product quality. Management took an initiative to introduce quality discussion groups and held lineside briefings on quality matters. Shop stewards and unions opposed these meetings. In response, management explained their needs and the trade unions in the Joint Works Committee (JWC) worked out a scheme for improving participation in the resolution of quality problems.[23] The unions' suggestion was that quality problems should be handled in a way consistent with the steps for grievance procedure.[24] The trade unions' approach was accepted for a period of six months. It was vivid in shop stewards' memories that once the union proposals were accepted, the company not only stopped the lineside meetings but also did not use the new procedure put forward by the unions. The conclusion drawn by the stewards was that the company intended to use its meetings to discuss quality matters mainly to undermine the unions. Thus, in 1984, the shop stewards succeeded in stopping the lineside meetings and curbed the development of the quality programme.

6 THE CRUCIAL ROLE OF WORK ORGANISATION

The preceding section underlines the more general question of how much management policy towards labour and work organisation matters in improving competitiveness. Ford management in both countries has pursued new approaches towards labour relations. It is recognised that flexibility in the organisation of work is essential to the improvement of competitiveness, but ways of achieving this have to be adapted to the national and plant context.

Management policies at Dagenham and São Bernardo have required very different kinds of reforms to achieve greater flexibility. This stems from different histories of industrial relations.

In both sites since the late 1970s the share of skilled workers in the labour force has increased and labour turnover rates declined. But these common trends arose from very different industrial relations policies. Both plant managements wanted to raise flexibility, productivity and quality. But at Dagenham, management's attention was mainly focused on changes in the job structure. At São Bernardo, management concentrated mainly on Employee Involvement to secure its goals.

Job structures differed significantly between the two plants. The distribution of the hourly paid workforce amongst the three manufacturing areas was similar but the job structure was different. There was a much higher proportion of direct workers to indirect workers in São Bernardo (see Tables 6.1 and 6.2). In Dagenham the general ratio of direct to indirect workers was 1.4:1, in São Bernardo it was 3.1:1. One consequence was fewer job titles at São Bernardo.[25]

How do these diverse job structures relate to productivity levels? Overall, the biggest difference in job structures was the employment of more direct workers at São Bernardo who did jobs which were done by indirect workers at Dagenham. Job demarcation lines in Britain secured levels of employment by specifying workers' tasks on the shopfloor. There was also greater union and shopfloor control over production matters at Dagenham. Work intensity and the rotation of tasks were not controlled by the union or Factory Committee at São Bernardo but at Dagenham they were matters of collective agreement. The same applied to shift patterns and overtime work. Consequently, in order to implement greater flexibility management at Dagenham was compelled first to deal with the more rigid job structure.

Since the application of the November 1985 agreement, Ford-Britain has moved towards an enormous reduction of job titles and more flexible deployment of labour. These were first steps towards the implementation of the sort of 'team work' and single job and wage classification schemes observed by Willman and Winch, and by Katz, in their studies of other car firms.[26]

The 1985 British Ford agreement dealt with many issues in the implementation of flexibility that would have been irrelevant in the Brazilian context because of differing legal frameworks or practices. The most interesting comparisons relate to the division of labour in the factories.

Craft demarcation lines have never existed in Brazil where the working class is more homogeneous than in Britain. The single trade union reflects this. In Britain jobs have belonged to the trade, and lines of demarcation between them have made changes in the division of labour at the shopfloor level much more difficult than in Brazil.

Unlike Britain, *additional worker training* has not been resisted in Brazil. The lack of job security and low wages have made Brazilian workers eager for new job opportunities and improvements.

The *versatility and flexibility* of Brazilian workers has been very wide. Since refusal to undertake tasks assigned by the foreman could result in dismissal for Just Cause (i.e. without compensation), versatility and flexibility have become intrinsic characteristics of jobs. The resistance of Brazilian employers to the 'professional wage' expresses their opposition to the introduction of 'rigidities' in the allocation of workers and changes that could lead to locking workers into defined groups or 'professions'.[27] Mobility inside the plants has been great because it has been customary to rotate jobs, moving workers to different machines, aiming at getting operators accustomed to all the shopfloor operations. However, less skilled operators have generally not been moved to machines operated by higher skilled workers. Job descriptions in Britain have had narrower task assignments.

Within job definitions in Brazil, *identification of defects, repairs and line feeding* have been taken for granted as a part of the job assignment. In Britain these tasks have been a matter of collective bargaining since November 1985.

The *implementation of process-control techniques*, such as Statistical Process Control (SPC), or completing monitor sheets or job cards, has been done in Brazil as part of the basic jobs of operators, maintenance workers and foremen. Brazilian skilled workers ordinarily have provided reports and records of actions to supervisors, whereas British skilled workers have negotiated about them.

Concerning *adjustments to equipment, cleaning and lubrication*, the minor tasks, which were introduced by negotiation in the British agreement of November 1985, have long been done by the operators themselves in Brazil. *Technical cleaning*, however, was done by an outside contractor at São Bernardo, whereas this was an in-house activity for Ford-Britain.

Ford-Britain has prioritised the reduction of indirect manning. But indirect manning was already low at São Bernardo (Tables 6.1 and 6.2).

The comparison of these practices shows that management problems were different in Britain and Brazil in the 1980s. In Brazil, provisions for moving workers between jobs were built into the existing system of labour relations. For instance, high rates of labour turnover were common during the 1970s.[28] The use of dismissal was seen as a major disciplinary tool and workers were trained to do as their bosses told them. Thus, one important task in the 1980s was to improve trust and involvement in the workplace. One way of enhancing commitment and participation was by implementing a policy of relative stabilisation of the labour force.

The stabilisation of labour turnover in Brazil reflected concerns of both unions and employers. Workers, for example, demanded the 'salário profissional'. In parallel Ford's president offered a deal for employment stability in return for a no-strike deal in 1984, and all the car companies in the area of São Bernardo offered unified wage and job structures in 1985. The motives were different but the various proposals reflected significant common concerns. The aim of the union was to increase its control over hiring and firing and to raise wage levels. The aim of management was to increase worker involvement and commitment to production.

Comparing Dagenham and São Bernardo we find that despite all the differences in conditions and strategies, recent developments show that their managements were clearly preoccupied with improving efficiency and quality in order to face the fierce competition. Given greater possibilities of intensification and flexibility in the actual use of labour in Brazil, we can conclude that management may have found it easier to improve competitiveness in the Brazilian site even if the British plant could automate faster.

7 TECHNOLOGICAL IMPERATIVES FOR COMPETITIVENESS? FINAL CONSIDERATIONS

A general conclusion from these case-studies is that other factors apart from technology are also responsible for productivity and quality. Together with automation, managerial capabilities and labour relations stand out. Organisational processes can be developed in many diverse ways and production strategies are differentiated in particular national contexts.

The questions analysed here concern the production of internationally saleable cars. The comparison has been between two sites where cars were made for the European market, and the scope of the investigation emerging from this is very wide. The competitive strategies developed by Ford in Brazil and Britain go far beyond the issues of the production process for car manufacturing described here.[29] This comparison can only deal with a few aspects of technology, labour and performance, but such international plant level comparisons are essential, though extremely rare.

Although technological innovation in the production process is central to competition it cannot be seen in isolation. Employment conditions and work organisation are very important for achieving higher product quality and productivity. Nevertheless, it is important to stress that the influence of technology may be great in certain processes. The importance of the modernisation of the Paint shops at both Dagenham and São Bernardo suggests the crucial role of technology in determining high levels of quality and efficiency. In this area management seemed to be able to do less to compensate for lower levels of technical modernisation. Moreover, modernisation in other areas at both plants in the early 1980s was essential if they were to play their designated role in the corporation's competitive strategies.

In other respects, the roles of management and labour are more crucial than that of technology: for example in the findings that the introduction of new techniques of Body Assembly in São Bernardo have reduced manning levels much more sharply than new automation at Dagenham. This resulted from more effective forms of shopfloor control in Britain. On the other hand, general employment levels in São Bernardo have increased owing to the export market, while large job cuts have occurred at Dagenham. This resulted from the success of corporate strategies for Brazilian sales in Europe, and the strategies for the implementation of changes in technology and labour organisation in Britain. The acceptance of the clauses for flexibility and versatility in the Ford-Britain 1985 agreement followed the closure of some shops, including the foundry at Dagenham. Workers were compelled to accept changes because of fear of bigger job losses.

How does this evidence relate to arguments stressing that a failure to develop and use the latest and most sophisticated technology is unacceptable and makes it impossible to compete internationally? The case-studies discussed here suggest that large-scale adoption of micro-electronics does not solve, *per se*, the problems of labour management and product quality. Micro-electronics may enhance productivity and product quality, but not alone. Micro-electronics are also likely to require new forms of labour management.[30] Nor are developing countries dependent on the *immediate* access to the *latest* automation technology in order to compete successfully in world markets. Countries like Brazil may be able to cope with the fierce international competition beyond the 1980s without increasing their dependency on technology and managerial skills from the developed world. This requires assessing where its present strength to compete lies, and managing, enhancing and making the best use of existing resources.

NOTES

1 I would like to thank the following for their helpful comments on a previous version of this chapter: Ruy Carvalho, José Cassiolato, Afonso Fleury, Tom Hewitt, Helena Hirata, Márcia Leite, Hubert Schmitz and Steven Tolliday.

2 For a detailed comparison of market, product and technological changes in these two factories, associated with changes in labour relations and work organisation, see Elizabeth Bortolaia Silva, 'Labour and technology in the car industry. Ford strategies in Britain and Brazil', Ph.D. thesis, University of London, 1988; and Elizabeth Bortolaia Silva, *Refazendo a Fábrica Fordista. Contrastes da Indústria Automobilística no Brazil e na Grã-Bretaha*, Hucitec, São Paulo, 1991.

3 See Alan Altshuler, Martin Anderson, Daniel Jones, Daniel Ross and James Womack, *The Future of the Automobile, the Report of MIT's International Automobile Program*, Allen & Unwin, London, 1985; and Davis Dyer, Malcolm Salter, Allan Webber, *Changing Alliances, The Harvard Business School Project on the Auto Industry and the American Economy*, Harvard Business School Press, Boston, 1987. The MIT study provides a broad overview of international competition centred on trade. The HBS analysis is more historically oriented with greater emphasis on national boundaries and social and economic relationships within nations. The accounts complement each other in their description of the present challenges facing the motor industry. See also the proceedings of the 'International Policy Forum of the International Motor Vehicle Program' of the Massachusetts Institute of Technology held in Italy in 1988, and in Mexico in 1989.

4 On the Japanese structure of production and work organisation in the car industry see Reiko Okayama, 'Industrial relations in the Japanese automobile industry, 1945–70: the case of Toyota', in Steven Tolliday and Jonathan Zeitlin (eds) *The Automobile Industry and its Workers*, Polity Press, Cambridge, 1986, pp. 168–89; Koichi Shimokawa, 'Product and labour strategies in Japan' in Tolliday and Zeitlin, *op. cit.*, pp. 224–43; and Kazutoshi Koshiro, 'Personnel planning, technological changes and outsourcing in the Japanese automobile industry', *Technology Management*, Vol. 2, Nos. 2 and 3/4, 1987. For a brief and broader account of the diverging views of the Japanese 'success' based on more recent literature see Elizabeth Bortolaia Silva, 'Remaking the Fordist Factory? Industrial relations in Brazil in the past and for the future', paper prepared for the Seminar 'Autour du Modèle Japonais ...' CNRS, Paris, Février 1990, pp. 28–45.

5 See Knuth Dohse, Ulrich Jurgens and Thomas Malsch, 'From "Fordism" to "Toyotism"? The social organisation of the labour process in the Japanese automobile industry', *Politics and Society*, Vol. 14, No. 2, 1985.

6 On the early history of the motor industry in Britain see Jean-Pierre Bardou, Jean-Jacques Chanaron, Patrick Fridenson, and James M. Laux, *The Automobile Revolution: The Impact of an Industry*, University of North Carolina Press, Chapel Hill, USA, 1982; George Maxcy and Aubrey Silberston, *The Motor Industry*, Allen & Unwin, London, 1959; and George Maxcy, *The Multinational Motor Industry*, Croom Helm, London, 1981.

7 On the early pattern of development of the motor industry in Brazil see Francisco de Oliveira and Maria A.T. Popoutchi, *Transnacionales en America Latina; El complejo automotor en Brasil*, Editorial Nueva Imagen, Mexico, 1979; Eduardo Augusto Guimarães, *Acumulação e Crescimento da Firma: um estudo sobre organização industrial*, Zahar, Rio de Janeiro, 1981; DIEESE, 'Indústria automobilística Brasileira', *Informe Econômico*, São Paulo, 1981; and Helen Shapiro, 'State intervention and industrialization: the origins of the Brazilian automotive industry', Ph.D. thesis, Yale University, 1988.

8 For details on FMC's corporate structure and strategy see Silva *op.cit.*, 1988, Ch. 4.

9 Sierra production was discontinued at the beginning of 1989, when it was transferred to Genk in Belgium.

10 Quotations are from fieldwork interviews carried out from 1985 to 1987 with employees from Ford-Brasil, Ford-Britain and Ford Europe.

11 Given the main concerns of this volume, a discussion of the importance of wages and other costs for competitiveness is not included in this chapter. For an account of such issues regarding São Bernardo and Dagenham see Silva, 1988, *op. cit.*, esp. pp. 106–7, 306 and 331–7.

12 For a discussion of approaches of these kinds see Silva, 1988, *op. cit.*, Ch. 1. For examples of the various approaches see: 1. On the technology's crucial role Joan F. Woodward, *Management and Technology*, HMSO, London, 1958; 2. A 'cultural' focus is provided by William H. Form, *Blue-Collar Stratification: Autoworkers in Four Countries*, Princeton University Press, 1976; 3. The Regulation approach, as developed in French literature, claims to integrate diverse dimensions of social

structures; see Michel Aglietta, *A Theory of Capitalist Regulation*, New Left Books, London, 1979.

13 There are various criteria for the definition of a car model. Sheriff argues that 'platform' is too broad and 'nameplate' too narrow. According to his suggested definition of a model as either 'a car with a completely unique sheetmetal or a car with a substantially unique sheetmetal and a revised track or wheelbase' there were four models at São Bernardo. See Antony Sheriff, 'The competitive product position of automobile manufacturers: performance and strategies', *International Motor Vehicle Program*, MIT, International Policy Forum, Italy, 1988, p. 2. In my previous work (Silva 1988, *op. cit.*, Ch. 5) I adopted a narrower definition of nameplate. A more complex definition is developed by John F. Krafcik, 'Complexity and flexibility in motor vehicle assembly: a worldwide perspective', *International Motor Vehicle Project*, MIT, International Policy Forum, Italy, May, 1988.

14 I use the term 'manning' only when the total workforce in the area is male. Women were employed only at Dagenham's Trim manufacture and in sealing operations at the Paint shop.

15 Source: For the Sierra: *The Industrial Robot*, December 1982; for the Escort: Ford-Brasil, 'Manufacturing engineering', n.d. For the complete welding distribution of the two car models see Silva 1988, *op. cit.*, Table 5.4.

16 Source: Silva 1988, *op. cit.*, Table 5.3.

17 Source: Silva 1988, *op. cit.*, Tables 5.5 and 5.6.

18 For a detailed account of these issues see Silva 1988, *op. cit.*, Ch. 9.

19 For details, see Silva 1988, *op. cit.*, pp. 209–14.

20 The criteria used by this Audit system were not disclosed in any of the interviews carried out either in Brazil or in Britain. In some cases the discussion of the data from these Audits, which I had obtained, made the interviewees uneasy.

21 Brazilian practice was based on a worldwide programme developed by Ford's Supply and Manufacturing Division and consisted of a series of actions to promote improvements in supplier relationships to improve product quality and reduce costs. These actions began to be implemented in Brazil in 1983. Similar changes in the relationship with suppliers were implemented in Britain following the launch of the 'After Japan' programme in 1981.

22 The 1986 Audit was the last one issued to senior employees worldwide.

23 The trade union structures in Britain and Brazil are very different. Ford-Britain dealt with eleven unions, Ford-Brazil with one. The JWC in Britain is a body of representatives of the company and of all the unions, in equal numbers. There is no such equivalent in Brazil, where a Factory Committee linked to the union existed for five years: from 1981 to 1986. After having been banned by the company in July 1986, elections for a new factory committee were held in November 1988.

24 The agreed process followed six steps, involving first the foreman and the local shop steward, and ultimately going to the Joint Works Committee (JWC). See 'Trade Union participation in quality matters', letter from management to PTA supervisors, Dagenham, 24 May 1984, also Silva 1988, *op. cit.*, pp. 367–9.

25 The job structures in both subsidiaries are discussed in detail in Silva 1988, *op. cit.*, Ch. 9.

26 See about British Leyland (nowadays Austin Rover) Paul Willman and Graham Winch, *Innovation and Management Control. Labour Relations at BL Cars*, Cambridge University Press, Cambridge, 1985. About General Motors in the United States see Harry Katz, *Shifting Gears,* MIT Press, Cambridge, Mass., 1985.

27 'Professional wage' (or salary) means the establishment of common pay levels for the same jobs. For instance, toolmakers, maintenance workers, welders and cleaners would be differentiated as job classifications but each individual would receive the standard pay for that job classification no matter in which company they were employed.

28 For accounts of labour management in the 1970s in Brazil see Elizabeth Bortolaia Silva, 'Política empresarial de controle da força de trabalho: rotatividade como dominação', MA thesis, Universidade de São Paulo, 1981; and John Humphrey, *Capitalist Control and Workers' Struggle in the Brazilian Auto Industry*, Princeton University Press, New Jersey, 1982.

29 The integration of the subsidiaries within the corporate plans was crucial in determining the strategies undertaken. Also, the competition pattern of the markets served by the subsidiaries was of great importance. For an account of these issues see Silva 1988, *op. cit.*, Chs 3 and 4.

30 For a discussion of this problem in relation to developing countries see Kurt Hoffman, 'Technological advance and organisational innovation in the engineering industry', report submitted to the World Bank, Brighton, Sussex Research Associates, 1988. For criticisms of this view see John Humphrey, 'New forms of work organisation in industry: their implications for labour use and control in Brazil', paper presented to the Conference on 'Padrões tecnológicos e políticas de gestão', São Paulo, 16–17 August 1989.

7 Employment and skills in the Brazilian electronics industry[1]

Tom Hewitt

1 INTRODUCTION

This chapter explores and documents the growth and nature of jobs in Brazil's electronics industry. It shows that this sector is in transition from labour-intensive assembly operations to more knowledge-intensive products and techniques. This transition, in turn, has a significant impact on the prospects for employment generation and skill formation. A driving force behind these changes has been active policy intervention. The focus is on the nationally-owned computer industry and, for the purposes of comparison, on the largely foreign-owned mainframe computer and consumer electronics industries.

Objectives and structure

The objective of the chapter is to examine the effects of policies in electronics on the industry's human resources. To this end, there is a brief discussion of the policy regimes for three sub-sectors of the electronics industry, followed by an overview of the growth of these sub-sectors over the last decade. The chapter then documents the changes in employment in the industry over this period and examines the qualitative aspects of employment: occupational composition, skills, training and ages. At the same time, the chapter discusses the obstacles which Brazil, as a developing country, faces in sustaining the growth of a skilled labour force in electronics which can keep pace with the expansion of the industry in certain sub-sectors.

Both changes in job and skill creation and the obstacles and timing of these changes vary within the industry. Therefore, the nationally-owned small system computer and peripherals industry (mini-computers and microcomputers) is compared, where applicable, to other sub-sectors, namely, large system computers (mainframes) and consumer electronics.

The argument is as follows. Electronics is a human resource based technology. As O'Connor has pointed out,

> The computer industry is essentially a knowledge-intensive industry wherein skilled, highly trained scientific, engineering and technical labour power is probably the single most important asset. Without such labour, even access to adequate financial resources and material inputs would not be a decisive advantage in a country's effort to develop an indigenous industry.
>
> (O'Connor 1985:325)

Until the end of the 1970s, multinational corporations carrying out labour-intensive assembly operations dominated the electronics industries of developing countries. The strategies of these firms, encouraged by host government policies, have strongly influenced employment patterns and the skill characteristics of the industry in developing countries. This resulted in the creation of jobs that for the most part were unstable and 'dead-end' in terms of the possibilities for upgrading skills or for fostering a skilled technical and engineering labour force. In addition, the possibilities for forward and backward linkages, in terms of technological learning, skill formation and skilled employment generation, seem to have been minimal.

In any economy, but particularly one which suffers from high unemployment, it is important to look for new sources of jobs. One purpose of this chapter is to examine the Brazilian policy regime for the electronics sector from the point of view of labour by asking first, have there been effective ways of creating jobs within national boundaries rather than importing products which would create jobs elsewhere? A second, more significant question concerns the means by which highly skilled technical and engineering employment can be generated.

The wider significance of these questions is summed up by Hoffman:

> the development of the electronics sector will assume the same role now normally attributed to the capital goods industry as a crucial source of skills and innovations that will have a wide degree of applicability throughout the economy and society as a whole.
>
> (Hoffman 1985:226)

In other words, the setting up of an electronics industry in Brazil, given its strategic position in the economy, is a way of internalising the economic and social benefits of electronics. As will be shown in the case of the computer industry, the combination of protective policies, national ownership of firms and an emphasis on developing technological capacity has produced a local demand for highly skilled labour which has been beneficial for all firms active in the sector (both foreign and national). This said, the initial phase of the industry has raised issues that are by no means settled. Despite the strategic concepts which guided the setting up of the national computer industry, it may well face tremendous pressure to change track and take up a more competitive stance than it has to date.

The industry faces a number of problems: the incompatibility of policy frameworks for different sub-sectors of the electronics industry; political opposition from home and abroad; economic pressure to become more price-competitive on the international market; and finally, growing pressures on the availability of scarce skilled labour due to the rapid expansion of the industry.

This study should therefore be treated as interim. Nevertheless, it does provide a detailed account of the 1980s from which lessons can be learnt on job and skill creation, some of which may have a wider policy application for other developing countries.

The chapter is organised as follows: Section 2 briefly maps out the differing and, at times, conflictual policy contexts of the three sub-sectors. It then shows the growth in output and markets for electronics goods.

Section 3 is a detailed presentation of data on employment and skills in Brazil's electronics industry. Beginning with the growth/ decline of absolute levels of employment in different sub-sectors, it goes on to examine the different skill composition, training practices and wages found in the industry. The information presents a picture of consistent variation between sub-sectors.

Section 4 concludes the chapter by returning to the issues raised at the beginning of this book. Namely, assessing the role of policy intervention in stimulating changes in the pattern of employment and the generation of skills in the electronics industry.

Sources

The data presented in this chapter are based on fieldwork carried out between May 1985 and March 1986 and the use of secondary sources published subsequently:[2]

1 Interviews in thirty-six firms covering three segments of electronics production, of which there were:
seventeen computer producers (computers and peripherals);[3]

thirteen consumer electronics goods producers (final and intermediate products);
six components producers (semiconductors and PCBs).

2 Interviews with specialists from:
universities;
government bodies;
professional associations;
technology suppliers.

3 Secondary sources:
specialist magazines;
newspapers;
academic and technical papers.

An important additional source to the information collected directly from firms in the sector was data from the Ministry of Labour's Annual Register of Social Information (RAIS). This provided a comprehensive data base on labour in the electronics industry which was used at two levels: at an aggregate sectoral level, with some disaggregation by sub-sector; and in small aggregate groups of companies representative of different sub-sectors of the electronics industry.[4]

2 POLICIES AND GROWTH IN THREE ELECTRONICS SUB-SECTORS

The main purpose of the chapter is to examine the evolution of employment and skills in three sub-sectors of the electronics industry in Brazil. This section provides brief comparisons of the policy contexts within which the sub-sectors operate and presents an overview of growth.

Policy contexts

The principal tool of policy in the *computer industry* has been the protection of certain segments of the computer market for nationally-owned firms; initially minicomputers, microcomputers and their peripherals. With the creation of the Special Secretariat for Informatics (SEI) in 1979, a protected market for national firms gradually spread to other areas, such as super minicomputers, automation equipment, micro-electronic components, teleinformatics, software, process control and instrumentation.

Some of the reserved market policy instruments used by SEI are:

the control of electronics imports; the concession of manufacturing licences to national firms; and the supervision of the demand for electronics goods and services by state organs and enterprises (Piragibe 1985:132). These provisions allowed SEI initially 'to attract domestic manufacturers into reserved market niches by slowing down the processing of import authorisations, which in combination with government procurement, created substantial pent-up demand to make domestic production by national firms a financially viable undertaking' (Frischtak 1986). As firms have become more established and as SEI has become more efficient, it has been possible to substantially reduce lead times for import licences and new projects (see Chapter 2).

Foreign computer firms located within the country (producing mainframes) also operate within the reserved market to the extent that the granting and withholding of import licences (which cannot contravene the reserved market for national firms), forces foreign firms to have growing indices of nationally produced inputs in their products (Piragibe 1985:132). Because of restrictions on market entry, foreign firms already located in Brazil – particularly IBM and Unysis – have no doubt benefited from the protection from import competition as a result of the informatics policy (Piragibe and Tigre 1989).

In the *consumer electronics* industry, in contrast to the experience of the computer industry, there have been no restraints to the inflow of foreign capital or foreign technology. Policy for consumer electronics has followed an international tendency in the industry. With the exception of Japan (and possibly France), government policy in consumer electronics internationally has been primarily *defensive* (Dicken 1986:334), using such measures as trade restrictions and voluntary export restraint agreements in an attempt to keep at bay the clear technological leadership of the Japanese.[5]

The consumer electronics industry in Brazil is also dominated by Japanese firms (and/or technology) which have production facilities within national boundaries in order to circumvent protective barriers. As a result, some observers have concluded that 'the profile of the consumer electronics industry reflects the absence of a regulatory policy for this sector' (SICCT 1985:137). This is not exactly the case even though the policy regime for the industry is quite different to that for computers.

The consumer electronics industry particularly has been influenced by regional development policies. The industry is now almost entirely located in the Manaus Free Zone (MFZ) and developments in the zone have meant that it has become virtually synonymous with the consumer electronics industry.

The economic attraction of the MFZ is a system of fiscal incentives. Firms located in the region are exempt from: some income tax, state tax on the entry and exit of goods, import tax after a predetermined index of product nationalisation has been reached, plus an array of other tax exemptions for specific products and individuals. The combined result of these incentives is a reduction in the cost of production of electronics goods of between 25 and 30 per cent relative to the rest of the country (Baptista 1985).

In sum, there have been no *direct* sectoral policies for the consumer electronics industry and regional policies have tended to create a hands-off attitude towards the industry. This has led to a certain conflict of interests within the electronics industry, in particular with the hands-on policy of the computer industry. The policy dilemma is summed up by Tigre:

> independently of their individual merits, the disarticulation of policies for computers, telecommunications and consumer electronics is a problem for the development of electronics in Brazil. Each sector has a distinct industrial structure which inhibits their integration into 'electronics conglomerates'. Thus, the opportunities of generating the economies of scale and scope inherent in diversified production are lost. Apart from this, the industrial and technological experience accumulated by one sector is rarely transmitted to the other, thus delaying the process of national capacity.
>
> (Tigre 1986:127)

Growth

This section examines the gowth of the electronics industry in the 1980s. Computer-related production accounts for the largest segment of the electronics market (Piragibe 1987; Piragibe and Tigre 1988; SEI 1989). Within computer production, it is the national firms which have undergone the fastest growth in output. Indeed, by 1986, national firms had passed a watershed by attaining 51 per cent of the total computer market (Piragibe 1987). By 1988, the ratio of output of the national computer industry relative to foreign computer firms was 2:1 (SEI 1989).

Various conditions set the context of and encouraged these changes in the computer industry. Computers were imported at annual growth rates of more than 30 per cent in the period 1969–76 (Piragibe 1985:115) and in 1974, they were the third largest item of manufactured imports in value terms (US$98.8 million) (Frischtak 1986:4). At the same time as this expansion of the computer market, the first oil crisis of 1973 and Brazil's growing balance-of-payments problems began to put a strain on import spending and the first steps were taken to produce computers within national boundaries.

Until 1975 there was a virtual absence of local production of computers. As Piragibe (1985:3) points out, the industry 'was born internationalised', based as it was on the import of final products or, as in the case of IBM and Burroughs (now Unisys), restricted to the final assembly in Brazil of products with high import content.

The dominance of foreign firms in the Brazilian computer market, based on ample technological, commercial and financial resources, imposed considerable entry barriers to local firms. In the opinion of Piragibe (1985:113) it also 'inhibited both the development of technology in the country and the creation of skilled employment in these activities'. Thus the state protected certain market segments from foreign capital competition.

A small number of newly-created nationally-owned firms were chosen to produce minicomputers, peripherals and, subsequently, microcomputers. From that time on the national computer industry underwent an impressive growth in sales, although this did not diminish the continued growth of foreign computer firms, whose sales increased threefold between 1980 and 1988 (Piragibe and Tigre 1988).

By 1986, the number of firms operating in the nationally-owned computer and peripherals market (excluding software houses) increased to nearly 300 (from thirty-seven firms in 1980) and there were thirty-one foreign computer firms in 1986, compared to four in 1980 (SEI, cited in *Senhor* 2 Sept. 1987). The size of installed capacity in computing places Brazil amongst the ten largest computer users in the world (Piragibe 1985:151).[6]

The number of firms in the *consumer electronics industry* has also increased. In the mid-1980s, there were thirty-nine electronics firms in the Manaus Free Zone accounting for 37 per cent of the region's total employment, although for only 13 per cent of fixed investment of all firms (*Info*, June 1985:62).[7] Since then a further twenty-four industrial projects in consumer electronics have been set up for intermediate and final products. Only some of these projects were to be taken on by already existing firms in the region. However, overall output increased only a little in the course of the 1980s.

The period 1972–74 is when the consumer electronics sector of the MFZ began to be consolidated. Such leading consumer electronics firms as Sharp, Philips, Sanyo, Evadin, CCE, Matsushita, National, and others installed factories. The attraction for these firms to set up in Manaus was a wide-ranging package of fiscal incentives. Apart from local tax exemption, one of the reasons for the migration of firms to the MFZ at the time was that there were no restrictions on imports – any import could come from abroad and there was no dollar limit to how much could be imported.

3 EMPLOYMENT AND SKILLS IN ELECTRONICS

The purpose of this section is to examine what impact the different policy regimes and growth patterns outlined above have had on the creation of jobs and the generation of skills in the three sub-sectors.[8] One of the main conclusions is that not only has the market reserve for national computer firms been a significant generator of new jobs in Brazil but also intervention has indirectly created skilled jobs which would not have been generated if the industry had been left to foreign firms. The difference between the routine assembly of goods and of understanding and modifying those goods lies in engineering and technical capacity. It is this human resource capacity which, in turn, sets the pattern for the future development and viability of the industry. As discussed in the introduction, electronics is a knowledge-intensive industry reliant on skilled engineering and technical labour. The fostering of this labour becomes a prerequisite for the sustainability of the industry.

Employment in electronics

Employment in the electro-electronics sector as a whole in Brazil has been declining since the beginning of the 1980s. This decline is roughly equivalent to the decline in manufacturing employment as a whole since the share of the sector in total manufacturing remained fairly stable. From a peak of 243,000 in 1980, the decline was steady until 1984 when employment grew 6 per cent relative to the previous year, reaching 200,000 people. According to information from CDI, the variation in employment reflects the variation in output of the sector (CDI 1984:5).

According to Ministry of Labour data (RAIS), in 1984 the electrical and electronics industry had 1.8 per cent (234,223) of Brazil's total registered employment (including agriculture and

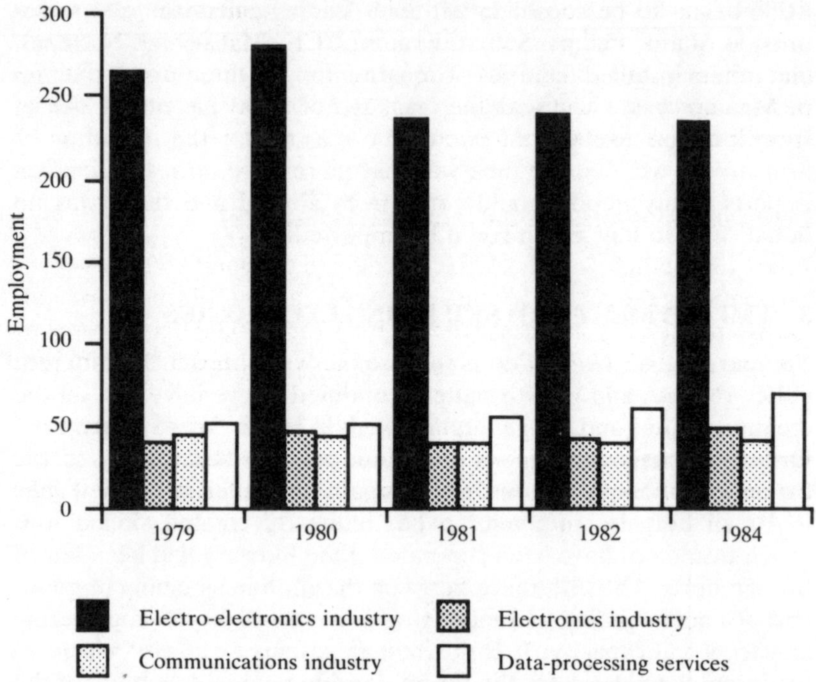

Figure 7.1 Employment in electronics and data-processing services 1979–84 (1,000s)
Source: Computed from data provided by the Ministry of Labour (RAIS)

services) and, within this, the electronics and communications subsectors accounted for 0.7 per cent (89,401) of Brazil's total. The employers' association of the electronics industry, ABINEE, estimates that the 200,000 employees of the electrical and electronics sector accounted for 5.1 per cent of total manufacturing employment in 1984.

Data from RAIS in Figure 7.1 show a slightly higher total employment in the electrical and electronics industry than the employers' association of the industry – ABINEE – estimates. The electronics equipment industry, which formed some 21 per cent of the whole sector in 1984, reached a peak of employment in 1980, then declined the following year. By 1984, the total employed in electronics had surpassed the 1980 level. Also notable is the level of and increase in employment in data-processing services.

Within the electro-electronics sector as a whole, however, the variation of both employment and of output is marked. Much of the decline can be put down to activities in the generation and transmission of electricity; telecommunications; and consumer electrical/electronics goods (CDI 1984). Consistently growing throughout this period, however, has been the computer industry. A closer look at specific sub-sectors bears this out.

Levels of employment in the computer industry

From 4,000 employees in 1979, according to the computer producers' association (ABICOMP), the national segment of the computer hardware industry alone accounted for nearly 30,000 employees by the end of 1985 (*Info*, Dec. 1985) and an estimated 55,000 in 1988 (SEI 1989).

Figure 7.2, based on a representative sample of 49 electronics firms, shows that the growth in employment is due wholly to the expansion of the national segment of the computer industry. While

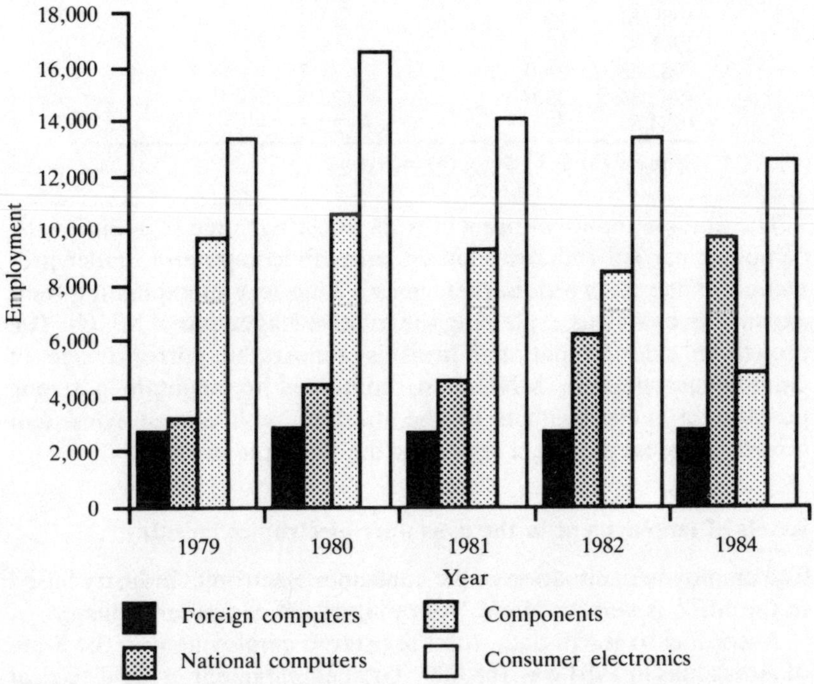

Figure 7.2 Employment in RAIS sample of electronics firms 1979–84
Source: Computed from data provided by the Ministry of Labour (RAIS)

the share of MNC computer firm employment has remained relatively stable since 1980 (SEI 1989), employment in national firms has grown considerably.

This aggregate growth is due not only to the entry of new firms into the sector but also to the continued expansion of existing firms, as shown in Table 7.1. The higher growth rates of the total of firms indicates the continued entry of new firms into the market and explains why the growth rates in the interview sample are consistently lower. Nevertheless, even in those firms which have been established longer (the sample) and which are amongst the largest in computers, the growth in employment continues to be impressive.

Table 7.1 Growth rates in employment in the national computer industry

	Nationally-owned computer firms	
Year	*Total*[a] *(%)*	*Sample*[b] *(%)*
1979/80	80.0	39.5
1980/81	14.0	13.2
1981/82	52.0	18.5
1982/83	25.0	21.7
1983/84	39.0	22.5
1984/85	30.0	20.5

Sources: (a) SEI (1986); (b) interviews

The entry of national firms into the most dynamic segment of the computer market (microcomputers and minicomputers), under protection of the reserved market, goes a long way in explaining their expansion, as well as explaining the relative stagnation of MNCs. The growth in sales in national firms is almost the mirror image of employment growth. MNCs also continued to maintain a strong position in their segment of the market, with almost consistent growth in sales, despite a small decline in employment.

Levels of employment in the consumer electronics industry

The employment situation in the consumer electronics industry based in the MFZ is very different from that of the computer industry.

According to RAIS data, total registered employment in the State of Amazonas in 1984 was 168,000. Total employment in the electrical and electronics industry was 15,451, or 9.2 per cent of the State's total employment. At least 82 per cent of this employment is in the

Table 7.2 MFZ: Number of firms and employment, 1980–4

	No. of firms			*No. of employees*		
Year	*Electronics (A)*	*Others (B)*	A/B %	*Electronics (A)*	*Others (B)*	A/B %
1980	31	153	20	19,905	29,658	67
1982	35	173	20	20,542	29,248	70
1983	42	186	23	20,094	30,006	70
1984	45	214	21	17.442	27,507	63
1980/84 % growth	45.2	39.9		−12.4	−7.3	

Source: SUFRAMA (cited in Baptista 1985:57)

consumer electronics industry itself. Table 7.2 shows employment estimates from SUFRAMA (Superintendency of the Free Zone of Manaus).

According to this information, the number of electronics firms in the MFZ in the period has been growing at a faster rate than in other industrial sectors (45.2 per cent and 39.9 per cent respectively). Despite this expansion, there has been a net loss of employment in the same period: −12.4 per cent for electronics firms and −7.3 per cent for other firms.

The data in Figure 7.2 show a steady decline in employment in the sample of thirteen consumer electronics firms in the Manaus Free Zone. These are the largest firms of the region, accounting for some 73 per cent of electronics employment in 1984. Despite the variation in sources, they indicate some consistency: employment in consumer electronics has been declining since the early 1980s, a time when the MFZ enjoyed relatively 'full employment'. The following years of economic crisis have meant that the MFZ has not yet reached this level again.

The solution, according to the technical director of FUCAPI (Fundação Centro de Anàlise de Produtos Industriais) 'is the constant attraction of new enterprises to the region' (*Data News*, 7 Jan. 1986, p. 20). New enterprises are going, but from the data presented in Table 7.2, this strategy is not succeeding in employing the growing numbers of those without work in Manaus. This is predominantly a semi-skilled female labour force with little or no alternative employment opportunities in Manaus. This situation has led one observer to conclude that in consumer electronics, the combination of changes in product and process technologies 'shows an inevitable tendency for the industry to lose its capacity to generate employment' (Baptista 1985:33).[9]

Labour force composition

The composition of the labour force is another indicator of the kind of social benefits which a national electronics capability can bring. An industry which is intensive in semi-skilled labour does generate much-needed employment. However, if a technological capacity in electronics is to be built up, what matters equally is the generation of skilled technical and engineering labour.[10]

By aggregating the RAIS data for the labour force composition into four broad occupational groups, we obtain a clear picture of the variations in occupational composition between sub-sectors as shown in Figure 7.3.

The presence of engineering, software and other professional skills as well as technicians is most marked in the computer industry (foreign and national). These data, however, do not tell us which activities these professionals carry out within the firms. Figure 7.3

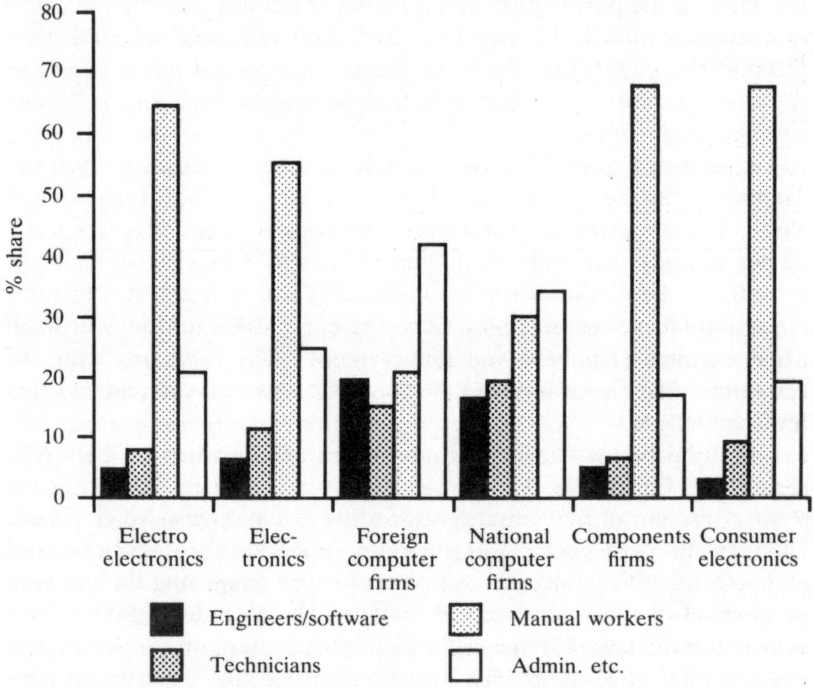

Figure 7.3 Labour force composition by occupation and sector (% share) 1984

Source: Computed from data provided by the Ministry of Labour (RAIS)

shows that foreign and national computer firms have comparable proportions of engineers and technicians.

Our research and other sources indicate that engineers in foreign computer firms are often involved in activities other than production, such as marketing, management and administration.[11] National computer firms, by contrast, employ much of their engineering staff in both production and, more importantly, in R&D (Piragibe *et al* 1983, SEI 1989).

More recent data in Table 7.3 show that the distribution of graduates has a similar pattern: MNCs employ few of their graduates in R&D (5.7 per cent in 1987) while national firms have a high concentration of their graduates in this activity (25.3 per cent in 1987).

Table 7.3 Proportion of graduates by activity in national and foreign computer firms (%), 1983–7

	1983		*1985*		*1987*	
	National	*Foreign*	*National*	*Foreign*	*National*	*Foreign*
Production	17.8	13.5	17.8	18.7	16.4	14.7
Sales	12.4	42.8	22.9	17.8	19.7	36.1
R&D	30.3	4.3	29.2	6.4	25.3	5.7
Maintenance	14.7	9.7	11.7	19.7	20.0	14.9
Administration	24.7	29.6	18.5	37.5	16.0	26.9
Training	–	–	–	–	2.6	1.7
Total	100.0	100.0	100.0	100.0	100.0	100.0

Source: SEI 1989

Despite these differences within the computer industry, a much greater contrast can be found between the computer industry and the consumer electronics industry. In the latter (and for the components industry) there is a very low incidence of engineers and technicians, even falling below the aggregate for the whole electro-electronics sector (see Figure 7.3).

In contrast to the situation for skilled labour, there are high concentrations of semi-skilled workers in consumer electronics and components (68.1 per cent and 68.0 per cent respectively). A large proportion of these workers carry out assembly tasks and there are very few skilled workers amongst them. In the computer industry, the proportion of semi-skilled workers is relatively low (20.6 per cent for foreign firms and 29.5 per cent for national firms). This is based on special tabulations of 1984 RAIS data.

Our research, carried out in 1985, tends to confirm the above trends and allows a direct comparison between the three sub-sectors. In the fourteen national computer firms for which there is sufficient information, the employment composition in 1984 was as follows. Product development accounted for some 15 per cent of the total employed.[12] Direct production in these national firms accounted for 30 per cent of the workforce, while indirect production involved a further 12 per cent of employment. The remaining 43 per cent in the sample were employed in administration, marketing, and post-sales maintenance and services.[13]

Of those involved in direct production in the fourteen national firms (30 per cent of total), more than half (52 per cent) carried out assembly work and a further third (21 per cent) were involved in product testing. The remaining 27 per cent of those in direct production carried out mechanical, metal working and packaging tasks. It is important to emphasise here that women accounted for 46 per cent of those employed in direct production. Direct observation on the shop floor suggests that they are concentrated almost exclusively in the assembly of printed circuit boards and wires and cables. That is, in semi-skilled activities. Assembly work for men, on the other hand, tends to be concentrated further down the process where the final product is assembled.

For eight firms on which there is sufficiently disaggregated information of the labour force composition, it has been possible to make a comparison between national and multinational computer firms and

Table 7.4 Percentage share of employment by skill category, 1985, in computers and consumer goods

Category of labour	*Computer firms*		*Consumer goods firms*
	National	*Foreign*	
R&D	12.5	3.7	1.9
Engineers	8.4	2.2	0.6
Technicians	4.1	1.5	1.3
Production	47.8	50.1	56.0
Engineers	2.1	2.7	0.9
Technicians	6.1	5.6	4.4
Maintenance	2.0	1.2	1.1
Supervisors	1.4	3.7	1.4
Assembly workers	36.2	36.9	48.2
Others	39.7	46.2	42.1
Total	100.0	100.0	100.0

Source: Interviews in 8 firms

consumer electronics firms. Table 7.4 summarises the relative share of the labour force by activity and by occupational category.

In R&D, a pattern similar to that already described is evident: national firms have a much greater percentage of their workforce here than either MNCs or consumer electronics firms.

Of the three sub-sectors, the consumer electronics industry has a higher proportion of its labour force (56 per cent) in production than either of the computer sub-sectors. In addition, the bulk of the production workforce in consumer electronics is in assembly (86 per cent of production), whereas assembly workers in the computer industry represent a lower proportion (about 75 per cent of production). Concomitantly, the share of skilled workers in production is higher in computers than in consumer electronics.

Table 7.4 clearly shows that the consumer electronics industry is a labour intensive assembly industry. Other sources confirm this. Calculating from a sample of 12,612 employees in the industry (drawn from RAIS data), skilled jobs make up only 11.6 per cent of total employment (0.3 per cent engineers, 8.6 per cent technicians and 2.7 per cent other professionals). By contrast, 68.1 per cent of the labour force are involved in manual work, of which 70 per cent are semi-skilled assembly workers and less than 1 per cent are skilled manual workers.

By contrast with computer firms, administrative and other services staff make up a low percentage of the total labour force in consumer electronics. This is explained in part by the location of head offices and sales personnel in the south of the country where final markets for consumer electronics goods are.

In sum, there is a pattern of employment which differs systematically between sub-sectors of the industry both in quantitative and qualitative terms. Most significant is the substantial growth in employment in the national segment of the computer industry and its high percentage of skilled engineering and technical labour. The build up in human expertise in electronics has, it appears, been aided by the market reserve for the national computer industry. Since firms in the sub-sector do not have 'parents' to turn to as a source of innovation the policy has forced them to accumulate their own technological experience.[14] The above data cannot prove the case, but it nevertheless suggests that such a fostering of talent and human resources would not have occurred if foreign firms had been allowed to enter this industry.[15]

Training, wages and labour shortage

The labour force composition has undergone a qualitative change in Brazil's electronics industry. The skilled technical and engineering component of the labour force has grown substantially. This has occurred under specific conditions, namely the setting up of a computer industry for which a principal goal has been to attain national technological capability. But the rapidly growing demand for skilled labour has resulted in shortages in the labour market. This has brought to the fore the question of training since it is here that the problem of the supply bottleneck will have to be resolved in the longer term. This section examines training practices for technicians and engineers and the extent to which the educational system copes with the demand for skilled labour. In addition, it discusses briefly wage structures in the industry and examines how wages are used to secure scarce labour.

Technicians are important in both production and R&D activities of the electronics industry. About 80 per cent of the demand for technicians is, obviously, in electronics skills. The majority of firms interviewed complained of the difficulty of recruiting electronics technicians with adequate and suitable training. The problem, however, is not the quantity of technicians available on the labour market. A study by the National Industrial Training Service (SENAI) on the demand for electronics technicians in the state of São Paulo shows quite clearly that the number of trained electronics technicians far exceeds the number employed: 21,000 as against 3,200 in the period 1979/83 (SENAI 1985:129). The problem is summarised by the study as '. . . a scarcity of technicians with *a professional training adequate to the necessities of the sector* . . . in other words, an inadequacy of supply relative to the sector's demand which is more a problem of qualitative, as opposed to quantitative demand' (ibid.:130).

Interviews with staff at SENAI suggested that the roots of this inadequacy lay in existing second grade courses in electronics. For example, of the 109 schools (public and private) in the state of São Paulo which offer courses for electronics technicians, only eleven were considered to be of adequate quality. Even in these schools, the third year of specialisation was considered to be inadequate through lack of resources for *practical* training.

SENAI (National Industrial Training Service) is an organisation linked to the Ministry of Labour but funded from employers' contributions. This gives industry a direct role in the training of technicians and arguably the opportunity for a greater control over the precise skills required of technicians. SENAI is beginning to respond to the problem by opening training schools which focus on the specific skill requirements in computer hardware and software (MCT 1988).

In the Manaus Free Zone, the problem of inadequate training is more serious, though the quantitative demand for electronics technicians is less. In this region steps are being taken by FUCAPI (Foundation for Industrial Support in Manaus), in conjunction with SENAI, to create a local supply of technicians. In 1985, for example, FUCAPI supplied seventeen courses in electronics to members of industry (engineers and technicians) and has recently created a second grade course to train 60 electronics technicians annually.[16]

To overcome labour scarcity in the computer industry, some of the larger national firms in particular are following the example of IBM and investing considerable resources in training. The strategy seems to be not only to fill the gaps in skills but to keep technology (embodied in employees) within the firms. This was a major pre-occupation of the managers and personnel officers interviewed. As a result, training programmes were often linked to other benefits such as finely-tuned salary scales, promotion possibilities, and so on. This desire to keep know-how within the firm is not only a result of inter-firm competition and the frequent migration of skilled employees between firms. It also occurred within firms where suitable technicians were in short supply. For example, some firms were concerned that their most experienced technicians attended engineering evening courses. As one engineer put it: 'what we want is good technicians, not bad engineers'.

'Good' *engineers* are in short supply. In an interview, the president of one of the larger national computer firms summed the situation up: 'at the level of human resources, specifically engineers, the situation is catastrophic. We are paying the price of the thoughtlessness and lack of foresight of the last ten years.' He estimated that the likely number of electronics engineers available to supply the industry in the next two or three years would be in the region of 2,000. Of these, his firm alone was planning to recruit some 600.

Research carried out by the Ministry of Science and Technology confirms this shortage and notes '. . . an enormous lack in the training and development of human resources for informatics in the country relative to that required to reach national technological capacity in the area' (MCT 1986:1). This problem has its roots in the lack of technical and financial resources in the principal centres of labour training which '. . . cannot meet the vigorous demand for "informatics" professionals adequate to the requirements of the labour market' (MCT 1986:3).

A report on human resources in electronics draws the conclusions of this situation:

> The market reserve is running serious risks ... there is a short term one which is external pressure, but there is another in the medium term which will come about by not perceiving that protectionism in itself is not sufficient. We urgently need aggressive measures for human resource training.
>
> (*Dados e Idéias*, August 1986:28)

In a situation where the public education and vocational training system cannot keep pace with the demand from an expanding industry, many firms have themselves taken on the task of training. In 1984, US$4.5 million[17] was spent by national computer and peripherals firms on the training of labour, with approximately 2 per cent of total employment devoted to training activities (SEI 1986). It is probably those firms which are investing in human resources training[18] – 'preparing for the end of the reserved market', as one engineer put it – which will prove to be the most competitive in the 1990s when, on present trends, the real effects and pressures on the local electronics industry will be felt.

In sum, the shortage of skilled labour in the computer industry becomes more acute as the industry expands. Despite the substantial increases in the supply of skilled labour, there still appears to be a shortage which public institutions are finding difficult to satisfy. The response of firms in the sector has been to take on the function of training. This contributes to the process of learning discussed in Chapter 2.

The scarcity of skilled labour is further illustrated by comparing the wage structure of the industry. Table 7.5 presents wage data for two occupations – engineers and technicians – in three sub-sectors of electronics. It summarises data from RAIS for the occupations and sub-sectors, as well as providing data for the electronics industry as a whole.[19]

There are a number of (sometimes stark) differences which deserve highlighting. Table 7.5 shows a differential between sub-sectors for *technicians'* wages.[20] In foreign computer firms, most earn over seven minimum wages, while in consumer electronics firms nearly 60 per cent of technicians earn less than five minimum wages. The national computer firms show more of a spread of wages, although more are paid above the median of five to seven minimum wages than below.

Table 7.5 Wage distribution for selected occupations in electronics (% shares of employees)

	Less than 15 MW	*15–20 MW*	*More than 20 MW*
ENGINEERS			
Electronics industry	48.0	28.2	23.9
Foreign computer firms	26.5	38.8	34.7
National computer industry	43.5	33.6	23.0
Consumer electronics firms	69.5	2.8	27.8
	Less than 5 MW	*5–7 MW*	*More than 7 MW*
TECHNICIANS			
Electronics industry	40.7	21.4	37.9
Foreign computer firms	1.4	10.3	88.4
National computer industry	33.0	26.7	40.2
Consumer electronics firms	59.9	22.7	17.4

Note: MW = Minimum wages.
Source: Computed from data provided by the Ministry of Labour (RAIS)

According to those interviewed in the national computer industry, the combination of labour scarcity and competitiveness between firms has produced a strategy of paying higher wages to keep technicians. In this category of labour, there is a tendency to recruit young, recently trained technicians, or, in some cases, to take on apprentices, training them on the job and giving them the possibility to be promoted to progressively more sophisticated and better paid tasks. The objective was to create a sense of identity with the firm.

In the case of engineers, Table 7.5 shows that there is a general tendency for their wages to be higher in foreign computer firms. National computer firms, again, show a spread of wages for engineers, while most engineers in the consumer electronics firms are below the median wage.

Comparing national computer firms and consumer electronics firms, we have seen a significantly higher ratio of engineers in the former than in the latter, owing to the greater emphasis on R&D activities which are engineer-intensive. The demand for highly qualified (and therefore higher paid) hardware and software engineers for R&D is a competitive battle that is mainly being fought between computer firms and not between computer and consumer electronics firms.

The competitiveness of firms in computers depends to a great extent on their access to hardware and software engineers both for production and, more importantly, for product development.[21] There are two sources of skilled labour for the industry. The first is from the university system which supplies both engineering graduates and experienced research engineers.[22] The low pay of the university system in Brazil cannot compete with the 'predatory demand for qualified professionals from industrial and service sectors' (MCT 1986:4).

The second source is from within the industry itself, which continually undergoes a circulation of highly skilled engineers and technicians. Nearly all of the firms interviewed expressed concern over the periodic loss of engineers, or even a whole team of engineers. Examples included: a software team of eight people; an R&D team of twelve people; and a whole production management team. Higher wages are used openly as a means of luring the best engineers to particular firms and to motivate them to stay.[23]

For the foreign computer firms, there are a significant number of engineers at the high end of the wage scale. This is not a function of skills needed for R&D activities, but more of the high value put on the MNCs' marketing and technical assistance activities, as well as the value placed by MNCs on the quality of their products which, because of their technological complexity, require qualified engineers to work in the area of quality control and equipment testing.[24]

The reasons for the observed wage differentials between sub-sectors may be summarised as follows. The scarcity of appropriate engineering and technical skills in electronics, in itself, puts an upward pressure on wages. Scarcity is felt most acutely by those firms which are involved in R&D activities, namely, national computer firms. The scarcity in the consumer electronics industry is caused by geographical remoteness from the main supply of labour and not by its demand for highly skilled labour. Scarcity of skilled labour has generated competition between firms for that labour. This occurs both between foreign and nationally-owned firms and between national firms themselves. Again, this puts an upward pressure on wages.

4 CONCLUSIONS

This chapter has focused on jobs and skills in the Brazilian electronics industry. The context in which the developments documented in this chapter are taking place is that of the Brazilian economy undergoing a process of technological modernisation in both industrial and service sectors. The social impact of this process, as evidenced by a growing body of research, may be detrimental in terms of job displacement.

In such a situation, it is important to know where new jobs are created. The national computer industry represents a new source of jobs to the extent that computers and peripherals are manufactured within national boundaries, rather than creating jobs abroad to produce goods that would then be imported to Brazil. In short, this is a question of securing jobs within the country rather than leaving it to the locational decisions of foreign electronics firms.

This said, conventional import-substitution measures (i.e. without a market reserve for nationally owned firms) would also have created jobs in the country. The case documented here, however, has had wider ramifications than previous industrial strategies in Brazil. The case of the consumer electronics industry indicates that without an element of control over the activities of foreign capital, there is a likelihood that the jobs created by conventional import substitution would be predominantly unskilled assembly work. Whereas, in the national computer industry we have observed a more skill-intensive labour force composition.

We cannot be categorical about this line of argument. First, because there are certain structural differences between the consumer electronics industry and the computer industry (internationally and in Brazil). Second, and more important, the evidence in this chapter has shown that foreign computer firms as well as nationally-owned ones have a highly skilled labour force. Thus, skill generation is not a simple equation about capital ownership.

We have, none the less, pinpointed some important differences between the skill composition of nationally-owned as opposed to foreign computer firms. By analysing the occupational structure of the electronics industry and identifying the importance of skilled engineering and technical labour for the future viability of Brazil's emergent computer industry, it has been possible to discuss a range of skilled occupations which, to date, have been considered 'uninteresting black boxes'.[25] That is, while skilled occupations are frequently taken as given, the importance of analysing the role of such occupations has become apparent in this study. If, as stated as a policy objective, technological capability is to be a key element in the future viability of the national computer industry, then knowledge of the precise nature of skilled labour requirements and the sort of activities in which that labour will be involved is a prerequisite.

There is clearly a difference in the use of skilled labour between foreign and national firms. The policy framework explicitly encouraged the development of local technological capability in electronics. There was a deliberate policy bias towards nationally-owned firms and a deliberate restriction on the areas of activity of foreign firms; this was a decision based on the infant-industry arguments elaborated in Chapter 2. The restrictions on foreign capital participation generated the need for skilled labour which could carry out product development and other research activities within the country. Foreign firms, for their part, already carried out R&D but outside Brazil. Given this background, some of the credit for the development of human resources in the computer industry must go to timely policy intervention, even though its impact on skills may only have been implicit.

It is difficult to say unequivocally that the policy regime for computers in Brazil was the only, or most effective, means of generating local capabilities, but it has indirectly encouraged the build-up of a substantial body of technical and engineering labour which forms the basis of such a capability. Despite this, there continues to be a severe shortage of skilled labour, exacerbated by the rapid expansion of the industry.

One final observation should be made. The period of research for this chapter was one of transition, when the national computer industry enjoyed a market reserve. A platform for further development has been built, but it is very difficult to project future changes in employment and skills. Nevertheless, the study has shown that the strategy which was behind setting up the national industry has had positive results in terms of job and skill generation. The timing of setting up the industry was crucial in this since there was then ample scope for employment creation and skill formation. If the industry is now coming under internal and external pressure to become more competitive than it has been to date, it is also in a much stronger position *vis-à-vis* the international technological leaders. What is important now is not so much the strict division between ownership of capital but that a Brazilian capability – embodied in engineers – has evolved. Brazil can afford to reopen the question of collaboration with foreign firms since there is now a capacity to negotiate more favourable terms of technology and skill transfer than a decade ago. It seems unlikely that this capacity would have been generated without a policy to foster the national industry.

NOTES

1 This paper has benefited from comments on earlier drafts from Hubert Schmitz, Helena Lastres, Mike Porteous and Elizabeth Bortolaia Silva.
2 For a detailed account of sources and research techniques see Hewitt (1988).

3 Three were foreign-owned computer producers and the rest were nationally owned producers of computers and peripherals.

4 For reasons of confidentiality, the Labour Ministry could not provide data on individual firms.

5 Such defensive polices have now extended to the East Asian NICs.

6 Although the market was only 10 per cent of the Japanese and 3 per cent of the US markets (Erber 1985:295). Nevertheless, in 1982, Brazil accounted for 41 per cent of Latin America's total installed capacity of computers. It had 1.7 times as many computers installed in this year as the next largest contender, Mexico (Nochteff 1985:5).

7 A reflection of the labour intensity of the industry.

8 The principal objective here is to examine the more skilled occupations in the industry. For a detailed examination of semi-skilled labour in Brazilian electronics, see Hewitt (1988).

9 We should note here the other large employer of semi-skilled female labour in electronics, the electronic components industry. The six firms shown in the components sector in Figure 7.2 carry out the final assembly of integrated circuits and are mainly located in the state of São Paulo. They show an even more dramatic decline in employment from 1980 to 1984 than in consumer electronics. In addition, this decline has taken place in a period of increasing demand for integrated circuits by the electronics industry.

10 The majority of assembly workers are 'classified' by the firms as semi-skilled. This includes both printed circuit board assembly and machine operators (drills, presses etc.). This classification relates to the occupation 'assembler' used by the Ministry of Labour. Others have used the label 'unskilled' for the same occupation (Humphrey and Hirata 1984:7), however, there appears to be no difference (in terms of tasks carried out and wages) between these two classifications.

11 This explains the extraordinarily high proportion of administrative and other service staff in mainframe producer firms in Figure 7.3.

12 The share corresponds to that found in earlier research in 1980 (Tigre 1983:95). This study emphasises the contrast with the human resources employed in R&D by MNCs and the fact that national firms have no 'parents' to turn to.

13 This high proportion of people in non-productive activities follows an international trend which has been noted by Soete (1985:67).

14 In absolute terms, in 1986 more than 5,000 professionals were employed in R&D by national firms and their R&D expenditure reached US$154.1 million which amounted to approximately 10.1 per cent of total sales (Piragibe 1987).

15 Apart from the example of the consumer electronics industry which has experienced two decades of the free market without showing any signs of the build-up of technological capabilities, the case of the Argentinian electronics industry is illustrative. What was once a burgeoning sector under national control is now virtually non-existent owing, it seems, to the opening up of the sector to uncontrolled imports of electronics goods (Tigre 1988).

16 FUCAPI is also promoting a technician-to-engineer conversion course, in conjunction with the University of Rio de Janeiro, for forty students.

17 Of this, 83.9 per cent was spent on training internal to the firm and the remainder for training in outside institutions.

18 See *Exame* (26 June 1985 and 19 February 1986) for two examples of human resources investment strategies.

19 Owing to the way wage data are processed in RAIS, it is only possible to give the proportion of those falling above, below or within a certain range of minimum wages.

20 The money value of the minimum wage became equal in January 1984. Before this, its value was 13.7 per cent higher in the South and South-East than in the rest of the country (*Boletim DIEESE*, January 1986). Note that *the minimum wage is not an adequate living wage*. By DIEESE calculations, the subsistence minimum for a family of two adults and two children is 4.9 minimum wages (*Boletim DIEESE*, January 1986).

21 Marketing capabilities are also a key factor in a firm's competitiveness.

22 Interviews in companies' personnel departments showed that, as supply becomes more constrained, there is an increasing tendency to employ the most 'promising' engineers before graduation.

23 For a detailed calculation of labour turnover for different occupations, see Hewitt (1988).

24 Thus, undoubtedly MNCs in Brazil are concerned and able to recruit highly qualified engineers.

25 To borrow Sayer and Morgan's phrase (1987:32).

BIBLIOGRAPHY

Baptista, M.A.C. (1985) 'A indústria eletrônica na Zona Franca de Manaus: consideraçoẽs sobre seu impacto na política nacional de informática e no processo de desenvolvimento na Amazônia', report prepared for ABICOMP, May.

Boletim DIEESE, Departamento Intersindical de Estatística e Estudos Sócio-Econômicos, São Paulo.

CDI (1984), *Perfil do Setor Eletroeletrônico*, Brasilia: Conselho de Desenvolvimento Industrial, Ministério da Indústria e Comércio.

Dados e Idéias, Gazeta Mercantil, São Paulo.

Dicken, P. (1986) *Global Shift: Industrial Change in a Turbulent World*, London: Harper and Row.

Erber, F.S. (1985) 'The development of the "electronics complex" and government policies in Brazil', *World Development*, Vol. 13, No. 3.

Exame, São Paulo: Editora Abril.

Frischtak, C. (1986) 'The informatics sector in Brazil: policies, institutions and the performance of the computer industry', Industrial Strategy and Policy Division, the World Bank, March, mimeo.

Hewitt, T. (1988) 'Employment and skills in the electronics industry: the case of Brazil', D.Phil. thesis, University of Sussex.

Hoffman, K. (1985) 'Microelectronics, international competition and development strategies: the unavoidable issues – editor's introduction', *World Development*, Vol. 13, No. 3.

Humphrey, J. and Hirata, H. (1984) 'Hidden inequalities: women and men in the labour process', mimeo.

Info, Rio de Janeiro: Editora Jornal do Brasil.

MCT (1986) 'Relatório final dos grupos de trabalho de recursos humanos em informática', Ministério de Ciência e Tecnologia, Brasília, October.

—— (1988) *Programa de Educação em Ciência e Tecnologia para o Trabalho*, Série Brasil Ciência, No. 8, Ministério de Ciência e Tecnologia, Brasília.

Nochteff, H. (1985) 'Government policies for the data processing industries in Argentina, Brazil and Mexico', Regional Meeting for the Initiation of a Regional Network for Microelectronics in the ECLAC Region, UNIDO, Caracas, June.

O'Connor, D. (1985) 'The computer industry in the third world: policy options and constraints', *World Development*, Vol. 13, No. 3.

Piragibe, C. (1985) *Indústria de Informática: Desenvolvimento Brasileiro e Mundial*, Rio de Janeiro: Editora Campus.

—— (1987) 'Policies towards the electronics complex in Brazil', Textos em Política Científica e Tecnológica, No. 28, CNPq, Brasília.

Piragibe, C. and Tigre, P. (1989) 'The Brazilian electronics industry: business strategies and role of the state', OECD Development Centre Project on *Technological Change and the Electronics Sector – Perspectives and Policy Options for Newly Industrialising Economies*, mimeo, Brasília.

Piragibe, C., Penna, M.V. and Tigre, P.B. (1983) 'Recursos humanos na indústria Brasileira de equipamentos de processamento de dados', Relatório de Pesquisa, Instituto de Economia Industrial, UFRJ, Rio de Janeiro.

Sayer, A. and Morgan, K. (1987) 'High technology industry and the international division of labour: the case of electronics', in M.J. Breheny and R. McQuaid (eds) *The Development of High Technology Industries: An International Survey*, London: Croom Helm.

SEI (1986) 'Panorama da indústria nacional – computadores e periféricos', Secretaria Especial de Informática, *Boletim Informativo*, Brasília, mimeo.

—— (1989) *Panorama do Setor de Informática*, Séries Estatistícas, Vol. 2, No. 1, Brasília, Secretaria Especial de Informática.

SENAI (1985) *Profissionais Vinculados à Area de Eletrônica – Demanda*, São Paulo: Mácia Helena de Lima, Divisão de Pesquisas, Estudos e Avaliação, SENAI, May.

Senhor, São Paulo: Editora Três.

SICCT (1985) *Proposta de uma Política Industrial no Brasil*, Campinas: Secretaria da Indústria, Comércio, Ciência e Tecnologia e Instituto de Economia, Universidade de Campinas, May.

Soete, L. (1985) 'Electronics', in L. Soete (ed), *Electronics and Communications: Technological Trends and Employment*, London: Science Policy Research Unit, Gower Street.

Tigre, P.B. (1983) *Technology and Competition in the Brazilian Computer Industry*, London: Frances Pinter.

—— (1986) 'Perspectivas da indústria Brasileira de computadores na segunda metade dos Anos 80', Instituto de Economia Industrial, Universidade Federal do Rio de Janeiro (IEI/UFRJ), mimeo.

—— (1988) 'How does Latin America fit high technology?', paper prepared for the International Symposium, *Technology Policy in the Americas*, Stanford University, December.

8 Revolution in a recession?

Advanced technologies and Brazil's machine tool sector in the crisis

Mike Porteous

1 INTRODUCTION

This chapter examines the introduction of numerically controlled (NC) and computerised numerically controlled (CNC) technologies in Brazil from the point of view of the machine tool manufacturers. Based on data and information gathered in the first half of the 1980s, it looks back at how the earlier patterns of growth shaped opportunities for machine tool firms to master these new technologies. It also looks forward, questioning the prospects for local firms to move on to more sophisticated areas of machinery production that are certain to be required in the 1990s. The central theme is that the opportunities for development in the area of NC and CNC machinery – as in other areas – was and will continue to be profoundly limited by much the same factors that constrained the local machine tool sector's earlier development. This perspective counters some of the more optimistic interpretations of indicators of growth and technological progress in Brazil's machine tool sector as well as the more positive accounts of the role of government policy in information technology in Brazil.

Up to 1980, the available aggregate indicators and detailed case studies suggested that some local firms were making significant inroads into this more advanced area of machinery manufacture, together with the foreign subsidiaries based in Brazil. Indeed the two largest domestic firms could point to experience with developing their own NC and CNC models as well as with using foreign licences to gain access to further advanced manufacturing technology. The Brazilian Government was also very much a player, with the Special Secretariat for Informatics (SEI) intervening to lay down the rules for the development of local suppliers of the electronics controls.

The evidence, however, from interviews with a sample of twenty-one local and foreign companies, representative of the different types of firms which made up the machine tool sector in 1984, suggested a more complex and less encouraging picture. The challenges of NC and CNC occurred at a time of profound economic recession. The impacts of the crisis on the machine tool sector at that time were so severe as to threaten seriously the very existence of a local capacity to develop and manufacture machinery. At the start of 1984, just three years or so into the economic recession, production levels were at around less than a third of those before the crisis. Half the sector's total workforce had been laid off and many firms, including some of the oldest and most reputable, had been forced into liquidation or had actually closed down. Although the research dealt with a period of profound change in the first half of the 1980s, it still appears that the implications of the continuing crisis for concerns with new technologies and government policy have yet to be given the attention they deserve. Brazil started the 1990s facing similarly turbulent conditions to those found in the early 1980s. Despite this, new technologies and the options for government policy are still discussed by many without consideration of the wider economic conditions.

The central question the article seeks to address is how Brazil's severe economic recession affected the relative strength of the indigenous and foreign machine tool firms at a time when NC and CNC posed new challenges to the industry. Did these and other related new advanced manufacturing technologies create openings for the local machine tool firms to overcome the worst effects of the crisis? Or were the potentially revolutionary impacts of the technology held back by these broader economic difficulties which lay beyond the control of individual firms or even of government technology policies? What, if any, impact could government policy in the area of advanced manufacturing technology have in the context of such a profound economic crisis?

After highlighting the concerns of existing literature, in Section 2, subsequent sections of this chapter turn to the evidence from the study of the impacts of the crisis on the machine tool sector in 1984. Section 3 details the collapse of demand and other aspects of the crisis for the machine tool sector. Some of the evidence on the introduction of electronically controlled machinery is presented in this context. Section 4 then turns to the role of government policy, both in the period leading up to the crisis and thereafter. Section 5 finally looks at the technological efforts of the sample firms which were attempting to enter the area of NC and CNC machinery. In each of these sections the evidence related to NC and CNC machinery is presented in the context of wider developments which influenced the nature and extent of opportunities created by these advanced technologies. Even though the empirical research mainly covers the recession years of the first half of the 1980s, the conclusions seem highly relevant at the beginning of the 1990s which threaten to be crisis years for the Brazilian economy. Some lessons for policy making are briefly suggested in the concluding Section 6.

2 PERSPECTIVES ON TECHNOLOGY AND THE CRISIS

There has been a longstanding and well developed interest within Brazil in the role of technologies and technical change in industrial growth, with a strong tradition of empirical studies doing much to inform policy-making (discussed in Porteous 1986). Curiously, little attention appears to have been given in recent years to the effects of the country's severe economic crisis on the opportunities for technological development. Two lines of work are particularly interesting in this respect and are briefly discussed here in order to set a background to the rest of the chapter.

Learning and ITC

The first line of existing research can be called the indigenous technological capabilities (ITC) literature. Through the 1970s, particularly, this line provided a wealth of studies of technological 'learning' inside firms in developing countries, challenging earlier views of dependence on foreign sources of technology. Rather than being caught in a vicious circle of dependence, many local firms showed signs of gradually accumulating technological experience to the point where they became no longer reliant on external sources of know-how in order to develop technologies of their own. In this way the studies also brought into focus dynamic aspects of technical change and innovation in developing countries, challenging conventional economic thinking on the comparative advantage of LDCs in technology trade (e.g. Lall 1982; Stewart and James 1982).

At the centre of certain countries' progress was said to be a process of learning-by-doing. This process was described as one of accumulating experience in an incremental manner, starting from carrying out adaptations and modifications of imported equipment to making use of local inputs, improving efficiency or making minor changes to suit local customers' preferences. Experience gained in these and other ways might be turned gradually to redesigning a product or process, possibly stretching the capacity of imported equipment far beyond its original specifications or even, in time, to the generation of completely new designs. The case of the machine tool sector was often of special interest in these respects as it was often emphasised that accumulated experience and tradition, built up through years of practical experience of machine making, was an important source of competitive strength for manufacturers. The early development of the more industrially advanced countries' sectors has been presented in terms of firms responding to a series of demands for increasingly specialised machinery which called for innovative efforts on the part of the pioneer firms (Rosenberg 1976). Any evidence from developing countries of such a gradual build-up of these kinds of problem-resolving innovative capabilities therefore was given special significance.

Following this tradition, in a set of case studies of the five largest Brazilian machine tool firms, Cruz *et al.* (1982) and Silva (1982) describe the gradual progress made from copying imported machinery to developing new machine tools, in two instances with NC and CNC controls. Erber (1982) and Teubal (1984) provided further studies of technological learning in the leading machine tool and other machine making enterprises in Brazil (though in some respects Erber's analysis went beyond the concerns of the learning school).

The ITC studies, however, typically paid little attention to three key 'limits to learning'. The first of these might be called historical or international limits to learning. To a large extent, the learning studies tended to presume the conditions of uninterrupted economic growth that had enabled firms gradually to accumulate and build up their technological skills and capabilities. The influence of the economic environment on the kinds of technological efforts made by firms was certainly recognised as important (Stewart 1984; Katz 1984). But major breaks in conditions were typically left out of view (for an exception see Erber 1983).

It will be argued that, in the case of Brazil, the conditions in which the machine tool sector developed were only sustainable in the shorter term. The government's overall strategy for industrial development and its responses to changes in the international environment placed opportunities for technological development on a precarious basis all too open to disruption. Furthermore the kinds of technological efforts required of firms in the crisis and problems faced once conditions changed were wholly different to those that previous periods of growth allowed.

A second 'limit to learning' which was rarely highlighted occurred at a sectoral level. The ITC literature's narrow focus on individual case studies of the leading firms often hid a picture of distorted and uneven opportunities for development within the rest of the particular sector. It will be argued that opportunities for technological development within the Brazilian machine tool sector were far more limited even in the periods of growth than would be typically supposed by the learning studies of the few success stories. Furthermore, the evidence in 1984 was that one impact of the introduction of NC and CNC machinery was to intensify such unequal opportunities for much of the local machine tool sector.

To complete the picture, a third limit to learning exists more at the level of the firm where most of the ITC learning studies have focused their attention. Here several ITC studies themselves have pointed to limits to what can be achieved by incremental learning alone. At certain points in their evolution, firms need to turn to outside sources of technical knowledge to supplement their experience. Investment in foreign technology may often be necessary. Contracting qualified personnel, investment in formal training or consultancies may also be ways of responding to the need to go beyond skills acquired by what has been called 'doing-based-learning' (Bell 1984; see also Fransman 1982). The design and production of NC and CNC machinery marks one such discontinuous point at which firms have to look to external sources of information and knowledge which internal 'learning' alone cannot provide. Design of CNC machinery calls for a fresh approach based on a knowledge of electronics and software as well as mechanical expertise. Apart from the electronic controls other critical components are typically provided by external suppliers rather than all being produced in-house. New approaches to marketing and sales services in turn are also required.

New technologies

In the course of the 1980s, the focus turned away from detailed case studies of learning within firms in developing countries to concentrate on new technologies. This literature needs less of an introduction here as it is discussed in other chapters of this book. However, a few points can be briefly raised.

First, the emergence of new advanced technologies whose impacts spread right across different industries clearly required a shift in focus from some of the more firm-specific issues of learning. How companies acquired the capabilities to master new technologies was still very much of interest but it was seen that investment in highly sophisticated process equipment and major R&D programmes were more centre-stage than the less tangible incremental accumulation of experience and learning. In this way, as attention shifted more to the advanced technologies, one limitation of the learning phenomenon became more apparent: that there are limits to what can be achieved within any firm by learning alone. But little attention appeared to be given to the two other important limitations on technological development which relate more to the wider economic conditions and to distorted opportunities for growth created within each sector.

Second, there seemed to be a tendency on the part of researchers and policy makers to focus first and foremost on the glamorous supplier sectors rather than on the users of new technologies. Within Brazil, the debate on micro-electronics and computers has tended to be couched in terms of the national importance of ensuring a local capability to produce computers and other strategic electronic goods versus the economic case for allowing a free trade in this and other areas. Comparatively little attention seems to have been given to the use of such technologies – whether imported or locally supplied – throughout the less glamorous areas of the industrial economy. As it is often argued that information technologies can play a rejuvenating role in many industries, the comparative lack of attention to user sectors is mystifying. In this respect machine tools are of special interest as the sector is both a user of micro-electronics and a key source of introducing automated manufacturing equipment into every other part of the industrial economy.

Finally it is worth highlighting that there are important peculiarities about the conditions of growth for many of the emerging hi-tech areas. In Brazil, at least, some of the key developments in micro-electronics have been in new sectors that even in the crisis were blessed with a special degree of government protection. This will be seen in the case of the suppliers of the NC and CNC units. The government's commitment to creating a local capability in certain areas of computers (Tigre 1984) and telecommunications (Hobday 1986) meant that these sectors remained cushioned from the destructive impacts of the changing conditions affecting most other, more traditional, sectors. As such, some of the new technology-supplying sectors were quite unrepresentative of the opportunities and problems faced by other industries.

3 TECHNOLOGICAL CHANGE, RAPID GROWTH AND DECLINE

In order to develop these points it is useful to start by looking at the available indicators of the machine tool sector's growth in Brazil up to the crisis and the extent of the impacts of both the economic recession and diffusion of NC and CNC on the sector's development.

The Brazilian machine tool sector: indicators of growth

Manufacture of machine tools in Brazil goes back to the beginning of the century. From the outset, the market has followed a cyclical pattern of growth largely in response to changes in international conditions of depression and growth. Since the 1950s these have been intensified by successive governments' policies of fostering rapid industrial and economic growth. The first of these, in the latter half of the 1950s, saw the Kubitschek government set out on the most rapid of import-substituting industrialisation programmes. As this ran into economic recession and political crisis, a military government took power, and pursued growth through the so-called 'miracle' years of the late 1960s and early 1970s. A further attempt to maintain the pace of change was made in the second half of the 1970s through debt-led growth.

Each period of rapid growth in turn generated major waves of demand for machine tools and other industrial equipment. Surveys carried out at the end of the first two major 'waves', after the automobile revolution in the late 1950s and at the end of the 'miracle years' (ECLA 1962; Vidossich 1974) showed a growth in number of local firms, their average size, employment and the range of products made. By the end of the 1970s there were over one hundred manufacturers of machine tools in Brazil. The available data showed a high domestic-supply ratio; in terms of the numbers of machines produced locally it reached 92 per cent in 1980. However in terms of value of machines the ratio was closer to 60 per cent. In 1979 Brazil was ranked as the world's thirteenth largest machine tool producer, ahead in terms of output of several more industrially advanced countries such as Spain and Yugoslavia. Its share in world gross output doubled from 1966/7 to 1979/80.

The aggregate statistics also suggested that the success of the sector did not rely only on internal demand. Exports of machine tools reached around 22 per cent of local production in 1979, surpassing both Korea (19 per cent) and India (15 per cent in 1978). The available data on destination of exports suggested that compared with other newly industrialised countries Brazil was not particularly adventurous in seeking a place in the international market. Inter-regional exports of machine tools within the Latin American countries as a whole actually increased from 80 to 85 per cent between 1970 and

1978 at the apparent cost of the share of exports to the more industrially advanced countries (UNIDO 1984). In contrast, over the same period Taiwan succeeded in penetrating into the more industrially advanced markets. Of the US$14.7 million imports of machine tools into the United States from developing countries, around 70 per cent was accounted for by Taiwanese machine tools alone (UNIDO 1984). As the predominant producer of machinery in Latin America, Brazil was able to take advantage of inter-regional trade agreements to supply a rapidly expanding market and had little incentive for looking to more advanced markets beyond the region.

One final indicator of the sector's comparative performance up to the crisis is of special interest, namely the production of NC and CNC machinery in Brazil and other NICs. The available figures do not allow accurate comparisons, but, taking figures for production of CNC lathes from Jacobsson's research (1984), the relative strengths of Korea and Taiwan stood out; both countries produced more than the estimated local demand. Jacobsson's figures for Brazil showed the sector's production of CNC lathes far behind that of Taiwan and Korea (36 in Brazil in 1980, 174 in Taiwan in 1981 and 222 in South Korea in 1982) but none the less indicated some progress into the more sophisticated areas of machinery production. Data gathered from the Brazilian Society for Numerical Control (SOBRACON) in the course of our research showed that by 1979 local production of NC machines had overtaken imports and was rising rapidly, from 40 NC machines in 1979, to 62 in 1980 and 120 in 1982 (SOBRACON 1984). By 1982 almost half the total stock of NC machinery installed in Brazil had been produced locally.

Collapse of internal and external demand

If the aggregate data showed rapid growth in production and export of machine tools and some signs of technological progress up to 1980, the picture thereafter was particularly bleak. The crisis that followed in 1980 was without precedent in terms of the scale of impacts upon the local machine tool producers.

From having served a domestic market in rapid expansion for some fifteen years, the machine tool sector suddenly faced a collapse in demand. The government's contractionist measures, eventually forced on it by the country's untenable foreign debt, began to have impacts on demand for industrially produced consumer durables, in turn passing on the effects to supplier industries. The autoparts industry, for instance, which was made up predominantly of a proliferation of small suppliers and represented the most important clients for the Brazilian machine tool firms, cut its total investment by more than half in only the first two years of the crisis. Many of the smaller autoparts enterprises closed down or turned to other activities.

The first years of the crisis show machine tools suffering from the greatest proportional falls in terms of receipts and employees of any of the other branches of machine manufacture. In just three years the sector suffered a loss of over half the total number of employees from 1980. According to official estimates, in 1983 only an average 22 per cent of capacity was utilised in the sector. While changes in sales over time are hard to measure, the available data suggest 1983 domestic sales fell to as little as 35 per cent of those in 1980. Production levels plummeted even lower, to 23.8 per cent and 31.8 per cent in 1982 and 1983 respectively of the number of machine tools produced in 1980 (Table 8.1).

As to exports, the picture was also alarming. Initially, as the economic crisis hit domestic sales, many local firms maintained production at previous levels, confident that exports to Mexico and other Latin American countries, at least, would continue. But in 1982

Table 8.1 Internal sales and domestic production of machine tools in quantities and real value, 1980–3

		1980	*1981*	*1982*	*1983*
Deforming machinery					
Number:	a.	8,638	6,968	4,336	7,866
	b.	10,316	8,930	4,479	7,852
Value:	a.	868	668	517	205
	b.	981	755	444	209
Cutting machinery					
Number:	a.	37,769	12,349	11,829	12,671
	b.	55,502	19,956	11,165	13,080
Value:	a.	2,580	2,396	1,255	915
	b.	3,636	2,953	1,315	1,129
Total					
Number:	a.	46,407	19,317	16,165	20,537
	b.	65,818	28,886	15,644	20,932
Value:	a.	3,448	3,064	1,772	1,121
	b.	4,617	3,707	1,759	1,338

a. = Internal sales; b. = Domestic production.
Value in Cr$ millions deflated using base year = 100.
Indices of machinery and equipment industry prices, Conjuntura Economica, Fundacao Getulio Varegas, various years.
Source: ABIMAQ/SINDIMAQ, various years

Mexico led several other Latin American countries in declaring a moratorium on its debts. The effect on these firms was devastating. Many were left not only with the effective closure of their most important remaining market but, in some cases, without payment for earlier sales. Such were the problems that several firms were forced into 'concordata' arrangements by which payment of debts is indefinitely suspended to stave off bankruptcy – but with obvious effects on the ability to secure any further credit. Not even the two largest, leading local firms escaped unharmed. Table 8.2 shows the machinery producers' trade association (ABIMAQ) figures of exports of certain selected machine tools from 1980 to 1983. The percentage falls in almost every category of machine tools are startling. Overall, the total number of machines sold in 1982 is barely 5 per cent of that in 1980. Notably it is the simpler, general-purpose types of machine tools which suffered the most dramatic falls in exports.

Diffusion of NC and CNC machinery

Against this alarming background of severe economic recession the introduction of NC and CNC machinery by 1984 was, not surprisingly, very restricted. NC and CNC machine tools was, in fact, the one area of machinery production to show any increase over the period. However, even with the fall in sales of conventional machine tools in

Table 8.2 External sales of machine tools, 1980–3: selected machinery in quantity

	1980	*1981*	*1982*	*1983*	*% falls*
Deforming machinery	1,446	1,343	237	83	94
Cutting machinery	14,476	5,396	696	2,316	84
– millers	111	68	22	5	95
– univ. millers	108	67	19	2	98
– drills	7,764	1,838	131	11	100
– plainers	698	66	22	22	97
– lathes	5,330	3,114	483	2,253	58
bench	1,162	1,034	155	1	100
up to 3,000 kg	4,047	2,007	297	2,202	46
over 3,000 kg	121	73	31	50	59
– saws	508	249	1	1	91
– grinders	11	7	–	1	91
– boring machines	3	–	15	–	100
TOTAL	15,922	6,739	933	2,399	85

Source: ABIMAQ/SINDIMAQ, various years

the crisis, total sales of NC and CNC machinery accounted for only 0.6 per cent of the number and 17 per cent of the value of total sales of cutting machinery in 1983. More recent data from Laplane (1989) shows this increased to only 2.5 per cent of the number and 31.6 per cent of the value of total sales of machine tools by 1987.

Most important, however, was that the diffusion of NC and CNC machine tools in user firms was very uneven. In 1980 just six enterprises possessed almost half of all the NC and CNC machines in operation in the country. The rest were divided between 146 user-firms, representing an average of only two per enterprise. Furthermore it was the large and foreign firms in Brazil that had the greatest share of such machinery (Tauile 1984). Two separate surveys at the time showed that around 60 per cent of users were foreign-owned. Almost 70 per cent and 80 per cent of these respective samples of user companies also had over 500 employees (Tauile 1984 and Rattner *et al.* n.d.). More recent studies provide data on the modest use of NC and CNC within machine tool firms but there does not seem to be any information on whether the limited take-up amongst user sectors has become less concentrated on the large foreign industries.

Certainly a restricted pattern of diffusion is to be expected in the conditions of recession prevalent throughout the 1980s. At the time of the research in 1984 most small and medium firms found the initial capital outlay beyond their means, no matter what the medium- and long-term cost advantages of the new technologies. In contrast impetus from parent companies, already well used to NC and CNC machinery, contributed to greater adoption by the foreign subsidiaries (indeed Tauile found evidence for this in his sample). However, it should be said that the second largest user of NC and CNC machinery was in fact the leading local machine tool producer. By 1983 the firm had installed over one hundred NC machines, most of which had been produced by the firm itself.

In conclusion, then, an important feature of the pattern of diffusion was that production of NC and CNC machinery was the only area to show a modest increase in the crisis. In the rest of the sector the picture could not have appeared more bleak. Furthermore, the market for NC and CNC machinery seemed to be heavily concentrated on large foreign subsidiaries. It will be seen that this was a market which the great majority of Brazilian machine tool firms had found particularly difficult to enter prior to the crisis. The government's general industrial policies were important here in creating conditions which distorted and limited the opportunities for local firms particularly. This point will be taken up in the next section.

4 CRISIS, UNEVEN OPPORTUNITIES AND GOVERNMENT POLICY

Following Freeman's pioneering work on long waves of economic development (e.g. in Freeman, Clark and Soete 1980; Freeman 1988) it has become fashionable in the literature to address the opportunities and needs for government policy in terms of the new wave of micro-electronic and information technology innovations (Perez 1983; Herrera 1986). In general terms this approach helps focus on the cyclical nature of growth. Broader social and institutional changes have also been emphasised as essential to the economic exploitation of fundamental technological changes (Freeman 1988). It was suggested that particularly the cyclical nature of growth opportunities has tended to be left out of view by the more narrowly focused learning studies.

As such, these perspectives on long waves accord with our own argument that the opportunities for the development of Brazilian machine tool sector and its entry into more advanced technology areas need to be seen as profoundly shaped by successive periods of growth and decline. However there is an important sense in which the cycles evident in Brazil had quite specific and concrete determinants. The importance of these in shaping opportunities for local firms can all too easily be lost from sight if the focus is left at the more general level of how some technologies can stimulate new waves of investment and growth. Government industrial policies and responses to changes in the economic environment were of particular importance in generating the conditions of rapid growth and ultimately severe recession. These in turn shaped the opportunities for development of the sector and are worth mentioning briefly before turning to the effect of the government's more narrowly defined policies aimed at fostering a local supply of the electronic controls.

The great leap into chaos

The causes of the severe recession and possible avenues for future growth have, of course, been the subject of intense debate in Brazil and need not be outlined at any length here. One line of debate is of special interest, however. In the crisis many Brazilian economists came to argue that the events after 1980 marked the end of an era for the Brazilian economy. One commentary, co-authored by one of the country's leading economists, Maria Conceição Tavares, captures the feeling in *The Great Leap Forward into Chaos*. A cycle of some fifteen years of external debt-led growth was started with the 'boom' and then maintained through the years in which adjustment to external shocks was put to one side. It took just two years for the real vulnerability of the economy to unfavourable international conditions to force itself to be recognised (Malan and Bonelli 1983; Malan 1983).

Looking back over the strategies pursued by successive governments it is hard to see how their ambitious plans for rapid growth could have been maintained beyond the short to medium term. First, going all out for the most rapid levels of growth inevitably would create internal pressures on the economy. This was perhaps clearest in the case of the Kubitschek government's highly ambitious 'Plano de Metas' (Plan of Goals) which aimed to achieve 'in five years what others did in fifty' (Gattas 1981; Lago *et al.* 1979). Inevitably, however, the inflationary costs of such a rapid import-substituting industrialisation and other prestigious government projects caught up once the first major industries had been established and expansion could no longer be maintained. Second, and more important than the internal pressures created by the government's own strategies was their reliance on a set of favourable external conditions all too open to change. Amongst the most important of these was the government's ability to make use of favourable international conditions through the 'miracle' years of 1968–72 and beyond. When the major Western economies felt the sudden impact of the first oil shock at the end of 1973, the Brazilian government found itself in a position of being able, in effect, to circumvent the crisis. Rather than adjusting the economy and absorbing the impact of the international economic recession, for some six years the government pursued a strategy of 'debt-led' growth, albeit at lower rates than in the 'industrial boom'. Not only did external international conditions allow this, but internally the military government's legitimacy rested on continued economic growth. Consequently Brazil's foreign debt rose from around US$12 billion at the end of 1973 to almost US$50 billion in 1979 (see Griffith-Jones 1983; Malan and Bonelli 1983).

In these ways, if successive governments' policies contributed directly to the phenomenal rates of economic growth up to 1980, they also placed industrial development in general on a precarious basis all too open to change and ultimately to crisis. With hindsight it can be seen that all the previous periods of rapid industrial expansion could only be viable in the short to medium term. In this way new technologies might well have been able to create new pockets of growth but ultimately the economic conditions in the country would not provide the foundation for lasting and more sustained development.

Structuring of opportunities

It would be wrong to imply that the difficulties facing machine tool manufacturers in Brazil were only related to general economic conditions not being right. For all the dramatic increases in demand for industrial machinery even in each period of rapid growth opportunities for the local machine tool manufacturers remained constrained to only certain segments of the market. To outline this we can return to some of the literature mentioned in Section 2.

It has been suggested that in many machine-making sectors in LDCs a duality of demand for machinery can exist (Cooper 1981). Demand for sophisticated machinery for the import-substituting industries may be met predominantly by imports or locally based foreign suppliers while local machinery producers' main markets are restricted to small-scale informal and less technologically demanding markets. Policies aimed at stimulating the general level of demand within the economy in this context would have only a limited impact on encouraging development. In addition, policy makers would need to consider how to encourage the more technologically dynamic sectors to use local as opposed to foreign sources of supply. At the same time, ways of overcoming the technological limitations of local machine making firms would also need to be addressed (Cooper *ibid*.:44). However the focus of most of the learning studies on the accumulation of skills in the few more successful firms has meant that it has not always been possible to draw lessons from existing work to begin answering these questions.

In Brazil the detailed case studies of the larger, more successful local machine tool firms certainly captured important instances of technological development at the firm level. But quite a different impression of the sector's overall technological development and the problems for policy making was given by the more representative samples and surveys carried out toward the end of each period of rapid expansion and by our own research (Vidossich 1974; Fleury and Muscat 1977; Porteous 1986). These showed few if any local machine tool producers making inroads into the supply of machinery for the multinational industries set up in Brazil. Indeed such was the pace with which the government sought to encourage industrial growth that it is difficult to see how local firms could have responded to the sudden increased sophistication of demand for machinery. Up to

1973, government subsidies for imports of machinery were even included in the package of measures to encourage foreign investment. Most local machine producers therefore were restricted to serving the proliferation of small local firms and workshops which grew up around the import-substituting sectors.

When protection of the capital goods sector finally became an objective of government policy in 1973, some measures at upgrading the technological skills of the local producers were set in place. Arguably, however, the most important effect of protection was to stimulate foreign suppliers to set up local subsidiaries in Brazil. This was clear from all of the foreign subsidiaries in our own sample which were set up in Brazil after 1973. The reasons they gave for establishing themselves in Brazil were either that they were invited by local subsidiaries of the same automobile assemblers their parent companies supplied in Europe or that they had been exporting to Brazil prior to the government's protection. All the subsidiaries benefited from total access to their parent firms' know-how. Such access, linked to the parent firms' positions in the world market gave the subsidiaries a special place in the newly protected domestic market.

In these ways, on the eve of the crisis in 1980, the great majority of indigenous producers faced a stratified and restricted market which offered relatively few stimuli for technological progress. The few successful cases appeared to be limited to areas where there were, as yet, no foreign subsidiaries, notably in precision grinding. For most other areas, however, there were even signs of a saturation of the local market, with many producers turning to exports to Mexico to make up for sales of their simpler, general-purpose machinery.

What impact then could policies aimed at encouraging the diffusion of NC and CNC and at strengthening local producers' capabilities in these areas have on local firms' opportunities for development? Would it have been possible for local firms to see their way through the crisis by taking steps to move into CNC machinery? The evidence already presented suggests that there would have been little room in the conditions of recession for local producers to attempt to enter this market. Demand for all types of machinery was very restricted and in the case of NC and CNC machinery was concentrated on the large foreign users. These were the markets which had previously demonstrated a certain discrimination against local technology and to which few local firms had succeeded in supplying before the crisis. Rather than opening up radically new opportunities, the diffusion of microelectronic controls in the machine tool sector appeared to be intensifying the separation of markets and duality in the sector that had previously limited opportunities for technological development in the earlier periods of growth.

Explicit government policy: SEI and other technology measures

It will already be clear that whatever the benefits of government policies in the area of the advanced technologies, in themselves they would not overcome the major problems facing the machine tool sector in the crisis. None the less, well targeted measures for helping local firms tackle the problems posed by design, development and production of NC and CNC machinery could conceivably have done much to help upgrade the local machinery-supplying sector.

The evidence, four years into the crisis, suggests that this opportunity was by and large being missed. The scope of SEI's operations and wider economic context provided little room for measures to encourage the use of electronics throughout the rest of the economy, not least in strategic areas such as NC and CNC machine tools. Furthermore the measures already existing for encouraging more general technological development in local firms had to all intents and purposes come to an end.

In 1982 SEI embarked on a policy of restricting the supply of foreign NC and CNC units to just four ostensibly local firms as part of its evolving plans to foster a capacity in micro-electronics within the country. Initially, these were given a period of four years in which they could use licences for selected foreign controls, free from competition of imports in order to 'master' the technology. The policy proved to be highly contentious.

In principle at least ensuring an indigenous supply of electronic components and controls seemed to have several advantages for local machinery manufacturers. Close links between electronic suppliers and machinery producers can be an important stimulus to innovation in the sector, with obvious benefits for all the machinery-using sectors in the economy. Such links would also be an important part of local firms' efforts to master the design and production of CNC machinery. Two of the local firms in our own sample had even sought to include production of the electronic controls as part of their moves into CNC machinery. One, the largest local machine tool firm, had plans to diversify in time into production and marketing of computer units. The other intended only to make controls for its own production of CNC precision grinders (for which hasty authorisation as a fifth producer using in-house rather than licensed technology was then required from SEI).

SEI's policy in the informatics area in general from its earliest days faced severe critics from inside and outside the country (as detailed by Tigre 1984, Fleury 1988, and Bastos in this volume). The questions at stake touch on fundamental issues such as the choice between either fostering key strategic technologies or selecting only those areas in which a country has relative strengths. Typical in this respect was that almost all of the foreign machine tool subsidiaries interviewed complained that the SEI policy restricted the choice of electronic control units available in Brazil to the four authorised sources regardless of whether these were the most advanced or whether firms had been used to other component suppliers. According to some of the respondents, the policy also indirectly raised the price of the controls as the suppliers either attempted to cover the cost of trying to master the technology or simply made use of the opportunity to increase profits behind the cover of SEI's authorisation.

The issues clearly demand more detailed analysis than is possible to give here. But a few points are worth highlighting. First, at the time of the research in 1984, even some of the firms authorised by SEI to sell licensed NC and CNC controls recognised that domestic demand was unlikely to make any one of the four operations cost-effective. Such diseconomies may have been justifiable on the grounds that protection would provide a breathing space for local firms to reach internationally acceptable levels. The technologies made available, however, were not the latest generation of NC and CNC equipment. In this respect Fleury (1988) found evidence of advanced machine tool designs having to be modified and compromised in order to be compatible with locally available electronic control units; this resulted in more expensive but less capable machinery.

If SEI's policies in the area of CNC supply appeared to be of little benefit to the machine tool sector, what of the other explicit government measures for promoting the sector's industrial and technological development? Two of the most important sets of measures were special financing for investments in capital goods, supplied by the agency FINAME, and the loans to encourage national firms to carry out development projects or add to their technological capability, by FINEP. Though both agencies continued to operate throughout the crisis, their impact on the machine tool sector was considerably reduced if not (in the case of FINEP) effectively brought to an end.

FINEP had made relatively low-cost loans available for development work. Although companies complained of the bureaucracy and length of time involved, many local firms made good use of FINEP's resources before the crisis. All this rapidly changed as nominal interest rates followed inflation to its record three-figure levels, effectively bringing to an end most companies' investments in improved plant or product development. Only two of the local sample firms at the time of the interviews claimed to have been able to make use of FINEP's resources after 1980. For the others either the new terms were prohibitive or any plans of developing new projects with or without subsidised funding were abandoned.

Similarly, although FINAME continued to provide financing for the sale and purchase of machinery, its operations were cut back after 1980. After 1980 the real value of transactions mediated through FINAME fell each year by around a third. The number of operations was reduced by 1983 to almost a half of those approved in 1980. To a certain extent, a fall in the use of financing for the sale and purchase of machinery is to be expected as plans for investment would have been put to one side in the crisis. However, the falls were also a consequence of the government's economic policies. FINAME's annual reports all speak of the restrictions placed on them by the Central Bank and the government in order to control the level of credit financing in the economy.

In conclusion, both before and during the crisis there were contradictory pressures on government policies which bore on the technological development of the machine tool sector and its response to NC and CNC. These do not seem to be readily captured in the narrow focus on the benefits of a local supply capability used by some commentators to justify the policy of a national reserve for key micro-electronic goods. If anything these perspectives should have directed attention at how effectively micro-electronics were being taken up and used throughout the economy.

At the time, however, the limits set on SEI (through wider economic pressures and policy decisions) meant the policy, at least in the area of computerised controls for machinery, focused only on supply. In terms of the influence of government on the development of the machine tool sector the most important factor was the more general industrial and economic strategies pursued before and during the crisis. These had by far the most profound impacts on the local sector, helping create a duality of demand and supply of machinery that limited most local firms' opportunities for technological development even in the periods of most rapid growth. In the crisis what measures had previously aimed to help upgrade the technological level of local firms now effectively came to an abrupt end. SEI's advanced technology measures in this context appeared to have little more impact beyond ensuring the prestige of local supply of the CNC electronic control units. Little was done by government to help the great majority of local machine tool firms or their users adopt the technology.

5 FIRMS' TECHNOLOGICAL EFFORTS

So far the focus of attention has been on the environment within which the Brazilian machine tool sector sought to respond to the challenge of CNC up to and then during the crisis. Government's general industrial strategies, themselves formed to a great extent in response to changing international conditions, have been given special emphasis. There is an important sense, however, in which firms create their environment and opportunities through their own strategies and past commitments to building up internal strengths and expertise. Clearly in this respect some local firms were making considerable efforts to move into production of CNC machinery, despite the growing economic uncertainty and apparent shortcomings of the government's more specific technology measures. It is therefore important to consider the significant part played by firms in mastering the development and design, production and marketing of NC and CNC machinery. As with the previous sections, the starting-point needs to be the context set by developments before the impacts of the recession and the radical changes of NC and CNC.

Technological development before the crisis

Here the contrast between the conclusions of the detailed case studies of learning in the few, apparently more successful firms and our own emphasis on the distorted and limited opportunities for development in the overall sector comes into view more clearly. Cruz *et al*. (1982) suggest that the case histories of the largest Brazilian firms demonstrate a significant 'technological jump' in the period from 1973 to 1980 when the sector finally received government protection. Their studies describe the progress made by these large firms, from initially copying imported machinery to developing original designs and, in two instances, culminating in moves into production of NC and CNC machinery.

However, the learning literature itself would suggest the need for some caution in drawing optimistic conclusions about the sector's state of development on the eve of the crisis. The major contribution made by Bell in covering a wide range of LDCs and NICs has been to show that learning is not an automatic 'something for nothing' process that takes place simply through the act of 'doing' or without any costs or efforts on the part of the firms. This means that cases of incremental learning in the largest firms need not necessarily have been repeated throughout a sector. Although examples of successful technological learning existed, counter examples also abounded, where enterprises showed little or no learning.

For Bell, two important factors would be stressed. First, conscious efforts on the part of firms and often significant costs are required in order to search out, collate, assimilate and make use of technological information which is generated inside the firm or is available from external sources. Second, as already mentioned, there are limits to what can be achieved by the internal 'doing-based-learning' alone. At certain critical points, firms have to turn to outside sources of technical knowledge to supplement their accumulated experience. Erber (1982) also showed that the kinds of capabilities built up within firms can be limited, for example, to detail rather than basic design skills.

We would want to extend this focus on the firm-level limits to learning by also emphasising the ways in which demand for different types of machinery gave rise to unequal opportunities for technological development within the sector. The broader historical and international conditions clearly set limits on the learning phenomena. In this way our sample of firms was chosen to capture the uneven pattern of development within the sector before the crisis rather than just concentrate on the more successful firms. Three distinctive groups emerged in relation to the strategies pursued up to 1980 and consequently to how opportunities to enter the NC and CNC area were viewed.

One interesting result to emerge is worth highlighting first. Basic elements of the management and organisation of the firms emerged as consistently related to improved chances for responding to the crisis and to NC and CNC technologies, despite the size of firms or apparent progression into different areas of machinery design – including even CNC. It is perhaps unsurprising that the best managed and organised firms would show the more encouraging prospects. However, the importance of such organisational aspects suggests that government measures to improve the flow of information and advice to companies in such basic areas as quality control and manufacturing best-practice could have very positive benefits – at least more effective than the high-profile measures to ensure a local supply of advanced electronic units. We will return to this after outlining the three groups of firms.

The first and largest group of local firms made up almost half the sector in 1980. It was restricted to producing relatively simple general-purpose machinery for workshops, small industrial enterprises and technical schools. For all their years of experience in machine manufacture (in our own sample ranging from twenty-five to forty-five years) little progress had been made beyond this market. Some models for relatively greater precision use in tool rooms had been developed in one or two cases through a mixture of copying and modifications. Typically, however, the machinery's suitability for universal, general purpose was the most important characteristic. Consequently little or no effort was made to develop more sophisticated models, beyond introducing occasional improvements. The common view was that there would always be a demand for the simple, universal machinery. A common strategy of many firms in this context was to diversify over a wide range of models or types of machine tools to guarantee sales, while remaining all the time in the more simple, universal segment of the market. NC and CNC was simply not a realistic option for any of these firms.

The strategy meant that if there was an accumulation of experience and learning it tended to relate to a few key individuals' expertise in modifying and redesigning machines. Little attention was given to improving the system of production or raising the skills of the workforce to improve efficiency. Production typically seemed to be organised on the basis of minimising the responsibility of the workforce, using the most routine and repetitive tasks possible and leaving such quality control as existed to final inspections by one or two trusted individuals.

A second, significantly smaller group of local firms was distinguishable by the progress made in terms of producing more sophisticated machinery. Inroads had also been made into the more demanding markets. Here examples of the kind of incremental progression, from simple models based on copies to the introduction of successive improvements through to original designs were evident. In the area of precision grinding, for instance, starting with the first universal models for tooling, many firms had subsequently progressed to the development and design of more complex and specialised automatic or semi-automatic models for production.

Competition with foreign subsidiaries and other Brazilian firms using licensed technology began to figure in one or two cases. Though significant as indicators of technological progress, the more complex machinery in very few cases represented the bulk of sales for such firms. Most firms still relied on lines of relatively simple machinery such as tool room millers, bench drills or universal grinders. Production machinery accounted for up to only 30 per cent of sales even in the more progressive firms. Use of foreign technology was also an important factor in cases of more advanced machinery sales. In the area of presses, for example, one of the leading Brazilian firms complained of a discrimination against local technology by the major foreign enterprises for presses above a certain weight and sophistication that was within what they believed to be their proved capability to design and produce. Several of the multinational automobile assemblers remained completely closed to the firm. Its main local rival, in contrast, used foreign licences to produce presses for the multinational industries, keeping its own models only for local industries.

By 1980 only two indigenous firms – the two largest in the sector – had progressed into the area of CNC lathes, using foreign licensing agreements and in-house designs. Their contrasting efforts, however, reveal an important feature of the technological development of nearly all the local firms. The largest of the two stood out in the sector as a development pole, setting the standards for production of, particularly, universal machine tools as well as leading the local sector in its efforts to acquire and master foreign technology in order to establish itself in the production of CNC lathes. It was also one of the few local firms to have established a well organised system of production control, introducing computerised production control and planning. It also had its own government-approved technical school and training programme. In contrast its slightly smaller rival was more typical of other local firms, in 1980 characterised by a deep-rooted lack of organisation and by personalised control quite unsuited to its size. All the product and production engineering activities were carried out in an informal, unsystematic fashion revolving around the individual efforts of a few key people in the firm. In this and other local firms, although formal design departments did exist, the initiation and development of new designs centred very much around the efforts of an experienced (but usually technically unqualified) director or owner. Such production control and planning as existed also seemed to depend on trusted foremen rather than preset methods and procedures.

The final group of producers which made up the sector in 1980, on the eve of the crisis, was the foreign subsidiaries. Their clients were almost exclusively the foreign multinational industries, which had of course provided the main stimulus for setting up operations in Brazil under the government's protection policies after 1973. Instead of gradually progressing into the development and production of increasingly sophisticated machinery, the subsidiaries stepped into their privileged position in the most sophisticated markets by virtue of having total access to their parent companies' designs and production know-how. 'Access to the most advanced technology through the parent company' was stressed in every case as the most important factor in competition. In contrast to all but a few of the Brazilian firms, production planning and control and rigorous systems of quality control were also the rule rather than the exception. In this respect, the biggest problem for the subsidiaries was producing their parent companies' machinery in the very different manufacturing conditions in Brazil where there was no reliable network of subcontractors. One respondent claimed that whereas in Europe the parent company would produce around 40 per cent of the final machinery, in Brazil they were forced to make 80 per cent, after the most strenuous of efforts to bring the level down from an original 90 per cent.

Compared with the pace of development in the international markets, the foreign subisidiaries were at first slow to introduce NC and CNC. But by 1980 they were already rapidly switching to offer their parent companies' CNC equipment. At the start of the crisis four foreign enterprises, along with the two leading Brazilian lathes manufacturers, included CNC equipment in their product lines (Cardosa, quoted in Rattner *et al.* n.d.). By the time of the interviews in 1984, all the foreign subisidiaries and joint-venture enterprises in our sample had come to offer such machinery. For them, the switch seemed to be relatively easy. When asked about sources of information used, requirements and problems encountered in entering the area, all the respondents emphasised their access to their parent companies' technical resources and experience. Training was undertaken with people from the parent firm. All the designs themselves were made fully and freely available. The only complaint was that SEI's policies limited the choice of electronic controls, regardless of which suppliers the firms had been used to working with.

Few local firms then were prepared by their earlier development to respond to the foreign subsidiaries' switch to CNC. Even the more innovative local firms tended to be characterised by a fairly rudimentary level of production planning and control ill-suited to producing NC and CNC equipment. The essential design and production engineering activities typically remained centred around the efforts of experienced individuals rather than more formal design teams and production systems. For the majority of local firms NC and

CNC was an even more remote possibility. Their past performance in producing only the simpler universal equipment provided few indications that any would be able to make the qualitative jump to NC and CNC machinery production.

Firms' responses to the crisis and radical technical change

It will be clear that once conditions changed and the sector began to feel the full impact of collapsed internal and external demand and the end of government assistance, little incremental learning could take place. With around half the sector's workforce laid off, much of the sector's accumulated experience was lost. What efforts could local firms, attempting to move into the only promising area of machinery production, make in this context?

Of course for the majority of local firms, NC and CNC machinery represented a jump way beyond their previously demonstrated abilities. Within the group of producers of simple, universal machinery, two strategies were evident (as demonstrated in Table 8.3).

Those that had resisted changes to their basic products and production methods in the previous periods of growth had few options open to them other than reducing their workforce and production to the point of closure. The slightly more innovative firms in this first group which had made efforts to broaden their product range and introduce new models on the basis of copies and modifications, but keeping all the time within the market for simpler general-purpose machinery, fared little better. The data from the interviews and the trade association show these firms suffering the severest contraction of sales within the sector. The most common responses were to diversify out of machine tools and into machining services, reductions to the barest minimum and then closure. In our own sample, 69 per cent of these firms' total workforce from 1980 had been laid off by the end of 1983.

The firms' own perceptions of the changes to their markets resulting from the crisis and the gradual diffusion of NC and CNC reflected their previous strategies. All expressed the view that what was involved was simply a sudden reduction in demand, forced by external factors beyond their own control. In this respect NC and CNC was not regarded as relevant. Though instances could be found of individual owners applying efforts to develop simple one-off custom-made machines or other products to order, in general most of these types of firms saw the crisis as purely one of demand. In their

Table 8.3 Overview of technological strategies before and during the crisis

Market segment	*Technological efforts before 1980*	*Changes to market after 1980*	*Av. % fall in employees*	*Changes forced on firms*
1 Simple general purpose machinery for workshops, maintenance depts etc.	(a) Minimal – little change to product/process	Greatest proportional falls in sales. End to exports	69%	Closure
	(b) Some development within simpler machine tools by copy and modification			Lay-offs to point of closure and/or diversify out of machine tools
2 More advanced 'finishing' and fine machining, small-batch production equipment Some sales to foreign industries	(a) More innovative strategies evolving from copy to modifications and original design. Production still informal	Substantial falls. CNC potentially displace some markets	56%	Diversification and product innovation. Formalisation of production eng. and PPC
	(b) Greater use of foreign licences. Production systems formal and routinised		45%	Product and process innovation. Some diversification
3 Exclusively production equipment for MNCs	Reliance on 'packaged' technology from parent company	'levelling off'. Wider market opened by CNC	35%	Accommodation. Switch to parents' CNC models

Source: Interviews

view there was little to be done other than sell at lower prices and wait for a general recovery of sales.

The second group of local firms, which prior to the crisis had made efforts to move into the more demanding segments of the market, showed more active strategies for responding to the new conditions after 1980. Their assessments of the changes taking place, including the slow but steady diffusion of CNC, were less fatalistic than was the case of the previous group. Despite also suffering severe losses of employees, there were far more instances of firms seeking to develop new models. Nearly all spoke of plans to move into NC and CNC machinery, though the extent of their efforts varied considerably.

Notably, those which prior to the crisis had only rather informal design and production control systems were now making their greatest effort in internal reorganisations. In some cases, studies to determine times for each stage of manufacture and assembly operations were carried out for the first time as a basis for introducing more systematic production control and for determining the minimum size of economic batches. Plans for entering NC and CNC machinery amongst these firms were, however, very hesitant and often were limited to modifying existing conventional models to NC. There was little evidence of the more radical changes in approach to design, production and marketing needed to realise the full potential flexibility and precision of CNC. Other firms in this category seemed to be watching to see how the market developed and where new areas of sales might appear. Some hesitancy was of course appropriate in the uncertain economic conditions and restricted market at the time. However, few firms seemed to recognise the extent to which CNC machinery would demand a wholly different approach to design, production and marketing.

In contrast (with one exception), the few local firms in our sample which had evolved more formal design and production control systems before the crisis had far bolder and more thorough-going responses to the CNC challenge. The case of the leading, largest firm has already been mentioned. Convinced that the firm's ability to maintain its leading position in the local sector would depend on having a strong capability in CNC machinery, the decision was made to manufacture the electronic controls in-house. Under authorisation of SEI, CNC units were imported and considerable efforts made to build up component suppliers to allow assembly of the hardware. Most impressive was its farsighted approach to software, involving bringing in a foreign expert to direct the development of an in-house team to ensure a foundation in basic software skills. New models of

CNC lathes were also designed and launched, alongside the range produced under licence. It also installed over a hundred of its own CNC machines into its plants in order to maximise the opportunities of learning at first hand the potential of CNC.

In many respects, however, this leading Brazilian firm was an exception, not least because of its sheer size and the resources available. An equally striking case of major innovative efforts came from a much smaller local firm. Prior to the crisis, the firm had been producing precision grinders under licence and a range of universal milling machines. As with many other local producers a large stock of unsold machinery was rapidly built up. Finally, production was stopped and the firm was forced to diversify to high-precision machining services and restoring old machinery. Around two-thirds of the workforce was laid off, a major reorganisation was undertaken in order to make up for lost personnel at the highest levels and the formal production planning and control was put to one side as so few machines were being produced. None the less, the firm was able to develop and launch a CNC precision grinder, developing in-house its own electronic controls.

The above cases of local firms applying considerable innovative efforts to acquire and master advanced technologies are worth highlighting if only to show what could be achieved. Nevertheless, in 1984 the general picture, in the midst of the economic crisis, was clearly one of NC and CNC machinery providing very few opportunities for local firms to overcome past constraints on their development. The foreign subsidiaries were already preparing to enter into competition for markets traditionally served, with some difficulty, by the few firms that had developed semi-automatic or precision machinery. At the same time some of the local firms making efforts to enter the NC and CNC area soon found that they had underestimated the difficulties and the radical changes needed in design and production.

6 CONCLUSION

Looking back at the data and research gathered in what has proved to be the worst period of a decade of severe recession what conclusions and general lessons can be drawn?

The first is the very uneven impacts of the slow but steady diffusion of the new advanced electronic control technologies for the machine tool sector in the crisis. Although promising instances of local firms making considerable innovative efforts in the area could be found, the general picture was one of the advanced technologies intensifying the distorted pattern of development of previous periods of growth.

In this context, government industrial policy at the time of the research had effectively come to an end as measures to manage the economy and foreign debt took priority. Although SEI's programme of measures ensured a high-profile 'indigenous' supply of electronic controls there was little scope for co-ordinating these with other areas of policy. No measures were in evidence to help either the machine tool producers or the ultimate users of NC and CNC machinery adopt the technology. At the same time the broader economic policies effectively worked against the sector, contracting demand and cutting back on previous assistance for sales of capital equipment. The existing government loans for development projects or improving firms' technological abilities were in effect no longer a viable option for almost all firms.

The lessons from the past were clearly not learnt. From 1985 to 1986 the transition government chose a course of going all out for 'booming' the country out of a crisis created by previous growth strategies. The net effect was an upsurge in imports, demand levels that local firms were unable to respond to after four years of lay-offs and idle capacity and, inevitably, deepened problems of inflation, foreign debt and instability.

Turning the focus to the decade ahead, what are the prospects for the machine tool sector to provide the advanced manufacturing technologies needed in any future industrial recuperation in the country? In this respect it should be mentioned that NC and CNC machine tools represent the more basic of advanced manufacturing technologies that are already widely accepted in other more industrially advanced countries as essential to achieving international competitiveness. Worldwide the trend in machine tool manufacture in recent years has been towards suppliers working with users to provide a more integrated solution to their specific manufacturing needs. In this way, many manufacturers of standard CNC machines now aim to offer more of a complete machining cell by adding on systems to feed in and pass on work. An ability to understand the customers' production problems and provide a solution rather than a discrete machine is essential. Few, in the West at least, see any prospect of competing on price with low-cost South-East Asian producers of basic CNC machinery.

As the inevitable pressures to open the economy build up, the options for the Brazilian machine tool firms are likely to be even more sharply divided. For those unwilling to make major changes the prospect would be to continue serving a declining domestic market with conventional workshop machines of low sophistication. Continued protection and isolation from international competition might allow some limited growth in this market. The real challenges and opportunities would be for those firms that see themselves as the key players in introducing international best-practice manufacturing technologies to their customers. Greater exposure to the advanced international market would be essential, if only to keep up to date with the key trends and developments in best-practice machine manufacture.

As to the role of government policy, it will be clear that the pressures on the government's more general economic strategy give very limited room for manoeuvre. Policy measures to stimulate demand might be attractive in the short term but are unlikely to be a realistic option. The past performance of the great majority of local firms also means that government intervention to encourage a greater take-up of NC and CNC would be likely to affect only those few firms already aware of the importance of the technology. Several specific policy measures do, nevertheless, suggest themselves as potential options for helping to upgrade local machine tool firms and, where appropriate, assist their response to CNC.

- *CNC awareness and promotion of best-practice measures for key user sectors*. Well targeted awareness programmes, concentrating on stimulating a better understanding of the importance of CNC and other advanced manufacturing technologies amongst the local engineering industries would potentially help raise the demands made by the local machine tool sector's key users such as the proliferation of smaller automotive component and parts suppliers. Two essential elements in any attempt to facilitate the adoption of CNC machinery in the local user industries would need to be emphasis on training and best management practice to ensure the effective use of the equipment. A promotional programme, using demonstrator firms, might help illustrate both the advantages of CNC and the all-important organisational and managerial changes needed to exploit fully the technology.
- *Assistance with CNC consultancies and feasibility studies for local machine tool producers*. A complementary, relatively low-cost, measure to the above would be provision of subsidies to cover a proportion of the costs of consultancy or feasibility studies for local firms seeking to make the switch from conventional to electronically controlled machinery. This would need to include an

assessment of the machine tool businesses as a whole, taking account of their existing production facilities and organisation, product mix and future plans and aimed at setting their sights on higher, internationally competitive practice. Above all the qualitative step changes needed in almost every area of firms' operations to switch to CNC would need to be a key theme.

- *Export assistance*. One of the few demand-related measures that had already proved feasible in the crisis was trade agreements with Latin American countries, reopening the export market on which many of the hardest hit local firms had previously depended. If government guarantees for export sales were targeted at the smaller producers and linked to attainment of basic standards of design and quality assurance, the effect upon local firms might help prompt an upgrading of activities.
- *Encouragement of science-base and industrial collaborations*. A final set of relatively low-cost measures relate to the role government can play in helping overcome some of the barriers to collaboration amongst otherwise competing local firms. University–industry programmes can provide one basis for the development of skills and understanding in advanced electronics applications to CNC, CAD/CAM, FMS and other areas of computerised integrated manufacturing. Such collaborative programmes can also be used to disseminate information on international best practice and developments in advanced manufacturing technologies, directing university research to more strategic and industrially relevant areas.

A principle in any such attempt to facilitate firms' transition to CNC machinery would have to be closer links with the industry itself than was evident in, particularly, SEI's interventions. As in other advanced technology areas, the most prestigious, high-profile programmes implemented from above without the participation or stimulus from the affected sectors inevitably have limited beneficial impact. By the same token, well targeted policies to encourage a raising of technological best practices would have little effect without the entrepreneurial drive and efforts of firms.

For the moment, however, a new Brazilian government started the decade without any industrial strategy or policy as such to pick up and build on, other than in isolated high-profile electronic technologies. Consequently, the response of the machine tool sector to the challenges of CNC in the crisis, like much of the sector's earlier development, looks set to be largely despite rather than with the assistance of government policy.

BIBLIOGRAPHY

Bell, M. (1984) 'Learning and the accumulation of industrial technological capacity in developing countries', in M. Fransman and K. King (eds) *Technological Capability in the Third World*, London: Macmillan.

Bell, M. *et al.* (1982), 'Limited learning in infant industry: a case study', in F. Stewart and J. James (eds) *The Economics of New Technologies in Developing Countries*, London: Frances Pinter.

—— (1984) 'Assessing the performance of infant industries', *World Bank Staff Working Paper*, No. 666, Washington, DC.

Cooper, C. (1981) 'Policy interventions for technological innovation in developing countries', *World Bank Staff Working Paper*, No. 441, Washington, DC.

Cruz, H.N. da *et al.* (1982) *Observações Sobre a Mudanca Tecnológica no Setor de Maquinas Ferramentas do Brasil*, IDB/ECLA/UNDP/IDRC Research Programme on Science and Technology, Working Paper 47, Buenos Aires.

ECLA (1962) *La Fabricacion de Maquinarias y Equipos Industrales en America Latina: II Las Maquinas-Herramientas en el Brasil*, New York: UN.

EMBRAMEC (1976) 'Estudo sobre maquinas ferramentas', Nucleo de Fomento, Rio de Janeiro.

Erber, F. (1982) *Technology Issues in the Capital Goods Sector: A Case Study of Leading Industrial Machinery Producers in Brazil*, Geneva: UNCTAD, TD/B/C.6/AC.7/6.

—— (1983) 'Technological dependence and learning revisited', Texto para Discussão, No. 34, Instituto de Economia Industrial, Universidade Federal do Rio de Janeiro.

Evans, P. (1979) *Dependent Development, The Alliance of Multinational, State and Local Capital in Brazil*, Princeton: Princeton University Press.

Fleury, A. (1988) 'The impacts of microelectronics on employment and income in the Brazilian metal-engineering industry', World Employment Programme Research Working Paper WEP2–22/WP 188, International Labour Office, Geneva.

Fleury, A.C.C. and Muscat, A.R.N. (1977) 'Analise a nivel de empresa dos problemas tecnologicas do setor de maquinas ferramenta', relatorio a FINEP, Sao Paulo: Fundação Carlos Alberto Vanzolini.

Fransman, M. (1982) 'Learning and the capital goods sector under free trade: the case of Hong Kong', *World Development*, Vol. 10, No. 11.

Fransman, M. and King, K. (eds) (1984) *Technological Capability in the Third World*, London: Macmillan.

Freeman, C. (1988) *Technology Policy and Economic Performance*, London: Pinter.

Freeman, C., Clark, J. and Soete, L. (1980) *Unemployment and Technical Innovation*, London: Pinter.

Gattas, R. (1981) *A Indústria Automobilística e a 2a Revolução Industrial no Brasil: Origens e Perspectivas*, São Paulo: Prelo Editora Itda.

Griffith-Jones, S. (1983) 'The changing international environment and its impact on developing countries', *IDS Discussion Paper* 188, Institute of Development Studies, University of Sussex.

Herrera, A. (1986) 'The new technological wave and the developing countries: problems and options', in R. Macleod (ed.) *Technology and the Human Prospect*, London: Pinter.

Hobday, M. (1986) 'Digital telecommunications technology and the third world: the theory, the challenge and the evidence from Brazil', D.Phil. thesis, University of Sussex.

Jacobsson, S. (1984) 'Technical change and technology policy – the case of computer numerically controlled lathes in Argentina, Korea and Taiwan', mimeo, University of Lund.

Katz, J. (1984) 'Technological innovation, industrial organisation and comparative advantages of Latin American metalworking industries', in M. Fransman and K. King (eds) *Technological Capability in the Third World*, London: Macmillan.

Lago, L.A.C. do *et al.* (1979) *A Indústria Brasileira de Bens de Capital: Origens, Situação Recente, Perspectivas*, Rio de Janeiro: FGV.

Lall, S. (1982) 'Technological learning in the third world: some implications of technology exports', in F. Stewart and J. James (eds) *The Economics of New Technologies in Developing Countries*, London: Frances Pinter.

Laplane, M. (1989) 'Setor de máquinas ferramenta: diagnóstico da situação atual', report for ABIMAQ/SINDIMAQ Política Industrial para a Indústria Mecânica Project; Instituto de Economia, UNICAMP.

Macedo, R. (1984) 'Brazilian children and the economic crisis: evidence from the state of São Paulo', *World Development*, Vol. 12, No. 3.

Malan, P. (1983) 'A questão externa' in *FMI x Brasil: a Armadilha da Recessão*, Rio de Janeiro: Forum Gazeta Mercantil.

Malan, P. and Bonelli, R. (1977) 'The Brazilian economy in the seventies: old and new developments', *World Development*, Vol. 5, No. 1/2.

—— (1983) 'Crescimento econômico, industrialização e balanço de pagamentos: O Brasil dos anos 70 aos anos 80', paper presented at FIPE, University of São Paulo.

Perez, C. (1983) 'Structural change and the assimilation of new technologies in the economic and social systems', *Futures*, Vol. 15, No. 5.

Porteous, M. (1984) 'Recessão e mudança tecnológica no setor de máquinas ferramentas do Brasil: resultados preliminares', relatório para CNPQ, University of São Paulo.

—— (1986) 'Recession and technical change in the Brazilian machine tool sector, D.Phil. thesis, Science Policy Research Unit, University of Sussex.

Rattner, H. *et al.* n.d. 'Produção e difusão de máquinas – ferramentas de comando numérico no Brasil', Research Report, No. 20, São Paulo: FGV.

Rosenberg, N. (1976) *Perspectives on Technology*, Cambridge: Cambridge University Press.

Silva, M.E. da (1984) *Inovação Tecnológica: Um Estudo de Caso*, São Paulo: Instituto de Pesquisas Economicas.

SOBRACON (1984) 'Situação atual do NC no Brasil', mimeo, São Paulo.

Stewart, F. (1984) 'Facilitating indigenous technical change in the third world countries', in M. Fransman and K. King (eds) *Technological Capability in the Third World*, London: Macmillan.

Stewart, F. and James, J. (eds) (1982) *The Economics of New Technologies in Developing Countries*, London: Frances Pinter.

Tauile, J.R. (1984) 'Microelectronics, automation and economic development:

the case of numerically controlled machine tools in Brazil', Ph.D. thesis, New School of Social Research, New York.

Teubal, M. (1984) 'The role of technological learning in the exports of manufactured goods: the case of selected capital goods in Brazil', *World Development*, Vol. 12, No. 8.

Tigre, P.B. (1984) 'Computadores Brasileiros: os desafios da crise', paper presented at FIPE, University of São Paulo.

UNIDO (1984) *World Non-Electrical Machinery: An Empirical Study of the Machine Tool Industry*, New York: UN.

Vidossich, F. (1974) *A Indústria de Máquinas Ferramentas do Brasil*, IPEA Serie Estudos Para o Planejamento, No. 8, Brasilia.

9 State policies and private interests

The struggle over information technology policy in Brazil

Maria Ines Bastos

1 INTRODUCTION

Brazil has managed to put national interests first in some areas of information technology (IT).[1] An 'informatics' policy was shaped during a period of deepening internationalisation of the Brazilian economy and against the interests of transnational corporations. Despite these countervailing factors, the IT policy has succeeded in reserving the domestic market for local capital. The aim of this chapter is to analyse the Brazilian state's room to manoeuvre in the formulation and implementation of the IT policy in face of both external and internal demands for change. It will not consider whether this policy was 'correct' (i.e. whether over time it provided a positive basis for continual growth), but will merely look at the ability of the state to withstand pressures to change policy.

Three phases can be distinguished in the development of the IT policy in relation to its scope, institutional support and political challenges.[2] The *setting-up phase* started with the creation of the Special Working Group (GTE) in 1972, covered the whole period of operation of the Coordinating Commission for Electronic Processing Activities (CAPRE) and ended in 1979 with the replacement of CAPRE by the Special Secretariat for Informatics (SEI). This was the time when the segment of mini- and microcomputers and their peripherals was defined as one to be reserved to local capital. A matter of restricted interest for the overall governmental structure, the policy remained insulated within the state apparatus and, protected by its support in the armed forces, was relatively safeguarded from interference by foreign firms and powerful state agencies. From a political standpoint, the policy's very existence depended mostly on the ability of its supporters to convince relevant players in the governmental structure of the policy's strategic opportunity.

An *institutionalising phase* started with the creation of the SEI and ended with the passing of the informatics law in 1984. The scope of the policy was enlarged to reserve the production of a comprehensive set of IT goods and services for local capital. The policy's success in establishing a locally-owned professional electronics industry and the potential of the Brazilian market for electronic products and services, contributed to the policy becoming more visible and, consequently, more fiercely challenged by its opponents.

The subsequent phase, from 1984 to the present, has been one of *consolidation of policy* instruments *and adaptation* of its goals and strategy to changing conditions both foreign and local. The current domestic debate on the policy's strategy and long-term objectives which involves local entrepreneurs acting in the electronics industry and other industrial areas, was triggered by local macro-economic determinants but occupied centre-stage throughout the US/Brazil conflict over the Brazilian IT policy. IT user firms and the potential victims of the US threat of retaliation are among the newcomers to the critics of the IT policy.

Different combinations of local political support have given the state some strength to face the frequent challenges to the policy since its launching. The IT policy has been supported by segments of the local intelligentsia and the armed forces, especially during its initial phase; by the national hardware producers group, as soon as the policy showed concrete results; by the majority of Brazilian representatives in the Congress, especially during the discussion and passing of the informatics law and during the preparation of the new constitution; and by national finance capital.

On the other hand, opposition came for some time mainly from foreign firms already acting in the Brazilian IT market but excluded from the local manufacturing of mini- and microcomputers. The antagonism of foreign companies toward the policy increased with the extension of the market reserve to a wider set of products and activities. A few local entrepreneurs and politicians who share economic and political interests with foreign firms gave internal support to the external challenges to the policy.

Recent developments have changed this scenario: new players, both foreign and native, joined the 'traditional' opposition. The US proceedings against Brazil under Section 301 of the Trade Act and the threats of retaliation to Brazilian exports were decisive in increasing external opposition and making clearer internal disagreement with the Brazilian IT policy. The significance of the initiative of the US government and the participation of US sectoral associations in the conflict brought strong allies to local subsidiaries of transnational corporations who have fought against the policy since its beginnings.

This chapter is organised as follows: Section 2 describes the shifts in the composition of the social alliances that supported the IT policy in its three phases and shows the linkages between these moves and economic and political factors.

Section 3 analyses the recent US/Brazil conflict over market reserve policy and the role this played first in revealing some of the contradictions among the social forces that supported the policy and second in giving the opposition new arguments.

The major outcomes of this conflict and the mutual concessions made throughout the negotiations are presented in Section 4.

Section 5 looks at some conclusions and policy implications. We conclude that the US/Brazil conflict, uniting external and local constraints upon the Brazilian state, was a turning-point for the autonomy of decision of the Brazilian government in this high-technology area. The Brazilian state's room to manoeuvre in IT policy has been diminishing and this has also meant the building-up of a negotiating capability among state agencies. This ability to learn has been a central condition for the policy's success and, in this sense, it can be seen not just as a case of successful development policy, but also as an example of good politics, i.e. of negotiation and bargaining.

2 THE SUPPORTIVE ALLIANCE

The IT policy was mostly developed during military rule in Brazil, and was indeed supported by military and intelligence sectors. However, in spite of these apparently exceptional political conditions for autonomous state actions, the policy has had a rough ride even within the Brazilian state apparatus itself, to say nothing of the obstacles of foreign origin. The policy's very existence and its success are due in large part to its supporters' political ability to make alliances and to negotiate.

Self-reliance in information technology was a central component of the nationalism that cemented different alliances between various social groups and mobilised internal support for the policy in the first two phases of its history. However strong the nationalist appeal of self-reliance in IT had been in the past, it was insufficient to overcome the centrifugal force of diverging interests among the allies that deepened during the 1980s owing to the Brazilian economic crisis and the dynamism of the electronics industry itself.

This section briefly describes the shift in the economic and political conditions for Brazilian state initiative in IT policy and the changes they have caused in the alliances that have supported the policy in the three phases it has gone through so far; namely, setting-up (1972–79); institutionalisation (1979–84) and consolidation/adaptation (1984 to the present).

Local support in the setting-up of the IT policy

The internal preconditions that facilitated the setting-up of the IT policy, in terms of technological capability and political awareness, were laid down almost a decade before as a result of the science and technology policy.[3]

The policy was established during a period in which the high growth rates of the Brazilian 'economic miracle' had already slowed down. Unfavourable international conditions, following the first oil shock in 1974, made the goal of promoting economic expansion dependent on greater foreign borrowing. Positive trade balances became of strategic relevance. Governmental policies were oriented to saving foreign exchange and to establishing long-term technological objectives.

The IT policy is one of the few successful technological policies launched in that period. It was set up in accordance with government concern over import control and incentives to import substitution. Import control of computers, the main IT policy tool, was also a powerful political instrument which the supporters of the policy used to gain the necessary blessing from other segments of the Brazilian state apparatus.[4]

The alliance that pushed the setting-up of the IT policy in Brazil was made out of segments of the intelligentsia within and outside the state apparatus and some navy officers.[5]

Members of the intelligentsia within the state agencies (also identified as the technical–scientific and planning community), especially in the National Economic Development Bank (BNDE), the Studies and Projects Financing Agency (FINEP) and in CAPRE, united with navy officers in the common objective of local technological development. Backed by the planning minister, they provided the 'within the state' bunker of IT policy, a position they used to convince the relevant players inside the state apparatus of the strategic opportunity for establishing the policy.

Outside the state, members of the science and technology community organised in the Brazilian Society for the Progress of Science (SBPC) and, later, in the Brazilian Computing Society (SBC) and the

Association of Data Processing Professionals (APPD) constituted a relatively small but highly mobilised force that pressed the government for consistency in IT policy. It was particularly during the Seminars of Computation at the University (SECOMU), sponsored by the Brazilian Council of Universities' Chancellors (CRUB) during the 1970s, that the intelligentsia expressed their nationalist strategy.[6] The participation of the scientific and technological community in the alliance was already explained by their ideological commitment to overcome dependence and their identification of the central role of local technological capability in setting the basis for a more autonomous development.[7]

The military adhesion to the alliance was pushed for two different reasons. At first, some Navy officers were moved to action by their concern with national security. Lack of local control over the new technology embodied in the shipping vessels would put national security at risk either by being unable to operate its own defence equipment or by being dependent on foreign maintenance. These strategic considerations were complemented, as soon as the first actions had been taken, by a nationalist reaction to the behaviour of subsidiaries of transnational corporations that affronted the decisions of the Brazilian state.[8]

The setting-up of the IT policy was favoured by the economic conditions that pushed Brazil to control imports and by the technical capability and political awareness that had been built up earlier. It was, however, the will and skill of a planning intelligentsia uniting the scientific community and segments of the military that was responsible for making room for the policy within and outside the government.

New supporting forces in the institutionalising phase

After the second oil crisis in 1979 and the rise of the interest rates in the international market, the deficit in the Brazilian balance of payments grew larger. This fact associated with rising inflation triggered the adoption of a recessionary policy which resulted in strict credit limits on public spending and cuts in direct public investment, as well as in spending by government agencies and in imports from public companies.

Growth rates fell. Between 1980 and 1983, the average annual growth rate of industrial production was −6 per cent and that of industrial employment −7.3 per cent (Carneiro 1986:170). To make the situation even worse, foreign direct investment started a sharp decrease in 1982. Worth US$2,782.8 million in that year, the net direct foreign investment (total foreign investment minus re-investment and repatriations) went down to US$1,104 million in 1983 and US$849.9 million in 1984 (Serra 1988:126).[9]

The professional electronics industry was an exception amid the recessionary environment. Computer production was not affected by the recessionary policy. It was already a burgeoning industrial segment with high growth rates pushed up in particular by demand for banking automation. The installed park of computers grew from around US$1,500 million in 1979 to about US$2,700 million in 1982 (Piragibe 1985:149–52). Locally-owned firms more than tripled their sales during 1979–84 and had a 49 per cent share of the US$1.7 billion total billings of the computer segment in 1984 (SEI 1987:8). National firms employed the majority of people working in the computer segment in 1983. University-trained personnel employed in the national sector more than doubled in 1979–83 and around one-third of these were working in R&D activities. Around 40 per cent of university-trained employees in the computer industry were absorbed by the subsidiaries of transnational corporations in 1983, and only 4 per cent of them were in R&D activities (Evans 1986:797).

Nevertheless, recessionary policies and financial crisis had negative effects upon science and technology policy.[10] Resources became scarce and the process of building up an institutional structure and financing mechanisms slowed down. This 'drought' stage of the Brazilian science and technology policy (Adler 1986:156) affected the confidence of the scientific and technological community in state actions in the area. However, it was only during the next phase of IT policy that the reduction of resources for science and technology, including the presidential veto to the R&D fund established in the informatics law, showed implications for intelligentsia disaffection from the supportive alliance of the IT policy. For the time being, its role was crucial in the policy's institutionalisation.

The gradual political liberalisation of the authoritarian regime that had begun under President Geisel's administration was carried on by Figueiredo's administration in spite of strong signs of non-conformism by the regime's hard-liners. The need for formalisation of the IT policy became a prime concern for its supporters who felt the policy yet too weak to survive in the instability of the transition to democracy. Pressures from foreign firms and even from the US government[11] had mounted throughout all the IT policy's still short existence. And in the beginning of the 1980s its supporters saw clear signs that the end of the market reserve was going to be pressed upon the Brazilian government. Deputy Cristina Tavares (PMDB) declared:

> a fact that awakened us for the need of urgent action was a meeting of the American Chamber of Commerce held in Brasilia but closed to Brazilian journalists. Afterwards, it transpired that they had been conspiring against the IT policy. The end of the market reserve had also been discussed during meetings between the US Department of State and Brazilian authorities. This kind of pressure convinced nationalist sectors of the Brazilian society either to institutionalise the market reserve into law or to risk the IT policy succumbing. If the policy remained based on directives and resolutions it would be easy to change everything with just another directive. A law would be much more difficult to change because this would have to be made through the Brazilian Congress.
> (Azevedo and Zago 1989:138; English translation: Bastos)

A strategy was then launched in order to make into law the set of norms that had embodied the IT policy so far. In June 1984 the Brazilian Executive introduced the Informatics Bill which was discussed, changed into a substitute and then approved in October in an emotional and tumultuous ten-hour session with over one hundred speeches, by 337 Deputies and 42 Senators with a solitary rejection by Senator Roberto Campos.

The IT policy's supportive alliance of the previous phase was strengthened with the adhesion of three new participants: the segment of army intelligence, the Brazilian Congress and the national producers of hardware.

The involvement of army intelligence was precipitated by Geisel's successor, Figueiredo, who was a former head of the National Intelligence Service (SNI).[12] It was from this area, and from foreign affairs, that Figueiredo took the constitutive members of a commission which he nominated to study the activities of CAPRE and to suggest what to do in the IT area during his administration. CAPRE's demise was suggested and its attributes passed on to a new agency, the Special Secretariat for Informatics (SEI), under the National Security Council (CSN).

Intelligence involvement was, at first, ambiguous. Besides the mutual suspicion that contaminated the relationship between intelligence officers, the planning intelligentsia and the scientific community, SEI's criticism of the narrow scope of the policy (because it was not yet concerned with micro-electronics and software) was also accompanied by the accusation that CAPRE was too liberal with transnational corporations.[13] SEI's pragmatic decisions[14] and intolerance towards the civil community who had thus far defined the IT policy, kept the intelligentsia in an opposing stance for some time. This situation changed when the process of turning the policy into law pushed the supporting forces to reunite.

National hardware producers entered the political arena having become a relatively important segment of the industrial sector. In August 1979, a group of Brazilian hardware manufacturers defected from the Brazilian Association for the Electric and Electronic Industry (ABINEE), where the foreign subsidiaries affiliates were preventing the Association from defining itself in favour of the IT policy, and created the Brazilian Association for Computers and Peripheral Equipment Industries (ABICOMP).

National manufacturers of hardware were a much more homogeneous group at this period than is the case nowadays. Many of these industrialists were genuinely supportive of the technological goal of self-reliance and were investing significant resources in R&D.[15] Reverse engineering was the usual way to learn the new technology, but some newcomers to the IT area, during the PC 'boom' of the mid-1980s in particular, made things easier by copying foreign models without much learning effort (Fregni 1988). Nevertheless, despite minor internal divergences mainly related to the process of technology absorption, the group of national IT producers was in this period much more cohesive than nowadays. The satisfaction with the segment's high growth rates and SEI's backing of their interests contributed to their cohesiveness.

The Brazilian Movement for Informatics (MBI) was formed while the informatics bill was being discussed and was decisive in pressing the Brazilian Congress to vote on the matter before the change of government.[16] The MBI brought together ABICOMP, SBC, APPD, SBPC, the Association of Data Processing Service Enterprises (ASSESPRO), the Society of Computer and Subsidiary Equipment Consumers (SUCESU) and the National Federation of Engineers (FNE) in the process of defending the IT policy principle of market reserve for local capital.

A smaller group comprised of ABICOMP, APPD, FNE, SBC and SBPC – the scientific and technological community that were combative members of the alliance in the setting-up phase plus the representation of the engineers and hardware industrialists – launched a national campaign to gain wider support for the informatics bill. Within a few weeks, 37 national-level and 300 state-level entities from twenty states had joined the movement (Azevedo and Zago

1989:139). The support of the scientific and technological community was crucial to the fight of the national industrialists for keeping the basic principles of the IT policy.

Pressed by foreign opposition and in the political juncture of transition from military rule to democracy, state actions towards institutionalising the IT policy developed within a situation of economic crisis, foreign indebtedness and IMF tutelage. The supportive alliance, that had been strengthened by new politically weighty participants, mobilised wide social support and gave the Congress the backing it needed to turn the IT policy into law.

Changing social support in the consolidation phase

Since the passing of the informatics law, high inflation, economic stagnation, political and institutional instability, the developmental problems of the electronic complex and US pressures severely weakened the alliance that supported the policy in the institutionalising phase.

Brazilian annual inflation rates which had stabilised around 200 per cent since 1983, began increasing by the end of military rule. The first civilian government in twenty years set up an innovative and unorthodox anti-inflationary programme, the Cruzado Plan, in February 1986. It brought about a brief economic surge expressed in higher rates of industrial growth and a decrease in unemployment figures. However, owing to mismanagement of the Plan, severe financial disequilibria reappeared, leading to the return of high rates of inflation. As a consequence, economic growth rates slowed down in 1987 and the economy stagnated in 1988. In February 1987, a drastic fall of Brazilian reserves of foreign exchange led to a unilateral moratorium on all interest payments of its debt to private banks.

The failure of the Cruzado Plan in curbing inflation added to the political instability of President Sarney's administration. The precarious political coalition of nationalists and some segments of the anti-nationalist forces that formed the basis of the Sarney administration broke up. After the second frustrated attempt at reducing inflation by unorthodox means, PMDB, the political party which had been in the opposition to the military government and which shared in Sarney's administration, was practically removed from government.

Domination of Sarney's administration by anti-nationalist forces had a negative influence on the IT policy. The Ministry for Science and Technology (MCT), created by the civilian government as a symbol of its concern with scientific and technological development and under which SEI was put, was extinguished in January 1989. Its tasks were transferred to the Ministry of Industry and Commerce, occupied at the time by an anti-nationalist. Pressure from PMDB representatives in Congress and from intelligence sectors resulted in the top of the scientific and technological institutional structure being restored in March 1989, this time as a Special Secretariat for Science and Technology (SECT) and directly subordinated to the President's Office. In December 1989 the Ministry for Science and Technology was brought back and the institutional reform made by the Collor de Melo administration in March 1990 revived the Special Secretariat for Science and Technology.

While, within the Executive, governmental actions were causing high instability in the institutional structure of IT policy, nationalist representatives in the Brazilian Congress were fighting a strenuous battle against the anti-nationalists to define constitutional principles that could give stronger backing to the policy. It is not at all clear, however, how much the IT policy supporters in the Congress have won. The new Brazilian constitution, approved in 1988, distinguishes 'Brazilian companies' from 'Brazilian national companies' to which some privileged treatment by the state is assured, including temporary protection (Art. 171), and defines the local market as part of the national endowment and, as such, a means to promote, among other developmental goals, technological autonomy of the country (Art. 219). There is no guarantee that the definition of effective control – that the majority of the voting shares should be owned by Brazilian residents – would in practice protect national capital.

The major political event that has marked the current phase of the IT policy so far has been the US proceedings against Brazil under the American Trade Act which opened in September 1985.[17] These proceedings developed in a situation of economic crisis and political instability when the alliance supporting the IT policy had weakened with the disaffection of the scientific and technological community and the military retreating back stage. Open support for the policy was reduced to some industrialists of the segment of professional electronics with the eventual help of a few institutions representing the technological community.

The feelings of the scientific and tecnological community towards the policy are well expressed in the statement by Prof. Nelson Maculan, the deputy director of the Engineering Graduate Programme at the Federal University of Rio de Janeiro. Commenting on the mistaken belief that an industrial policy would automatically bring about scientific and technological development, he said:

> This mistake has been made in the computer area: we have a market reserve for national industrialists, but the entrepreneurs themselves do not have the necessary understanding of the strategic role of human resources which, however, are the only means to sustain and to speed up the sector's development.
>
> (Mineiro 1987:55; English translation: Bastos)

A study of human resources in IT prepared for the Ministry for Science and Technology concluded that bad technical and financial conditions in Brazilian universities were the main causes of poor university training in IT. Universities were equipped with few computers and most of these were more than ten years old. For example, CTI, a research and development centre established in the mid-1980s for developing computer-related technology, has been understaffed and poorly financed. The study suggested urgent actions in order to give universities and research centres the resources they needed (MCT 1986).

The development of the Brazilian electronic complex had brought to light some features and problems that affected the cohesiveness of the supportive alliance of IT policy and altered its relationship with governmental agencies in charge of the policy. Disarticulation of the different policies towards the various segments of the electronic complex has aggravated inefficient inter-industrial relations within the complex, especially those involving the production of components and the rest of the complex. Fragmentation in the industry of professional electronics has resulted in small-scale production affecting price performance[18] and technology strategies (Tigre 1988:117). The availability of financial resources for investment in capital and technology, which was not problematic during the market growth, became risky and uncertain under conditions of slowing-down growth and high inflation. This has moved state agencies to encourage medium-sized and smaller firms to merge, which has happened recently in favour of the great local conglomerates[19], and has increased in many other smaller firms the willingness to collaborate with foreign capital.

Brazilian industrialists in the protected area are now being pushed to improve their international competitiveness. This is particularly true for those moving towards an export orientation stimulated by the instability and uncertainty of the local market and by limits to import substitution. Moreover, the pressures by the US government and Brazilian users for easing import restrictions bring closer the time when Brazilian firms will have to face higher foreign competition in the local market. This time has been brought closer by the spread of the idea that market reserve is due to end in 1992.[20]

In any case, Brazilian firms anticipate a higher competition either within or outside the local market in the near future. Cost-effectiveness and access to rapidly changing technologies are more than ever major concerns to the Brazilian IT industrialists.

Conglomerates on the one hand, and niche firms, on the other, are in a better position to face the competitive challenge. They both have built in, at different scales, a potential for achieving higher cost-effectiveness and for absorbing technological developments. They would benefit, for instance, from a quicker process of technology acquisition. In the case of the conglomerates, some analysts think that the former's export orientation would be facilitated by the establishment of alliances with foreign capital which would provide them with international credibility and an overseas distribution and service network (Tigre 1988). Thus, conglomerate entrepreneurs and niche industrialists are more relaxed about the possibility of a selective change of the Brazilian IT policy: conglomerate businessmen, because their relative strength gives them bargaining power; niche-firm industrialists, because their knowledge of users' needs and the diminutive market size may be efficient shields against predatory competition.

The bulk of the professional electronics firms are in a more fragile situation because of their small-scale production, lack of resources for investment in capital and technology and low research tradition. Some of these industrialists, especially those who fought for the establishment of the policy, are still its combative supporters. For them, changes in the policy may put their own existence at risk. They would prefer an open debate on the alternative paths to the policy where they could try to mobilise social support once more for the goal of technological self-reliance. Most of the manufacturers in the segment of professional electronics, however, are willing to have unconditional permission to associate with foreign capital (*Ciencia Hoje* 1989:75). From their point of view, the perspective of sharing potential profits in the near future more than compensates for the riskiness of their present investments. The quest for technological autonomy is not their concern.

The position of local industrialists on the matter of software, especially operating systems, the difficulty ABICOMP has found in discussing with its affiliates their investments in the medium and long run, and its vain attempts at defining an ethical code for industrialists' actions are signs of the shifts in their support for the policy. The majority of industrialists favour foreign-developed operating systems instead of locally-developed equivalents. Some of them fear to reveal their willingness to become associates of foreign capital and, consequently, they are hesitant to commit themselves to an ethical code that could bind them to local technology (*Ciencia Hoje* 1989:74–6).

The analysis of the supportive alliances of the IT policy made in this section shows how the building up of social backing by relevant social forces had helped the Brazilian state in setting up and institutionalising a hi-tech industrial policy. It also shows how this support was undermined by economic crisis, political instability and problems stemming from the development of the electronic complex. The pressures of the US government reinforced these disrupting effects of internal conditions because the conflict brought to light and fuelled an unprecedented widespread local criticism to the policy.

3 THE US/BRAZIL CONFLICT AND LOCAL OPPOSITION TO THE IT POLICY

In September 1985, the US government announced the start of an investigation into the Brazilian IT policy and, under Section 301 of the Trade Act, the possibility of economic retaliation if discriminatory or unfair trade practices were found against US interests in Brazil.

The US/Brazil conflict over IT policy is an example of the toughness of international relations in this new technological era. The position of developing countries in the international arena does not facilitate their negotiations in these issues, but some of them, like Brazil, are not so weak as to jettison their policies. Besides their own markets being a highly valued resource, the internationalisation already reached by their economies gives them bargaining power and makes the threats of economic sanctions almost a rhetorical exercise. Foreign debt, however, is a restriction of the state's room to manoeuvre in defence of its policies in a situation of international threat of economic sanctions. In any case, resistance to foreign pressures, with consequent major or minor concessions, is a subject that also depends on the dynamic of internal social forces.

Section 2 described the shifts in local support for the Brazilian IT policy. This section describes the evolving opposition and the rise of a new criticism of IT policy and explores the effects the US/Brazil conflict had upon them.

The conflict, developed throughout 1985–89[21], provided the opportunity for the local opposition to show itself without being blamed for defending anti-Brazilian interests. On the contrary, the conflict made room for the IT policy's criticism to be linked to a defence of broader Brazilian interests. The Brazilian IT policy's nationalist appeal to *self-reliance in information technology* gained a powerful counterpart in the struggle over *Brazilian integration in the international market*.

Isolated claims from American firms against the Brazilian IT policy can be traced back to the setting-up phase of the policy. As already mentioned, IBM started actions in 1976 to make and sell a mini-computer model in Brazil in open confrontation with CAPRE's Resolution No. 1. The Brazilian state agencies involved had to act strategically in order to curb the IBM initiative.[22] Nevertheless, subsequent pressures from Carter's administration led Brazil to compromise and approve local production of, and trade in, two medium-sized IBM computers. This decision was made by CAPRE in its last meeting before being superseded by the SEI.

From that time onwards, claims by American firms against IT policy have been taken up by the US government which has demanded changes in the Brazilian policy. President Reagan insisted on the IT policy being included among trade and financial subjects under negotiation between the two countries when he visited Brazil in 1982. In March 1983, a USTR (United States Trade Representative) report accused Brazil of being not merely protectionist but biased against transnational corporations. In April 1983, in a meeting with the US Treasury Secretary, the Brazilian Finance Minister and other authorities, the US Trade Secretary declared that IT policy had become a compulsory topic in any negotiation between the USA and Brazil. IT policy entered the 'hidden agenda' where political issues in the interest of the US have been made part of the negotiation on debt rescheduling and new money (Griffith-Jones 1989:22).

The US/Brazil conflict was a culmination of all these past pressures. Its initiation was a result of a US Congress call for action to decrease the US trade deficit and of demands by the American Electronics Association and the US Computer and Business Equipment Manufacturers Association for easing Brazilian import restrictions and for establishing local regulation on intellectual property rights. The USTR proceedings against the Brazilian IT policy were initiated under Section 301 of the US Trade Act which allows economic retaliations for discriminatory or unfair trade against American interests.[23] A history of the evolution of the conflict is presented in the Appendix at the end of this chapter.

Of particular importance is how this conflict affected the internal opposition. Remarkable shifts in its composition have been the progressive involvement of various agencies within the state itself, of IT user firms and of the potential victims of US retaliation. The arguments of the opposition have changed accordingly, going from the traditional and aggressive criticism against any restriction of foreign capital to more moderate and consequent terms in which it has been quite close to the national IT companies' own vision of the necessary changes in the policy.

Among the natural allies of the foreign IT firms have been the domestic IT user firms, especially in the industrial sector where the larger companies are foreign subsidiaries. The US/Brazil conflict added to this group local exporters of orange juice, shoes, textiles, plywood, steel products and planes, who were the most directly affected by the threat of economic retaliations.

Local opposition to IT policy grew within the state apparatus. At first it was reduced to the realm of the Ministry of Communication whose relationship with CAPRE and SEI had been marked by deep differences in conception and action strategies. The Finance Ministry's and the Foreign Relations Ministry's negotiation of foreign debt and the US/Brazil conflict itself moved them to a position of criticism of the IT policy. The legal advisory body of the President's Office has been the most recent addition to the opponents of the IT policy. It has been searching for a possible unconstitutionality of the Informatics Law (No. 7232) and for evidence of improper definition as 'national' given to some of the IT firms that went public.

The criticism of local opposition has been centred on the policy's two pillars, the market reserve and its justification in terms of local technological development. The US/Brazil conflict sharpened the opposition's arguments so that it gathered higher mobilising capability than it had previously had.

The criticism that had prevailed from the policy's beginning until the early 1980s was couched in emotional terms and expressed as a rejection of any restriction to foreign capital. The IT policy was accused of being xenophobic and of embodying a kind of economic 'racial' prejudice, as only pure national capital is allowed to enter the reserved market. More than that, the opposition stressed that the market was not reserved to local capital in general, but to some few local capitalists, sometimes to a single entrepreneur. The policy has been viewed since then as protecting a local fiefdom.[24] The liberal Brazilian policy towards foreign capital in all industrial areas and the foreign firms' domination at the upper end of the IT market were, of course, not taken into account by the critics.

In contrast to the above reasoning, the current dominant criticism presents more convincing arguments. They are rooted in a paper by the Federation of São Paulo Industries (FIESP) in which it acknowledged the market reserve's merit as a policy orientation but pointed to some corrections to be made to the policy (FIESP 1983). One such improvement would be granting minority participation of foreign capital in local firms within the reserved market as a way of facilitating their access to foreign technology. Law 7232/84 (Art. 12) and Decree-Law 2203/84 incorporated this suggestion.

The current debate over joint ventures with foreign capital has begun with the US/Brazil conflict. The case of GSI, a joint venture of a local industrial steel-making group interested in informatics and IBM, approved during the US/Brazil conflict after long hesitation by SEI, gave the opposition the chance to go further in demanding a flexible interpretation of national control as defined in the law. If local technological control is not required of joint ventures, then Law 7232 will have lost its main link between an industrial and a technological policy.

It is precisely in the area of IT policy's approach to technology development that the views of traditional and current opposition differ. For the traditional opposition, the emphasis on local technological development is as irrational an aim as it would have been to have set 'the reinvention of the wheel' as a national goal. The significance local technology development has in the policy is viewed as a distortion of the real local need, that is, to have the best and most efficient technology regardless of where it was developed.

The opposition's new focus on the technological issue is well represented in FIESP's paper. As stated in the document, the entrepreneurs' motivation to discuss IT policy was twofold. One was their concern with the modernisation of industrial production and the other was their acknowledgement that an IT-based industrial revolution was not only unavoidable, but also impossible to reject and inadvisable to postpone. In addition, despite Brazil's struggle to balance local technological development and the absorption of more dynamic foreign technology, corrections were needed in the form of governmental action in order to avoid the pitfall of extremist positions. One such position would lead to the isolation of Brazilian technological development from what is going on in the rest of the world.

The US/Brazil conflict reinforced FIESP's earlier arguments. From that period onwards, the policy has been accused of aggravating the whole Brazilian industrial sector's isolation from the international market, an idea that has spread to manufacturers, including those in agribusiness, and even consumers.

Thus the conflict set, for the first time in the IT policy's still short history, the local IT sector against the whole Brazilian economy. 'Will Brazil defend an island policy in order to save the informatics sector and to condemn all local exports to stagnation?', asked an important newspaper lead article (*Jornal do Brasil*, 10 September 1985). Referring to the conflict, another leading newspaper article commented: 'the problems Sarney's administration faces are related to the whole country. Their solution depends on the technological evolution of the industrial sector, the international competitiveness of Brazilian manufacture and the openness of the local economy' (*O Estado de São Paulo*, 16 February 1986).

Local technological capability in IT, the major justification for a protectionist industrial policy, was made responsible for the lack of modernisation of the Brazilian industrial sector and for the potential loss of Brazilian exporters. Moreover, local R&D in information technology itself has also been censured.

> The road that leads to the independence of Brazil includes cooperation with other countries and giving high priority to technological development. It is essential for the country to invest in areas such as research on tropical diseases, on deep-water oil extraction, on selected seeds and on genetic adaptation of cattle to tropical conditions. We cannot let the idea of technological advance be reduced to electronics, computers, software.
>
> (*Jornal do Brasil*, 23 August 1986)

In fact, this reduction has never been present in Brazilian science and technology policy. Here, the use of the concept was intended to counterblast the principle of local technological development that justifies the IT policy.

This section has described the changes in opposition to the Brazilian IT policy that resulted from the US/Brazil conflict. Related to its composition, the main shifts were, on the foreign front, the decisive involvement of the US government, and on the local front, the progressive compliance of several governmental agencies, IT-user industrialists and the exporters potentially affected by the threat of economic sanctions. Regarding their criticism, the conflict gave the opposition the opportunity to mobilise social support in the name of the international isolation the policy was imposing on the whole Brazilian economy.

The Brazilian state's room to manoeuvre in a hi-tech policy was heavily affected by the US/Brazil conflict. A crumbling supportive alliance, a strengthened opposition, high political instability, increasing disarticulation within the state and the general economic crisis narrowed the limits for the state agencies' actions, including negotiating a solution for the conflict. Even under such unfavourable conditions, the Brazilian state had the ability to negotiate what was probably a second-best solution. The results of the conflict are discussed in the following section.

4 IMPLICATIONS FOR THE POLICY'S SCOPE AND IMPLEMENTATION

The central US demand had been for the policy to be made more flexible, and there is no doubt that this was satisfied. Nevertheless, this does not mean that the Brazilian state surrendered to all US demands because, as will be presented in this section, the main outcomes of the conflict resulted from concessions made on both sides.

The strong nationalist reaction may have caused a shift in the US strategy, moving it from its initial demand of changing the Informatics Law. Despite being strengthened during the conflict, local opposition was not yet strong enough to face local reaction to any attempt at altering that law, which can only be made through the Brazilian Congress.

The firm Brazilian position of not negotiating the Informatics Law, but only discussing specific cases related to its implementation, led the US to change its strategy and the negotiations to a more realistic field. If it was difficult to change the law, the US could obtain a more favourable implementation of it instead and, moreover, it could influence the making of subsequent legislation. From the other side, for Brazil to succeed in its IT industrial policy, which was under both foreign and local pressure for change, it could accept some flexibility in the implementation of the law and could even sacrifice some sectors not yet fully regulated.

Respecting the strategy of getting better results from the implementation of the policy, the target for US attack was the SEI. The SEI was accused of being too strict in interpreting the law and very inefficient in making any decision within schedule. This criticism gained local support from some of the local subsidiaries whose import requests had been rejected by SEI with the justification that the items would be locally made in the near future. Supporting these firms' complaints, the Brazilian Foreign Minister declared that 'SEI cannot be an immobilizing agency' (*Jornal do Brasil*, 20 July 86). In the same month and year, the local press published news of the existence of a paper by the National Intelligence Service (SNI) and the Army in which the authors express their doubts about the benefits that could have resulted from SEI toughness toward foreign capital. They concluded that, in spite of the market reserve being effective in establishing a local IT sector, the restrictiveness in its implementation was already causing local technological backwardness and, consequently, threatening national security.

How to prevent SEI being so strict in its activities? During the development of the conflict, various attempts were made to look for a way to reduce SEI's powers. The possibility of dividing its tasks among related governmental agencies was considered. This alternative would increase the policy's fragmentation and make it easier for interested parties to influence particular decisions.

No such change occurred to SEI within the time span of the conflict. It remained the executive secretary to the National Council for Informatics (CONIN) and gained new responsibilities from the software law. The only changes related to SEI at that time were internal reforms announced in August 1986. They consisted of hiring new specialised people, giving the regional SEI offices some autonomy of decision and establishing an internal data-processing system to speed up decision-making. With this internal reform, SEI defined new schedules for itself: for common cases of import and local production requests, the decision would be made within sixty to ninety days. Given the difficulties of restricting SEI's institutional powers at that time, its opponents concentrated on demoralising the agency. It was accused of manipulating CONIN meetings and decisions and was practically humiliated when CONIN approved local trade in the operational system MS-DOS version 3.3 after SEI rejected the authorisation for trade in the Brazilian market of MS-DOS because of the existence of a local similar product. However, after the result of this particular conflict, President Collor's administration has recently changed the institutional status and the attributions of the once central agency in charge of IT policy.[25]

In response to US demand for a list of non-reserved items, the Brazilian government issued a new CACEX Communication No. 171/86 that specifies the equipment the import of which is controlled by SEI. All non-digital items were excluded from this new list and twelve categories of equipment eliminated. One new category (optical fibres) was added and the remaining ones were better defined. A new list was prepared by a special commission created by President Collor's administration and ratified by CONIN in late October 1990. The new list now excludes several goods the imports of which depended on previous authorisation by the former SEI. In accordance with Decree No. 99541, of 21 September 1990, the goods not included in the new list are free to be imported or manufactured within the country.[26]

The US request for the adoption of standards resulted in the Brazilian Code for Information Exchange, prepared by the Brazilian Association for Technical Standards (ABNT) and presented to CONIN in 1986. However, the resistance larger nationally-owned firms had shown to adopting locally-developed operating systems can be interpreted in the light of US pressures for standardisation in the Brazilian market.

In August 1986, SEI gave its approval to GSI, an association between the Brazilian group Gerdau and IBM. Despite having a high-speed network of data communications connecting its data-processing centre in Rio to its industrial plant in Campinas and to all its affiliates in diverse Brazilian cities, IBM declared that this network would not be used by the new firm. Its activities would be limited to the use of application software already available in the IBM data-processing centres. New software would eventually be developed in the future, depending on the customer's needs.[27]

It was, however, in the area of software regulation, left open in Law No. 7232/84, that the most outstanding result of the US/Brazil conflict can be found. Even in this case, negotiations brought about a kind of compromise between the contenders. First formulations of the software bill were much closer to the US demands than the final software law. The idea of allowing the establishment of 'consortiums' of foreign and local firms was later excluded from the bill sent to the House of Representatives. The software bill established copyright protection for twenty-five years (the US demanded fifty years), excluded foreign firms from local trade in software, unless it would be used in the hardware made locally by foreign subsidiaries, and made previous filing at SEI a condition for local trade in any software. This filing would only be accepted by SEI under the condition of non-existence of a local equivalent.

Opposition to this Bill was made by the Brazilian Association of Software Houses (ABES), which wanted no limits for local trade of foreign software, SUCESU and ASSESPRO, especially the latter, which assessed the Bill as giving little support to locally-developed products. The Senate made some important changes in the text approved in the House of Representatives. Some of these were in order to limit SEI powers,[28] but the most important was related to ASSESPRO's demand for stronger protection of the locally-developed product. The substitute bill established that a fee should be paid for all imported software in order to stimulate either local development or the use of readily available local products. The fee would be collected to form a Special Fund for Informatics and Automation that would support technological development in the area.

When President Sarney vetoed the fee for imported software, ASSESPRO and national entrepreneurs reacted by saying that this decision rendered the Software Law useless as a protection for local products. Considering that Brazil had only very recently started to develop software, there were few local functional equivalents to the foreign product. Among these few, SISNE (a MS/DOS-like) and SOX (a UNIX-like) were already being undermined by other mechanisms.[29] So, the only protection given by the Software Law is to local traders of foreign software that have no research activities in the area. Imported software will be able to enter the Brazilian market, which will certainly discourage local initiatives in this area. The principle of technological development orienting the market reserve policy was, in this case, abandoned.

5 CONCLUSIONS

The purpose of this chapter was to analyse the Brazilian state's room to manoeuvre in the formulation and implementation of IT policy in the face of external and internal demands for change. The discussion on the varying economic and political conditions for state action during the three phases which the policy has gone through so far, the shifts in domestic social support it has received and the escalating foreign pressure against it, have shown diminishing room to manoeuvre for state autonomy of action in the area. The US/Brazil conflict was a turning-point in this process as it expressed in a direct form previous indirect foreign pressures for liberalisation of the IT market in Brazil and articulated local discontentment with the policy.

The launching of the policy was an act of state autonomy in conditions of large room to manoeuvre. The initiative started with support from the military, benefited from the need to control imports and from the availability of human resources with the technical knowledge required, and was directed to a segment of the market still unexplored by the foreign firms that dominated the local market. Its insulation within the state apparatus was another condition that facilitated autonomous action. In this situation, the policy's formulation depended exclusively on what state officials (military and civil) thought desirable and convenient. It resulted from a strategic appraisal of the important role of IT in the economy and its implications for national sovereignty. But this was not a conclusion reached by industrialists individually or via their sectoral associations and seems not to have been among the concerns of the local industrial sector with its future perspectives of growth. It was not pressed on the state by interested parties nor did it reflect interests rooted in the productive process.

Nevertheless, even in this phase of greater autonomy for state action, and its corollary – greater room to manoeuvre – it did not mean the relief of negotiating support for their acts for state agencies. They fought for the backing of relevant players within the government structure and had to deal with the combative segment of the intelligentsia which was responsible for putting IT policy onto the national agenda. Conforming to the governmental goal of controlling imports was also used as a way to get support from within the state. Accepting the anti-dependence arguments of the scientific and technological community was not only a sign of some ideological convergence between the forces supporting the policy, but a powerful although risky way for state agencies of a right-wing military regime to gain societal support for their policies. State officials in charge of the policy took this risk and the subsequent efficiency of social mobilisation around the concepts of technological self-reliance and national control over high technologies showed the choice had been correct.[30]

State agencies also had to keep the pressures from foreign capital under control in order to avoid turning the policy into a *cause célèbre*. This was done by the strategy of market segmentation, by involving other more powerful state institutions such as CDE and by making some concessions.

The ending of military rule, mounting direct foreign pressures and the economic fragility of a yet burgeoning industrial segment explain the decision to institutionalise the policy into law. This decision shows in itself the recognition of the state agencies that their room to manoeuvre was becoming so narrow that the whole policy could be in jeopardy unless it could be put under the protection of Congress. With the liberalisation of IT policy in the hidden agenda of foreign debt rescheduling, the direct criticism of the USTR and the demands of domestic hardware producers for continuing protection, state agencies were pushed to harder negotiation. The strategy of involving wider interest groups in the preparation of the Informatics Bill was in accordance with the previous pattern of action of these agencies.

Mobilisation of societal support gained momentum and influenced congressional decision on the informatics law.

Negotiations were mainly established throughout the discussion of the Informatics Bill, despite the more controversial subjects such as the wide definition of the area protected and the concept of 'national company' coming out of any possible deal. Most of the changes to the Bill, including the creation of a fund to be used for research and development in this field, were later vetoed by the President. Subject areas around which a compromise among interested parties failed to be achieved, such as software regulation, transborder data-flow and data-banks, were left open for future legislation.

The US/Brazil conflict developed at a particularly sensitive moment for the Brazilian government. Political instability in the transition from military rule to democracy within economic crisis and during creative but unsuccessful attempts at controlling inflation were conditions that weakened the whole Sarney administration. The agencies involved with IT policy were among the few segments of the state apparatus that kept some legitimacy in view of their competence. Even so, and probably because of this, SEI was a central target for attacks by the US government and, when it proved difficult to reduce its powers, it turned to demoralisation. It was, however, due to the resistance and negotiating ability still remaining in the institutional structure of IT policy (supported by the Congress in the front- and the military in the back-stage) that the policy was not dismantled.

The state agencies in charge of IT policy seem to have always been aware that the large autonomy of action stemming from the authoritarian state to which they belonged was not enough to make their actions valid in the eyes of those subject to it. They have made a real effort to justify their position and actions, to make them acceptable, and thus make the policy legitimate. In the launch phase, this drive towards legitimation was accomplished by mobilising general societal support via the scientific and technological community. The legitimation the IT agencies generated has been, in subsequent phases, sustained by policy achievements. Furthermore, the policy's rules and directives were made wholly legal and legitimate through discussion in the Brazilian Congress and the passing of the Informatics Law. However, it was mostly the claims and aspirations of the national hardware producers that were considered by the policy. This provoked disaffection of the intelligentsia outside the state and became a hindrance to the policy's legitimacy among other segments of the industrial sector. On top of this, problems in the development of the electronic complex have brought about new challenges to the state apparatus affecting the social image of its efficiency. The US/Brazil conflict was decisive in bringing to light and fuelling the belief that the interests of the majority of the Brazilian industrial sector were not being served by the policy and that it was, in fact, contrary to these common interests. In a nutshell, the conflict jeopardised the policy's legitimacy.

The conflict had two important political results. One was, as already pointed out, to broaden the group of domestic critics of the policy, thus affecting the balance of internal support to it and, consequently, reducing the policy's legitimacy. The other was to restrict the Brazilian government initiatives for consolidating the policy, as was the case with software regulation.[31]

This reduction in IT policy legitimacy was also a sign that the state agencies in charge had abandoned the strategy or been ineffective in their legitimation effort, after the passing of the informatics law. Whatever may have been the explanation for the IT institutional structure lagging behind the dynamism of Brazilian society, a strategic mistake it had never before been blamed for, its acceptance of the need to review some of its previous orientations was an indication that the ability to adapt to new circumstances was also learnt during all those years of negotiation and compromise. Therefore what is surprising in the analysis of the development of the conflict is not the extent of what the Brazilian negotiators were compelled to give up, but what they have managed to keep.

Changes in the Brazilian IT policy are, in any case, an expected outcome of its current phase. The central issue now is no longer whether to keep the policy as it has been implemented so far, because it has well acknowledged limitations, nor when to adjust it to the changing market conditions, because new internal and external situations are already evident, but how to change it in such a way that its positive achievements are not thrown away. Some liberalisation of the policy is expected to happen[32] in order either to speed up the process of industrial modernisation, to expose the nowadays unafraid national IT producers to a beneficial stronger competition, or to deal with the resources for investment in capital and technology that domestic producers are seeking and some foreign firms are now willing to negotiate.

One of the tasks of the agencies in charge of the policy and, in fact, of the Brazilian government as a whole is, then, to build up a new legitimate order that relies neither on defensive nationalism nor on blind liberalism. In this respect, it is crucial to consider the continuing protectionism which can be observed around the world despite all the liberal rhetoric. It is within this context that a new direction will have to be found for the Brazilian IT policy.

APPENDIX: HISTORY OF THE US/BRAZIL IT POLICY CONFLICT

7 September 1985 – Announcement of the USTR proceedings.

January 1986 – The US Department of State asks the Brazilian Ministry of Foreign Relations for formal explanation on the Brazilian Law No. 7232, the Informatics Law.

4 February 1986 – Caracas: consultative meeting. Deputy USTR heads the American delegation and the Executive Secretary for Informatics, the Brazilian delegation. Brazilian delegation discusses the IT policy under GATT rule.

17 April 1986 – Brazilian Congress votes on Law No. 7463, the First Informatics Plan (I PLANIN).
April 1986 – Letter from the US Secretary of State to the Brazilian Minister of Foreign Relations suggests they should personally devote their attention to the case.

9–10 May 1986 – Seminar, in the US, organised by ABICOMP and the American Electronics Association (AEA), for Brazilian and American IT businessmen.
14 May 1986 – The US Economic Policy Council decides that the Brazilian IT Policy constitutes unfair trade practice and retaliation is justified.
May 1986 – Brazilian Ambassador to GATT informs the institution of the government's intention to appeal to GATT if there is retaliation against Brazilian exports to the US.
– Brazilian Senator Severo Gomes (PMDB-SP) introduces Bill No. 99/86, the National Self-Defence Bill, establishing several sanctions to foreign interests in Brazil in the case of restriction to Brazilian exports to foreign markets.
May 1986 – Brazilian Minister for Commerce and Industry declares being himself favourable to an extension of the market reserve principle to the area of pharmaceutical products.
– Microsoft Corporation announces its intention to sue four Brazilian companies for infringement of copyright on its MS-DOS operating system.
– The US Deputy Secretary of State visits Brazil.

2 July 1986 – Paris: second meeting. The US delegation, headed by the USTR, presents a list of twenty-six items, organised in five sections, expressing their demands. The sections are: trade and investment, intellectual property rights, transborder data flow, subsidies and technical standards. The Brazilian delegation was headed by the Brazilian Secretary General of the Ministry of Foreign Relations.
July 1986 – The Chairman of the SEI visits Japan and meets potential suppliers of electronic components to Brazilian IT companies.

11 August 1986 – Paris: third meeting. The Brazilian delegation discusses some of the twenty-six items. The US delegation demands positive actions towards more flexibility in the implementation of the policy.
17 August 1986 – The SEI announces internal reforms.
26 August 1986 – The CONIN approves the regime of copyright for software.

10 September 1986 – The Brazilian President visits the US.
September 1986 – The SEI approves the IBM–Gerdau association for data processing.
– Brazil announces the intention of publishing a list of non-reserved items.
– The US government postpones the conclusion of the proceedings to January 1987.

November 1986 – The SEI begins to institute administrative reforms designed to speed up its processing of company applications.
– A US-Brazil 'ad hoc' group is set up to review individual complaints by American companies.
– The CACEX Communication No. 171 is issued to narrow the range of products for which imports are restricted and eliminating SEI's powers to control the imports of unspecified products as long as they contain commands or controls or other systems with digital logic of the numerical control type.

December 1986 – The software bill is sent to the Brazilian House of Representatives.
13–14 December 1986 – Brussels: fourth meeting. Brazilian delegation declares its mission complete as all information required is given. The US delegation asks for further discussion on the software bill.
23 December 1986 – The USTR declares no trade retaliation related to the IT conflict until July 1987.

26 December 1986 – The USTR announces the exclusion of thirty Brazilian products from the GSP list of concessions.
30 December 1986 – The US government announces the suspension of the part of the Brazilian 301 case which dealt with administrative procedures. The parts referring to software and investment were retained and action on those parts postponed until July 1987.

12 March 1987 – The USTR public hearing on intellectual property rights for software in Brazil, on the rules for software trade in Brazil and on restrictions to foreign investment in the Brazilian IT sector.
17 June 1987 – Mexico City: discussion on the changes made in the software bill by the Brazilian House of Representatives.
24 June 1987 – The Brazilian House votes on the software bill.
30 June 1987 – The US government suspends the part of the Brazilian 301 case dealing with intellectual property rights and asks the USTR to continue to pursue the investment part, but without deadline.

15 July 1987 – Hearings before the US Subcommittee on Commerce, Consumer Protection, and Competitiveness of the Committee on Energy and Commerce, House of Representatives.
July 1987 – The SEI turns down a request by Brazilian companies to license Microsoft's MS-DOS on the grounds that a local similar product had already been developed.

6 November 1987 – Some Senators from the US Democrat Party introduce a resolution in the US Senate recommending the White House to start sanctions against Brazil because of the IT policy.
13 November 1987 – The US government announces retaliation by imposing punitive tariffs on US$105 million worth of Brazilian exports to that market. The list of the products affected was not announced.
November 1987 – Brazilian authorities call local subsidiaries of foreign companies to discuss the imposition of punitive tariffs on Brazilian products.
18 November 1987 – The Brazilian Senate votes on the Senate substitute for the software bill approved by the Brazilian House.

4 December 1987 – The Brazilian House votes on the Senate substitute and approves the Software Law No. 7646.

January 1988 – The CONIN discuss SEI's previous decision against local trade of Microsoft's MS-DOS. The decision is to allow local trade of the version 3.3 of that operating system.
– The American Electronics Association (AEA) asks the USTR to postpone the economic sanctions for 120 days.

February 1988 – The US government announces the suspension of the retaliatory measures.

13 May 1988 – The procedures for implementation of the Brazilian Software Law are published.

6 June 1988 – The Brazilian President José Sarney goes to the US for the UN General Assembly's opening session. In a press conference he criticised the Brazilian IT policy and declared he would not vote for it again as he had done in 1984.

January 1989 – The Ministry for Science and Technology, under which the SEI was placed, is closed down and its tasks transferred to the Ministry of Industry and Trade, at the time headed by a notorious opponent of the IT policy.

March 1989 – Creation of the Special Secretariat for Science and Technology directly subordinate to the President's Office.

October 1989 – The US government drops the Brazilian 301 case.

NOTES

1 This has not been so for the whole of the electronics industry. The nationalist policy towards the lower end of the computer market and most of the professional electronics industry has not had the same positive results in the semiconductor industry. The development of the latter was hindered by the liberal government measures towards the foreign-dominated consumer electronics industry. The local telecommunications industry has also developed through a policy oriented by a particular concept of national business that has not affected foreign control over companies. Various segments of the Brazilian electronics industry have been subject to policies implemented in different and sometimes conflicting ways by independent state agencies.

2 For a detailed description of the development of the IT policy see Helena 1980, Piragibe 1985, Adler 1986 and Dantas 1988.

3 The effort to create an institutional structure and an explicit policy for science and technology, rooted in the 1950s and early 1960s, benefited from the prosperity of the economic 'miracle' and resulted in a significant amount of resources for higher education, graduate courses at home and abroad and research grants (Langer 1989). A considerable segment of highly trained individuals was formed and, partly because of the difficulties in getting a job compatible with their ability, many of them developed a critical opinion of the poor results of a science and technology policy which had no strong links to industrial development. In the IT area they were called the 'frustrated nationalist technicians' (Evans 1986:792).

4 The contribution of IT policy to global macro-economic goals was substantial. Computer imports in 1969 were worth US$13.3 million. In

1974 this grew to US$99.8 million and became the third highest import item among manufactured goods. In 1975 Brazil spent US$111.9 million on computer imports (Piragibe 1985:121 n. 12).

5 For a detailed account of the participation of the Brazilian intelligentsia, the planning community and the Navy in this initial phase of the IT policy, see Adler 1987 and Dantas 1988.

6 One of the 1974 SECOMU's conclusions was that the domestic IT sector should be linked to the local endeavour of mastering the whole technological knowledge of computer manufacturing, from the central processing unit to electronic components. A recommendation was issued to government agencies stating that special incentives should only be given to nationally controlled firms that gave proof of using genuine national technology. The 1976 SECOMU proposed the reformulation of the IT policy in order to exclude the association to foreign capital and suggested direct state control over an industrial firm.

7 See Katz 1981, Westman 1985, Adler 1986, Evans 1986, Botelho 1988.

8 CAPRE's Resolution No. 1, of July 1976, that started the market reserve for local capital in the area of mini- and microcomputers and their peripherals, was openly affronted by IBM's decision to enter the local market with its minicomputer /32. The foreign subsidiary rejected CAPRE's suggestion for it to export the whole /32 production and, taking advantage of its privileges in obtaining generic import licences, imported the /32 components, started a marketing campaign and even collected 400 statements of interest. It was only in January 1977 that a decision from the Brazilian Council for Economic Development (CDE) gave CAPRE the authority and strength to curb IBM's initiatives.

9 This decrease was sharper than the fall in production and also reflected the changes in US income tax legislation that stimulated capital repatriation. Net foreign investment in Brazil continued to decline during the years of production growth: in 1985 it was US$765.1 million and in 1986 it was US$203.6 million (Serra 1988:126).

10 See Porteous, Chapter 8 of this book.

11 See Section 3 of this chapter for details.

12 SNI's National Intelligence School (ESNI) had been doing experiments, since the early 1970s, to develop a cryptograph that could enhance information safety. From this experience resulted the conviction that the IT policy's goals would never be achieved unless the development of micro-electronics was stimulated also.

13 The group in charge of investigating the activities of CAPRE, two members of which later became chairman of SEI, was negatively impressed by the fact that, in 1977–78, the central agencies in charge of IT policy considered the convenience of a joint enterprise between Fujitsu and the Federal Data Processing Service (SERPRO) as a means of allowing a foreign firm other than IBM to act in the local market of medium computers, competing with the Big Blue subsidiary. The group also judged as too liberal the decision made by the planning minister at the end of the Geisel administration to allow local production of IBM medium-sized computers.

14 One of the IBM projects for entering the Brazilian market with medium-sized computers, rejected by CAPRE at its last meeting, was presented to

SEI which approved its local production and trade in very strict conditions. The same occurred to a Hewlett-Packard desktop microcomputer. SEI's decisions were justified by pragmatic reasons. Local demand for such computers could, at that time, only be met by imports.

15 Domestic computer firms were investing a significant amount of their sales in R&D, reaching an average share of 11 per cent in the period 1979–82. Although impressive in Brazilian terms, research investment was still minimal by international standards. In 1982, COBRA, the only state-owned and then the biggest computer manufacturer in Brazil, had a research budget of US$10 million, an amount representing only 12.5 per cent of Data General spending in R&D during that year (Peritore 1988:33).

16 The military regime candidate, Paulo Maluf, who declared himself against the market reserve, had at that time a chance of being elected President of the Republic in the indirect election of 15 November 1984. The bill restoring direct presidential election had been rejected and the possibility of the candidate of the opposition winning in an indirect election, despite being acclaimed in popular rallies, was not yet clear.

17 For a summary of negotiations between the US and Brazil governments, see Section 3 of this chapter.

18 Price performance in the reserved segment of the Brazilian electronic complex seems to have been an argument largely used in the struggle over the IT policy, with opposing sides using different data as evidence for their positions. More balanced assessments have come to light very recently and make a positive mark for the policy's supporters. See Chapters 2 and 10 of this book.

19 There are five electronic conglomerates in Brazil, namely, Itau, Machline (SID), Elebra Electronica, ABC Sistemas and Cataguases Leopoldina. The first two are the biggest, with significant participation in the professional and consumer electronics, components and telecommunications segments. The Itau conglomerate includes the second largest Brazilian private bank, Itau Bank, whose strategy is to increase its investment in the electronics industry in order to change its present 75–25 per cent between financial and industrial assets to a 50–50 basis (Tigre 1988:125). Recently it acquired Philco with which it has entered the segment of consumer electronics. The Machline conglomerate is linked to the largest Brazilian private bank, Bradesco Bank, through SID Informatica which is its professional electronics subsidiary. The merger of medium-sized firms with the conglomerates is exemplified by the case of Scopus, one of the innovative national firms in the professional electronics industry, which sold 70 per cent of its voting shares to Digilab (of the Bradesco group) in April 1989. One year before, Digilab joined Itautec (from the Itau conglomerate) as a shareholder in Sistema/Rima, a firm manufacturing printers.

20 In fact, what Law No. 7232/84 establishes is the end of import control by SEI in 1992. After that year, import control can be made by CACEX, the foreign trade department of Banco do Brasil, or left to SEI for another period. In this case a new bill would have to be introduced in the Brazilian Congress. Nevertheless, the duration of the restriction to local production by other than nationally-owned firms is not defined by the law (see Art. 9 *vis-à-vis* Art. 8 VI).

21 The US/Brazil conflict, also called by the USTR 'the Brazilian 301 case', was dropped in October 1989 when two other cases against Brazil were already proceeding. In July 1988, the USTR had opened a case initiated by American pharmaceuticals companies because of lack of patents rights in the Brazilian market. By mid-June 1989, the USTR initiated investigations on 'quantitative restrictions and import licensing of agricultural and manufactured products' in Brazil, one of the three 'priority countries' for the US to negotiate the liberalisation of their policies. The objective and conditions for these negotiations were established in accordance with the 'Super' and 'Special' Section 301 of the Omnibus Trade and Competitiveness Act of 1988. The other two countries are Japan and India. Brazil has also headed the 'Priority Watch List' with respect to intellectual property protection. In November 1989, the USTR started the revision of the status of each of the eight countries in the list to identify the extent of their denial of adequate and effective protection of intellectual property rights in order to orient the bilateral negotiations for changing the situation considered harmful to the economic interests of the United States.

22 The Brazilian Council for Economic Development (CDE) was led to establish that a proportion of local capital and the use of local technology were conditions for approval of IT industrial projects. Covered by CDE conditions, CAPRE selected nationally-owned firms in a bid for minicomputer production in Brazil. Data General, a firm that had no significant position in the Brazilian market, wrote to the USTR complaining about Brazilian restrictions on the import of minicomputers and saying that the IT policy was a bad example to other countries.

23 An analysis of alternative explanations of the attitude of the US government towards Brazil and the implications of the self-initiation of the Brazilian trade case by the President of the US can be found in Evans 1989.

24 These arguments were expressed in very aggressive terms in dozens of speeches, newspaper articles and books. See, for instance, Campos 1985.

25 The present Department for Informatics and Automation Policy (DEPIN), into which the SEI was transformed, is part of the Secretariat for Science and Technology (SCT). Its attributes are not yet clear as SEI's tasks were transferred to the SCT and not to the DEPIN and therefore are formally in the hands of the Secretary of the SCT. These changes were established by the Provisional Measure No. 222, of 11 September 1990, and its corrections published on 13 September 1990.

26 Among the items excluded are line printers, keyboards, electronic telex terminals, stored programme control for telex and telephone exchanges, digital control for brakes and anti-swerving, digital systems for the manufacture of paper, and industrial robots.

27 This decision was made under strong protest by ASSESPRO. ASSESPRO claimed that SEI violated Article 12 of Law 7232 because, in this case, the local partner obviously has no control of the technology to be used by the new firm.

28 The filing of any software would be automatically renewed after three year's time independent of SEI's decision, and the agency would have to make a decision about all filing within 120 days' time.

29 On 13 September 1990, the Brazilian Finance Minister and the Secretary for Science and Technology issued a joint decision, 'Portaria' No. 536, to create a special interministerial commission which would have thirty days to propose improvements to the Law No. 7646/87, the Software Law. Article 4 of the 'Portaria' defines the directives for the preparation of the changes of the law: (I) elimination of any restriction to the distribution of software within the country, (II) replacement of the local equivalent concept by other instruments, (III) assessment of the present system of registration of software.

30 There was, however, a weak link in the chain that bound the military and the intelligentsia within and outside the state apparatus. This was the latter group's left-wing discourse that left open a path for the policy's opponents to attack the otherwise untouchable military nationalism. Within a right-wing military administration, the IT policy backing by left-wing researchers and university teachers gave the antagonists of the policy, especially those linked to the interests of foreign capital, a wedge to try to break the consistency of the policy. This sensitive point was explored in the transition from CAPRE to SEI.

31 The conflict also inhibited the Brazilian state initiative from extending to other hi-tech areas, such as biotechnology and fine chemicals, the same political orientation as in the IT sector.

32 Such change is incorporated in the Second Plan for the Development of Informatics and Automation, the II PLANIN, due to be raised in the Brazilian Congress. In fact, this has been postponed since April 1989, when the I PLANIN was extended for another six months. It did not begin in October of that year probably because of the proximity of major political changes expected to be brought about by either one of the two stronger candidates for the presidential election. The new Collor administration's II PLANIN defines some changes in the IT legislation (Informatics Law and Software Law), reduction of the protection given to the industrial sector and the selection of the micro-electronics and software segments as national priorities. Before the discussion of the II PLANIN by the Brazilian Congress, however, important changes had already been made in the IT policy directives. The revision of the Software Law, the Bill defining 29 October 1992 as the date for the end of the control of local manufacturing of IT goods and replacing the concept of national company defined in the Informatics Law by the one defined in the Brazilian new constitution: all these changes were launched before the plan that proposes them was approved.

BIBLIOGRAPHY

Adler, E. (1986) 'Ideological guerrillas and quest for technological autonomy: Brazil's domestic computer industry', *International Organization*, Vol. 40, No. 3, 673–705.

Azevedo, C. and Zago, G., Jr (1989) *Do Tear ao Computador. As Lutas pela Industrialização no Brasil*, São Paulo: Politica Editora.

Botelho, A.J.J. (1988) 'The political economy of professionalization: the birth of the Brazilian informatics policy', paper presented to the 4S/EASST Joint Meeting, Amsterdam, 16–19 November.

Campos, R. (1985) *Além do Cotidiano*, São Paulo: Record.

Carneiro, D.D. (1986) 'Perspectivas del endeudamiento externo Brasileno 1986–1990', in Ricardo Ffrench-Davis and Richard E. Feinberg (eds) *Mas Alla dela Crisis de la Deuda. Bases para un Novo Enfoque*, Santiago do Chile: Ediciones CIEPLAN.

Ciencia Hoje (1989), Vol. 9, No. 51.

Colson, F. (1985) 'New perspectives on the Brazilian computer industry: 1985 and beyond', *Multinational Business*, No. 4: 1–9, UK, EIU.

Dados e Idéias (1987), Vol. 2, No. 111.

Dantas, V. (1988) *Guerrilha Tecnológica. A Verdadeira História da Política Nacional de Informática*, Rio de Janeiro: Livros Tecnicos e Científicos Editora.

Evans, P.B. (1986) 'State, capital, and the transformation of dependence: the Brazilian computer case', *World Development*, Vol. 14, No. 7, 791–808.

—— (1989) 'Declining hegemony and assertive industrialization: US–Brazil conflicts in the computer industry', *International Organization*, Vol. 43, No. 2, 207–38, Spring.

Fregni, E. (1988) 'Política de informática. É preciso privilegiar o essencial', paper presented to the Seminar on Informatics: Present and Perspectives, organised by APPD, 29 October, Rio de Janeiro.

Griffith-Jones, S. (1989) 'A history of debt crisis management', in S. Griffith-Jones (ed.) *Third World Debt: Managing the Consequences*, London: IFR Publishing Ltd.

Helena, S. (1980) 'A indústria de computadores. Evolução das decisões governamentais', *Revista de Administração Publica*, Vol. 40, No. 3, 673–705.

Katz, R.L. (1981) *Nationalism and Computer Technology Transfer: The Brazilian Case*, M.Sc. dissertation, Massachusetts Institute of Technology, Cambridge, Mass.

Langer, E.D. (1989) 'Generations of scientists and engineers: origins of the computer industry in Brazil', *Latin American Research Review*, Vol. 24, No. 2, 95–111.

MCT (1986) *Relatório dos Grupos de Trabalho de Recursos Humanos em Informática*, Ministério de Ciencia e Tecnologia, October, Brasilia.

Mineiro, P. (1987) 'A guerra da tecnologia', *Caderno do Terceiro Mundo*, No. 101, August.

Peritore, P.N. (1988) 'Hi-tech import substitution: Brazil's computer policy and the possibilities for social transformation', *Alternatives*, Vol. 13, No. 1, 27–54, January.

Piragibe, C. (1985) *Indústria da Informática. Desenvolvimento Brasileiro e Mundial*, Rio de Janeiro: Editora Campus.

SEI (1987) *Panorama do Setor de Informática*, Brasilia.

—— (1989) *Panorama do Setor de Informática*, Brasilia.

Serra, J. (1988) 'Conversão da dívida: realidade e alternativas', *Revista de Economia Politica*, Vol. 8, No. 1, 124–31, Jan./Mar.

Suzigan, W. (1989) 'Restruturação industrial e competitividade nos países avançados e nos NICs Asiáticos: lições para o Brasil', in W. Suzigan *et al.*, *Restruturação Industrial e Competitividade Internacional*, 7–32, São Paulo: SEADE.

Tigre, P.B. (1988) 'Business strategies in the Brazilian electronics industry', in C. Piragibe (ed.) *Electronics Industry in Brazil. Current Status, Perspectives and Policy Options*, Brasilia: Ministry for Science and Technology, Centre of Studies in Scientific and Technological Policy, November.

Westman, J. (1985) 'Modern dependency: a crucial case study of Brazilian government policy in the minicomputer industry', *Studies in Comparative International Development*, Vol. 10, No. 2, 25–4.

10 Learning in industry and government

Achievements, failures and lessons[1]

José Cassiolato
Tom Hewitt
Hubert Schmitz

1 INTRODUCTION

Hi-tech does not offer quick fixes for the problems of industrial development. Using and producing it efficiently means embarking on a difficult learning path. Indeed, learning is the unifying theme that runs through the preceding chapters of this book. The purpose of this final chapter is to pull together the observed learning experiences in industry and to draw conclusions about how they were fostered or inhibited by government policy.

So as to bring out the general policy relevance of this discussion on learning, it is organised around the following questions:

- Should government policy in LDCs target producers or users of hi-tech?
- At what stage in the development of new technology should LDC governments seek to foster local supply of that technology?
- What criteria can help subsectoral targeting in building up a domestic supply capability?
- What is the relative effectiveness of explicit and implicit policies for generating skills in industry?
- What is the most effective way of making foreign firms contribute to the domestic accumulation of skills and know-how?
- What explains the observed learning in government and what is its significance for the industrial policy debate?
- What is the political room to manoeuvre for sustaining policies which restrict imports and the participation of foreign firms?
- What are the implications of short-run stabilisation measures for long-run technology policy?

This is an ambitious if not unwieldy list. Our objective is not to give each question an exhaustive treatment, but merely to pull together explicit or implicit answers that can be derived from the preceding chapters. In some cases, additional empirical information is provided in order to develop the argument further. Where appropriate we also draw on the experience of other countries.[2]

This chapter remains, however, anchored primarily in the Brazilian experience, which means that there are some limitations to the lessons one can draw. These arise from the size of the Brazilian economy and its stage of industrial development. Clearly few LDCs have as large an internal market as Brazil, which can have implications for policy. For example, it is easier to have both a market reserve and competition between producers. The relevance of most of the policy discussion in this chapter is not however conditional upon large internal markets.[3]

The stage of industrial development plays a more important role. In fact, the discussion in this chapter is geared to countries with a minimum threshold of industrial development, human resources and infrastructural support. Without these, hi-tech equipment can rarely be produced or used efficiently. In practice, this means that this chapter – as previous ones – is of relevance primarily for semi-industrialised countries.

2 POLICIES FOR PRODUCERS OR USERS? A FALSE DICHOTOMY

Controversy has been the hallmark of Brazil's policy for informatics since its inception. It has been argued consistently that the producer orientation of the policy has worked to the neglect and disadvantage of users.[4]

The analysis of preceding chapters leads us to suggest that the division between a policy for users and a policy for producers is a false dichotomy. This becomes clear when we unbundle the issue into static and dynamic concerns. Static concerns have concentrated on the high price charged by local producers which are said to have an adverse effect on users' competitiveness and on the diffusion of informatics goods. Let us recapture the evidence.

Chapter 2 by Schmitz and Hewitt includes an extensive discussion of differentials between local and international prices. It shows that the price of locally made equipment tends to be double the US price, but that opening up the Brazilian market to foreign firms would not necessarily reduce the cost to the user. Mainframe computers and other electronic equipment, not included in the market reserve for national firms, show similar price differences as compared with the US market.

There is further evidence on price differentials which is worth referring to. A recent study by Teixeira (1990) provides the most thorough account of how prices have developed over the period 1984–90, both for protected Brazilian products and for their IBM equivalents in the US. Her study shows average real price declines for a variety of Brazilian PCs ranging between 6.1 per cent and 9.4 per cent per quarter. These price reductions match, and in some cases slightly surpass, those in the US. Nevertheless, given the difference in starting positions, Brazilian prices have remained above US levels.

One of the reasons is the high price of inputs as shown in Chapter 2. A further issue to be explored is whether oligopolistic practices in a protected market are responsible for the high prices charged by national firms. Two facts are relevant here. First, there has been a gradual but constant deconcentration in the industry: in 1984 the five largest IT firms were responsible for 48.18 per cent of total sales of computers and peripherals covered by the market reserve; in 1988 the five largest were responsible for 38.46 per cent. In microcomputers, in 1988, no firm was responsible for more than 20 per cent of total market (both in value and units sold) (SEI 1989). Second, recent research suggests that price is not the main competitive weapon in the Brazilian IT industry. Baptista *et al.* (1990:191–212) conclude that firms producing microcomputers put more emphasis on improving technical services for users and reliability of the product. As regards pricing policy, it is true that most firms in the sector operate with a high mark-up. However, far from being a special characteristic of the sector, such high mark-ups over costs have been a general feature of Brazilian industry throughout the 1980s.[5] Hence, the point to be made is that such price-setting behaviour of IT firms is indeed partly responsible for the US–Brazil price differential. But rather than a specific consequence of the market reserve, this behaviour is a general response of Brazilian industry to inflation and economic crisis.

In comparing Brazilian and US prices, it is also worth recalling that the latter do not represent international prices. For example, according to our calculations, UK prices in 1989 were between 30 and 60 per cent above US prices (see Chapter 2). Similarly, Meyer-Stamer (1990) suggests that prices in Germany are substantially above those in the US. Thus while comparisons with US prices are important to monitor Brazilian prices, the price gap is a misleading measure of the actual cost of the existing policy regime to the consumer.

This becomes even more apparent in comparisons with other Third World countries where a market reserve for national firms does not exist and where price differentials are similar to, if not greater than, those between Brazil and the US. Tigre (1988) quotes a study which shows that PCs manufactured by IBM in Mexico are sold at a price which is 267 per cent higher than that of the same IBM product in the US. The Argentinian example is a very interesting one. In the absence of locally manufactured products, Cline (1987) found that in 1986 prices of IBM microcomputers were 3.7 times higher than in the US and suggested that the existence of a 100 per cent tariff was the main cause of the discrepancy. Azpiazu *et al.* (1990), however, give several examples where the same disparities were found with a much lower level of tariffs, suggesting oligopolistic practices as the main reason for the sharp differentials. In fact they point out that price differences were reduced considerably in 1987 when a promotion scheme was introduced and local manufacturers started to produce IBM clones (p. 26).

Oligopolistic practices by subsidiaries of international firms also seem to occur in the area of software (see chapters by Gaio and Bastos). A survey of price differences of 14 major general-purpose software products which can be freely imported in Brazil found that all are significantly overpriced in Brazil. As shown in Table 10.1 Brazil–US price differentials range between 2.5 (Supercalc and Framework) and 4.1 (Wordstar 2000).[6]

Static price considerations are frequently combined with the criticism that locally produced equipment is of low quality and

Table 10.1 Software products: differences in prices, Brazil/US, September 1989

Software	*US price (US$)*	*Brazil price (US$)*	*Brazil price/US price*
MS-Word 4.0	209	740	3.5
Word Perfect 4.2	219	715	3.2
Wordstar 2000	248	1,036	4.1
Dataflex	409	945	2.3
dBase IV	453	1,261	2.7
Paradox	429	1,647	3.9
Lotus 1–2–3	299	1,135	3.7
Quattro	149	570	3.8
Supercalc	305	780	2.5
Chartmaster	205	600	2.9
Freelance Plus	320	1,055	3.3
NS-Chart	255	794	3.1
Framework	429	1,081	2.5
Symphony	440	1,473	3.3

Source: Porto, Silva and Laplane (1990)

therefore diffuses slowly. Even when real price differentials are taken into account, the diffusion of computers produced by national firms has been very rapid, much higher than in other developing countries where free imports are allowed (see Table 3.1 in Chapter 3). According to the US Department of Commerce (1987) not only was the Brazilian market for microcomputers and software one of the fastest growing in the world throughout the mid-eighties, but also, as Figure 10.1 shows, it represented one of the highest percentage shares of Industrial Product in LDCs. Thus, by 1987, the Brazilian market for microcomputers represented almost 1 per cent of its industrial product, more than most other countries in Europe and the Far East.[7]

So where does this leave users? The international comparison of price and diffusion levels does not add up to a picture in which Brazilian users are the victims of a producer-oriented policy. Our

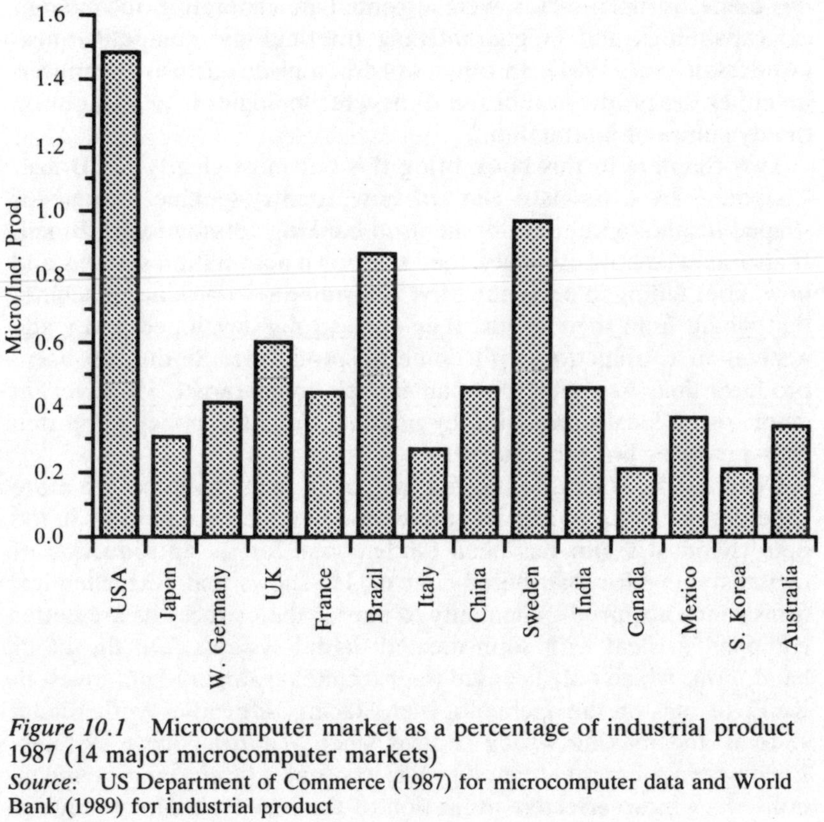

Figure 10.1 Microcomputer market as a percentage of industrial product 1987 (14 major microcomputer markets)
Source: US Department of Commerce (1987) for microcomputer data and World Bank (1989) for industrial product

main point, however, is that a dichotomy between producer and user concerns is unhelpful. Interaction between producers and users seems essential. This is most obvious for maintenance and technical assistance. Carvalho, for example, shows in Chapter 5 that the higher cost of locally produced equipment used in the petrochemical industry is offset by better maintenance services for such equipment.

Maintenance is only part of a dynamic relationship between producers and users. Learning by interaction is seen in various chapters as essential for attaining competitiveness. Such learning rarely develops if equipment is imported from far away. Local production facilitates user–producer loops. Case studies carried out in the context of Nordic countries confirm this. For instance, Denmark's specialisation in dairy machinery, Sweden's in metal-working and wood cutting technology and Norway's in fishery technology cannot be credited merely to their factor-endowments. Closer interaction between producers and a competent and demanding domestic user sector were essential in generating technological capabilities and in guaranteeing international competitiveness (Anderson *et al.* 1981). In other words, to place exclusive emphasis on either the production or use of new technologies is to lose out on the dynamics of interaction.[8]

Two chapters in this book bring this out most clearly for Brazil. Chapter 3 by Cassiolato showed how country-specific user needs shaped technological developments in banking automation in Brazil. It also showed how users designed their own automation systems and how, after failing to persuade MNC subsidiaries to produce machines that would fulfil their needs, they successfully developed and made systems in conjunction with domestic producers. Setting up user–producer links has never been an explicit policy target. However, in emphasising local production by national firms the policy permitted user–producer links to emerge.

Chapter 6 by Carvalho showed that, even in the absence of a more coherent policy for industrial automation, competence of users in the petrochemical sector has been fundamental for the introduction of digital systems for distributed control. He shows that petrochemical firms which acquired a capability to master their processes are better equipped to deal with sophisticated digital systems. On the other hand, firms which only licensed their technology and did not invest in R&D to master the technology are facing difficulties with digital systems and making wrong choices when acquiring these systems. Even where the right technology was chosen, it often remains under-utilised. A more effective utilisation of the new technology requires that producers develop user-specific solutions. To do so, close interaction between users and suppliers of both hardware and software is essential. Carvalho's policy point is therefore not that the build-up of a local hardware industry is mistaken. His critique is that the market reserve, which is oriented towards suppliers, is not enough, that it does not feed into a general industrial policy which stimulates synergies between users and producers. This issue of learning by interaction is pursued further in the next section.[9]

3 LEARNING AND STAGE OF ENTRY

This section draws attention to the special significance of interaction between users and producers at an *early stage* in the development of new technologies.

A crucial question faced by policy makers in the newly industrialised countries is at what stage (if any) they should seek to foster the local supply of new technologies. Would it make more sense, in the early stages, to rely on imports of such goods or to enter hi-tech production? What would be the relative costs and benefits of both strategies? There are two competing approaches to this question, which were set out in the introductory chapter but are worth recalling.

Neo-Schumpeterians would argue that periods of technological transition are the most appropriate for newly industrialising countries to try to catch up with the leaders. An earlier and fuller assimilation of the new technologies combined with crucial institutional innovations in firm organisation, financial and credit systems and the development of qualified human resources would be essential parts of a 'national system of innovation' which is fundamental for the establishment of the international competitiveness of any economy.

By contrast, neo-liberals and other followers of product cycle theory argue that entry at the early stages of a technological breakthrough implies considerable cost-related inefficiencies. In other words, policies that fostered early entry represent inefficient attempts to 'reinvent the wheel'. Efforts to localise production of hi-tech goods in the Third World would merely reproduce previous vintages already developed in more advanced economies.

Thus, for neo-liberals, lack of externalities militates against an early entry into new areas such as information technology, biotechnology and new materials, while for neo-Schumpeterians it is precisely the entry at this stage that would generate country-specific externalities, particularly through interactions between users and producers.[10] Here we will examine the implications of early entry.

The absorption process of a new technology is itself a key determinant of the direction in which technologies evolve. In other words, the diffusion of a technology and its development are simultaneously determined within any competitive market economy (Metcalfe 1986).[11] The environment shapes the pattern of technical evolution which in turn redefines the environment. Different diffusion environments generate different patterns of technical advance (Georghiou *et al.* 1986). This means that there are country/region specificities in technological design.

The fact that technical problems arising during the innovation/diffusion process require joint collaborations between producers and users has some major implications, particularly that competitiveness of the producer becomes structurally linked to the competitiveness of the user (Perez 1988).

This is why the early stages of the development of a technology are so important for Third World economies according to neo-Schumpeterians (Freeman 1982). The proximity of the users to the suppliers of an innovation plays an important role in the design and de-bugging of new products, especially hi-tech products. For this process to be successful, or even possible, the supplier needs ready access to a reasonably large constituency of potential users with the complete range of skills necessary to apply the innovation successfully or the expertise necessary to formulate problems for the innovator to solve (Walsh 1988).

What is significant for the new technologies is that as the process of technological innovation, particularly in its early stages, comprises a significant degree of uncertainty both at a technical and at an economic level, users may be positioned in such a way as to actually shape the subsequent paths of technical development if they possess the minimum necessary level of economic and technical knowledge and skills.

The geographical proximity between groups of users and producers constitutes a comparative advantage, which increases with the complexity of technical links to be established between them and with the degree of non-standardisation of production. This type of advantage has a more significant effect the more radical the innovation and the earlier in its life-cycle the industry concerned (Lundvall 1983, 1985, 1986).

These issues have been explored very little for the case of Brazil, even though its informatics industry has received much attention.

This is where the chapter by Cassiolato makes an original contribution to the debate. There it is argued that the success in designing banking automation products and systems geared to the specificity of the financial system of Brazil was achieved precisely because entry was made when the sector was in its early stages internationally. The unwillingness of subsidiaries of multinational corporations to tailor their products to the conditions of the local market led data processing departments of the most powerful banks to design their own automation processes. Later on these banks commissioned the production of hardware and software for their specific requirements. Decision to enter in the early stages was crucial to the competitive status of the industry in Brazil. But entry has to be accompanied by a sufficient level of scientific and technical capabilities not only in producers, as Chapter 3 on banking automation shows, but also in users. These capabilities are essential not only in the hi-tech sector but also in the other sectors of the economy, as stressed by Carvalho in Chapter 5 on the petrochemical sector.

The importance of users' competence in mastering their own technologies to speed up the introduction of new technologies locally developed is further reiterated in Chapter 4 by Gaio. She discusses the argument that software could be an optimal point of entry for developing countries into the IT complex, given its labour-intensive nature and relative maturity in some segments. Analysing the Brazilian example, she shows how in the more mature, general-purpose packages market foreign suppliers just set distribution channels in Brazil with limited externalities to other software activities. Local firms find it very difficult to enter this market segment. This contradicts the argument that developing countries hold comparative advantages in mature industries.

She also emphasises some of the problems occurring in the more dedicated software segment. In this market packages are written to meet individual user requirements. Recent studies have suggested the existence of a significant degree of heterogeneity in technological mastery across the Brazilian industrial sector at the end of the 1980s (Ferraz 1989; Coutinho and Suzigan 1990). In one way such variety helps the establishment of local independent software houses which can reap the benefits of proximity to users in pursuing dedicated software solutions. However, on the other hand, as Gaio points out, such diversity on the level of technological capabilities leads to differentiated and fragmented user–producer connections.

In sum, the dynamics of technological change are considerably enhanced by interaction between users and producers at an early stage of the development process. The nature of this interaction encourages the development of location-specific technologies which are more responsive to local conditions than imported technologies. The process of development through interaction at an early stage also encourages a build-up of capabilities which are essential for further rounds of innovation.

4 BACKWARD LINKAGES AND LEARNING

In the last two sections we have been concerned with the learning effects of forward linkages which arise in the interaction between users and producers. Here we make a similar argument for learning from backward linkages. Since the latter are explored very little in accounts of the Brazilian informatics industry, this section includes some empirical observations not available elsewhere. Our findings (presented below) lead us to conclude that there has been an increasing division of labour between producers of end products and suppliers. Specialisation and de-verticalisation of production have been significant. The process is far from complete but the speed at which backward linkages have developed is remarkable.

These conclusions have to be seen against the frequent observation that industrial firms in LDCs tend to verticalise their production since the industrial structure does not include supplier networks. As Katz (1987) pointed out, the empirical evidence shows that manufacturing firms in LDCs make much less use of subcontracting than their counterparts in advanced countries.

The first period of the development of the IT sector in Brazil shows a pattern of excessive verticalisation. Some authors suggest that such a trend is responsible for the inefficiency of the sector (Frischtak 1989). In fact, it was necessary to verticalise production in the early stages of the industry in Brazil owing to scale and tolerance requirements for inputs. In other words, firms which supplied mechanical parts, plastics and other inputs to other industrial sectors, such as the auto industry, were not interested in supplying to the infant IT sector or could not attain their more demanding specification level. As a result, at the beginning, IT firms had to produce even plastic boxes and most mechanical parts of their computers and peripherals. This had changed substantially by the end of the 1980s. A recent survey of suppliers to the IT sector shows a network of approximately 2,000 specialised firms supplying parts and components to the IT firms.[12]

What seems remarkable is that subcontracting in IT in Brazil has been accomplished in less than a decade, a much shorter period than was needed in Latin America in other less technology-intensive industries (such as the engineering sector) where technological capabilities were acquired primarily through licensing or joint ventures (Castaño *et al.* 1980; Katz 1987). In fact, Katz (1987) points out that the degree of subcontracting in LDCs attained under these circumstances seems to have increased at a slow pace, taking almost two decades to be achieved.

Research, carried out in March 1990 by the authors of this chapter, on the eight major Brazilian producers of computers and peripherals established that all went through a process of deverticalisation.[13] Specialisation of suppliers was achieved mainly through learning-by-doing since most of the new producers of parts and components were offspring of the first vintage of IT firms. The interviews showed that the new successful Brazilian IT firms and the new specialised suppliers of parts and components came about as spin-offs of the old ones. One national firm producing peripherals cited, among several others, the production of printed circuit boards which had recently been deverticalised and estimated that such external sourcing reduced output costs by 10 per cent.

Particularly telling is the example of a computer firm which entered the market relatively late. It was set up in 1985[14] and was able to start less verticalised than the old firms, because by then suppliers of mechanical and plastic inputs could be found. But, competing in the higher segment of the microcomputer market, this firm found that some parts were not yet available from outside suppliers. In particular, parts such as high-performance video terminals, which entail scope economies and require advanced design and engineering, still have to be produced internally.

These observations point to a pattern of collective specialisation whereby production subjected to scale economies has progressively become the realm of specialised firms and production subjected to scope economies is still closer to R&D-intensive firms. As a consequence, two waves of verticalisation can be observed. The first one, which has probably been completed, refers to the creation of hundreds of small firms which specialised in products and processes which are not technology-intensive such as assembling circuit boards and producing mechanical parts, plastics and so on. The second one, which is already beginning, refers to products which are more technology-intensive and which reap scope economies.

It should be pointed out that the foreign computer firms have contributed to this build-up of specialised suppliers. Not voluntarily, though, but because government policy ('law of similars') forced them to buy locally. Thus the two bigger multinational subsidiaries making mainframes in Brazil have around four hundred different internal suppliers of parts and components. Such parts integrate products which are exported and have had their technical quality certified by agencies and associations responsible for technical standards in the US and Europe (Eicher 1989:21–2; Proença and Caulliraux 1989:100). Multinational subsidiaries are also increasingly relying on national firms to produce OEM[15] equipment, such as monitors, disks, tapes and small computers for their systems. One of the subsidiaries interviewed by the authors acknowledged that during the early stages of the IT industry in Brazil quality and delivery time used to be a problem but it no longer is. This firm also mentioned one example whereby modifications specific to the Brazilian market were introduced. A national firm supplies this subsidiary with approximately 7,000 terminals per year on an OEM basis and the logic of the terminal was jointly developed according to specific requirements of the internal market.

Having emphasised the achievements in deverticalisation and specialisation, we also need to mention a major problem: these advances do not always translate into lower input prices. The main cause is that in *some* of the domestic input markets there is little competition. As mentioned in Chapter 2, there are cases of suppliers (some of which are multinational companies) abusing their monopoly or cartelised positions. Liberalisation of imports might therefore be essential in some product lines – as part of an overall industrial policy.

In conclusion, this final chapter is concerned with drawing together the various learning processes in the IT sector and drawing lessons from how they have been 'engineered'. This section drew attention to the learning which occurred in the course of deverticalisation of the IT sector. The observed specialisation and inter-firm division of labour is merely a recent example of age-old economic principles asserting themselves. This is what industrialisation is about. It would not have occurred in such depth, however, without an infant-industry policy for the national informatics producers and without local-content regulations for foreign computer firms. The limitations of this subsectoral strategy also become apparent. The creation of backward linkages includes supply industries outside the informatics sector governed by remnants of previous outdated government rules. Problems and lessons emerging from this lack of a more comprehensive industrial strategy are addressed in a later section.

5 EXPLICIT VERSUS IMPLICIT POLICIES FOR HUMAN RESOURCES

In this chapter the stress has been on processes of learning. The key question now is to examine the mechanisms responsible for these processes. The argument in this section is that implicit measures for the training of human resources have been more significant than explicit ones.

Implicit measures do not necessarily aim at human resources training as such. They consist of incentives (in particular trade, tax and monetary policy) which influence the production decisions of private firms. State investments in the education, training and research system are explicit measures. The two types of measure are both important and complementary but the combinations can vary. In the 1980s, few LDCs have been able significantly to increase explicit support; equally few have the resources to expand such expenditure in the 1990s. Brazil is no exception (see section 10 of this chapter). Short-term crisis management and the squeeze on public expenditure make a strong argument for additional reliance on the private sector. This is why we focus in this section on implicit measures. Brazil's own experience in the 1980s suggests that trade policy can be a very powerful instrument to bring about the training of human resources. Let us elaborate.

The human resources issue is particularly pertinent in the electronics industry. As O'Connor (1985:325) has argued, 'The computer industry is essentially a knowledge-intensive industry wherein skilled, highly trained scientific, engineering and technical labour power is probably the single most important asset.' Thus, the industry is characterised by an enormous demand for skilled labour. The shortage of supply of such labour has been well documented for the advanced industrial countries, particularly for software and hardware skills but also in areas of management (Ernst 1985; Soete 1985; Kaplinsky 1988).

The skill intensity is less clear for LDCs. The traditional view of the electronics industries of LDCs has been one based on direct foreign investment in offshore export platforms: foreign subsidiaries and their subcontractors used mainly cheap, unskilled female labour to carry out repetitive assembly tasks. Particularly the semiconductor and consumer electronics industry were the focus of the widely disseminated literature of the late 1970s and early 1980s on the 'new international division of labour'[16] (Fröbel *et al*. 1980).

The old pattern of labour use can still be found in certain regions and sub-sectors of electronics in LDCs, but there has also been a clear transition to a new pattern, documented in Hewitt (1988). It may be characterised as using a much wider range of skills and, in particular, a high proportion of skilled engineering and technical labour. The origin of this increased demand for skills is the attempt by LDCs to establish a local hi-tech sector and to stimulate a local capability in new technologies.

The employment and skill implications of such attempts in the case of Brazil are presented in Chapter 7 by Tom Hewitt. It reinforces the argument made in the book's introduction that labour in hi-tech needs to be viewed as a *resource* as opposed to a *cost*. Chapter 7 shows that in the last decade there has been a substantial increase in the skill base of the electronics industry within the protected segments of computer production in Brazil. This increase is not only due to the expansion of the industry but also to the way that the industry has developed: generating in-house capabilities to both design and produce computers and peripherals for a range of market segments.

This is not to argue that Brazil has no scarcity in informatics skills. Far from it. A study carried out by the Ministry of Science and Technology notes 'an enormous lack in the training and development of human resources for informatics in the country' (MCT 1986:1). The root is seen in the under-resourced public education, training and research system. While both the scarcity and the under-funding are real, it is not sufficiently recognised that a massive growth in human resources has taken place in spite of inadequate state provision. The bulk of the learning has not taken place in universities or technical colleges but in the computer firms themselves.

The training intensity of informatics firms is confirmed by M.T. Fleury.[17] Tax incentives for training (Law 6297) contributed to this, particularly to the formalised training. In our assessment, the unstructured training has been even more important, notably the acquisition of skills in the course of learning-by-doing. The market reserve created the space for this to occur. Firms which wanted to occupy this space and expand had little alternative but to take on new challenges with the labour that was available. It is true that some firms managed to make short-term profits by pursuing a 'screwdriver strategy' – mere assembly of copied products. Sustained success however required investment in human resources. It is worth recalling survey findings which show that local informatics firms spent approximately 10 per cent of their sales in R&D and engaged a higher percentage of their workforce in R&D than international firms operating in mainframe computers (for details see Chapter 2).

Indeed, not even the critics of the market reserve deny that skill levels were raised dramatically in the course of the 1980s.

What lessons can we draw from this? It would be wrong to conclude that public educational provision does not matter. Educational institutions and private industry complement each other. In the East Asian NICs, the state set favourable incentives for the private electronics sector and invested heavily in technical education.[18] The latter is difficult in most LDCs because of the shortage of adequate financial resources (O'Connor 1985; *Microelectronics Monitor*, October 1989). This is why the Brazilian experience of an implicit human resources policy is particularly interesting.

The essence of such an implicit policy is that the state structures earning opportunities in such a way that firms need to provide training in order to succeed. That structure was provided through the instruments of protection. As discussed elsewhere in this book, this protection has not been without cost and we have shown some products to have a higher price than on the world market. To some extent this higher price can be seen as users paying for the training necessary for producers to be able to put these goods on the market. Given the difficulties of establishing an effective and equitable tax system in LDCs, this seems an appropriate way of recovering the cost of training. It also runs counter to the argument of critics of protection who, as seen in Chapter 2, use the high price for users as proof of the inefficiency of protection.

Our emphasis on implicit measures has extra relevance where – as in Brazil in the 1980s – public finance for explicit measures is not available. But the point itself is not new. It merely reinforces project findings from cross-country comparisons on how government policies have affected the achievement of 'technological mastery' (IDRC 1978; see also Dahlman and Westphal 1982). The main conclusion was that implicit as opposed to explicit measures had more impact on technological capabilities, particularly in the early development of industry.

6 LEARNING IN GOVERNMENT

The need to address this issue arose in response to the influential neo-liberal position that there are problems inherent to government intervention such as corruption and inefficiency. Chapter 2 concludes that the Brazilian informatics policy does not support this. While errors were made, they were outweighed by a number of positive features in government performance. There was no corruption, administrative delays which were initially long were subsequently reduced to minimal levels, a capacity to negotiate with powerful internal and external opponents was built up (see also Chapter 9), there was a capacity to adapt and work out modifications to the policy regime, particularly in the late 1980s.

These achievements deserve special attention for a number of reasons: first, the common perception of government performance in this area is overly influenced by accounts of the messiness and conflictual way in which such a controversial policy is (inevitably) conducted. Second, such performance, since it is not common, demands explanation. For this purpose it would probably be helpful to draw on Arturo Israel's (1988) analysis of institutional performance. He suggests that 'specificity' and 'competition' are key determinants of how government agencies perform. The experience of SEI seems to support this, since it had a clear and specific mandate and was under constant pressure from opposing forces which can be considered a competition surrogate. This latter point is dealt with further in Section 9.

Third, the notion of learning in government is absent from large parts of the industrialisation debate. The importance of learning in industry is accepted by all schools of thought and an integral part of the infant-industry argument. Concern with learning to raise infants, however, is rarely on the agenda. An exception is the recent neo-Schumpeterian work on the national system of innovation by Lundvall and collaborators (see introductory chapter). Their work is concerned explicitly with institutional learning, but as a body of research it is still at a relatively early stage and has focused so far on the OECD countries. In the mainstream development debate, the issue has surfaced in Pack and Westphal's (1986) analysis of South Korea's industrial strategy. The success of that strategy is linked to a continuous process of government agencies accumulating industry-specific knowledge and adapting policy to changing circumstances.

In conclusion, there is a need to make 'learning in government' an explicit research issue. This applies to research on industrial policy in general and particularly so in the hi-tech field where the pace of change is so rapid. More than a research issue it is also a question of frame of mind. The international policy debate is overly geared towards prescription and blueprints. Creating a capacity to take informed decisions and implement them seems a more fruitful goal. Particularly in areas with a very fast rate of technical change, development opportunities are a moving target and government intervention needs constant rethinking. Otherwise good policies turn sour and inflexibility turns good results into brittle achievements.

7 PICKING WINNERS

One of the most difficult decisions is which sectors or subsectors to foster. Picking winners is particularly hard since one of the main characteristics of the new information technology is its pervasiveness. Carrier branches are therefore hard to select 'ex ante'. The difficulty is underlined by Brazil's own experience. Systems for banking automation turned out to be the most competitive subsector (see Chapter 3), not as a result of foresight but as something that emerged in the course of protection for a fairly wide range of electronics products. Most economies are smaller and have to focus more narrowly; even Brazil may have to do so in the 1990s. Such subsectoral targeting is inevitably a political process, but there are economic criteria which can usefully guide decisions.

The first relates to the size of the technological gap. The larger the distance between the national and international frontier the more difficult it is for any country to efficiently produce the goods in question. This is clearest where the bridging of this gap would require huge financial resources.[19]

The second relates to the degree of standardisation. The more standardised the hi-tech product, the less a 'self-reliance' type of policy would be necessary.

> The more easily a product can be bought off the shelf and serviced by a standard maintenance technician, the less it matters where or by whom it was produced. The more custom-specific the product, the greater is the importance of being able to count on direct service from the manufacturers themselves. Thus the custom and semi-custom ends of the range with the relevant services are the safer markets in the longer term.
>
> (Perez 1990:24).

Taking the first and second criteria together, it follows that imports tend to be a better solution for homogeneous goods which require high investments for their development and production. The clearest example is standardised microchips which involve high R&D expenditures and huge investments in automated plants: it is estimated that in 1990 a wafer fabrication facility would cost US$300 million and in 1985 alone, American and Japanese companies were spending around US$3 billion in R&D, all of which pose a severe barrier to entry for newcomers (Hobday 1990:22–4). There is little point in LDCs engaging in such production.

The third criterion follows from the previous discussion on interactive learning. The more user–producer connections matter in the development and mastery of the technology, the more advisable it is for a country to produce it internally. What seems to be realistic is to choose user sectors where some previous technological capability exists. The success of small developed countries – such as the Scandinavian ones – in developing their electronics industries was due to the fact that, instead of trying to compete in commodity products, they specialised in specific requirements of their user industries. Their strategy involved buying in fairly standard components or modules and assembling them into products and systems with a large amount of design and software involved (Lundvall 1989). As there are, in such countries, advanced user firms demanding applications of electronics technologies to specialised tasks, suppliers have been able to compete in their niches on product features, performance and service rather than price.

Recent trends seem to have increased the scope for such types of strategy even in microchips. This is the so-called shift to ASICs (application-specific integrated circuits). The traditionally integrated spheres of product design and fabrication of ICs appear to be decoupling and such a trend may well reduce the economic, technological and financial barriers to entry in the design and application of IC technology (Hobday 1990). The emergence of an independent design sector for ASICs could well mean that user capabilities will be essential for a rapid diffusion of integrated circuits. As a result, at policy level commodity 'semiconductor production should be discouraged but ASIC design capability would be encouraged' (Perez 1990:20).

In conclusion, the more important user–producer relationships are during the learning process, the more advisable it is for a country to start technological development and production. The relative gap in this case may not be so crucial since imports could not be the solution, because they involve some degree of 'inappropriateness', or are not backed up with specialised services, or because they simply would not occur.

The fourth criterion relates to the executive capacity of governments. 'The successful use of aggressive selective intervention requires the technical and administrative skill to implement desired strategies intelligently and the political ability to terminate intervention if there is no movement toward the desired goal' (Pack and Westphal

1986:124). Where such capacity does not exist, sectoral targeting makes little sense, even if a systematic view of technology suggests strategic intervention. Here we must however recall our earlier argument on learning in government. An executive capacity (capability to take informed decisions and implement them) does not fall from heaven. Building it up takes time and probably a fair amount of learning-by-doing.

8 LEARNING TO NEGOTIATE WITH FOREIGN CAPITAL

Access to foreign technology on favourable terms is an important way of speeding up the process of technological learning for late industrialisers. But negotiation with foreign firms over technology transfer is itself a learning process. In this section we will focus on this process by asking the question: What are effective means of making foreign firms negotiate technology transfer and other investment decisions on terms favourable to LDCs?

We will argue that the phasing of policy is important. Policies which are restrictive towards foreign capital in the early phase have the advantage of giving LDC industrialists and government the time and space to learn what they need and on what terms. At the same time, these policies force foreign firms with an interest in the local market to negotiate less restrictive terms of access. Such restrictions appear to have a pay-off when, in a later phase, it is possible to open up access gradually and selectively.

The relation between governments, foreign corporations and local capital is complex and has formed the centre of much of the debate around development strategies and policies.[20] In the electronics industry, large corporations from the US, Japan and, to a lesser extent, Western Europe dominate the technology and the markets for electronics goods and services. For LDCs then, interaction with foreign corporations is necessary and inevitable. However, for LDCs to be able to attain what they want from direct foreign investment, as opposed to foreign corporations simply achieving their own corporate goals, the former have to be in a position to negotiate.

Let us step back and ask why it is necessary to negotiate and set the terms with foreign capital. The answer can be found in the experiences of several countries in the past.

O'Brien (1985), for example, examined the extent to which multinational subsidiaries can, over the longer term, adequately substitute for building an indigenous capability. After extensively discussing the experience of countries which adopted a policy of attracting MNCs as the main means of developing the electronics sector, he concluded that 'this approach does not appear to be a viable one for effecting major improvements in these countries' capabilities for developing technologies' (O'Brien 1985:209–10).[21] The reasons for such failure have been: (a) MNC subsidiaries tend to carry out little R&D in LDCs; (b) MNCs have had little impact on the technical capabilities of indigenous firms since they have largely been 'offshore' manufacturing platforms for US and Japanese firms; and (c) because of the limited scope of their activities there has been little 'spin-off' effect, no perceivable movement of skilled personnel and little creation of technology linkages from MNC subsidiaries.

Even in larger European countries an OECD study found that governments had to rely on measures such as the use of real or threatened trade barriers and access to public purchasing (particularly in computers, telecommunications and semiconductor components) for getting MNCs to perform R&D locally (OECD 1980).

By contrast, Freeman (1987) compared these methods with the Japanese policy of rejecting foreign investment with local firms assimilating and improving upon imported technology. He observed that in this latter case, as development, production and marketing went ahead, the whole organisation was committed to the new products and processes in a way that was relatively uncommon in other countries.[22]

Apart from Brazil, the developing countries which have followed the most restrictive policies towards foreign capital in electronics are South Korea and India. In South Korea, according to Mody (1987), there has been a close collaboration between national political and industrial interests in electronics which gave a strong basis for restriction of foreign capital. As a result the Koreans have been able to 'negotiate much superior technology transfer deals' (1987:28). This is not a situation peculiar to the electronics industry in South Korea. Luedde-Neurath documents the extensive (formal and informal) controls over direct foreign investment in the economy as a whole which included: tight investment screening; extensive government interference in corporate decisions; and extensive reporting requirements (Luedde-Neurath 1988:85–8).

The Indian experience is more mixed. A study which examines the Indian computer industry up to the late 1970s concluded that the policy of exclusion of foreign firms had opened up opportunities for local development of industry (Grieco 1984). Similarly, Heeks (1991) argues that the restrictions on foreign capital in the 1970s helped to establish a local hardware industry with some internal capacity. In

1984 the policy regime was liberalised in a number of ways including some removal of import restrictions and duties and opening up to foreign capital. This liberalisation seems to have had a twofold effect. New firms established since then have relied mainly on screwdriver technology, that is the assembly of copied products, often licensed from foreign firms. In contrast, those national firms established in the 1970s appear to have negotiated agreements with foreign firms which further their internal technological capabilities (Heeks 1991).

Thus, the two countries besides Brazil which have had the most restrictive practices towards foreign capital push us towards the conclusion that negotiating power needs to be built up not only at the political level but also at the economic and technological level. How do these experiences compare with those of Brazil? In order to set this out, it is useful to recall the course of events.

In the early days of the informatics industry, foreign capital was uninterested in collaborating with Brazilian capital. By the end of the 1980s both foreign firms that already had production facilities in Brazil and those wishing to enter the market were more amenable to negotiate with Brazilian firms and the government over collaboration.

The details of the past inflexibility of foreign corporations are well documented (Tigre 1983; Piragibe 1985). Initially, the large computer firms were unwilling to license technology to Brazilian firms. Nevertheless, they were interested in operating within the Brazilian market as evidenced by the applications to the government by foreign firms to manufacture minicomputers in Brazil. National firms, in their turn, were forced to turn to smaller foreign computer firms as a source of technology. Even then, some of the licensors imposed marketing restrictions on the licensee (Tigre 1983:16 and 154).[23] Indeed, at the time of setting up the market reserve, the issue at stake was not the exclusion of foreign capital but access to technology and know-how on terms favourable to the development of capabilities locally. If the big firms had negotiated at all, a reserved market might not have been necessary. The informatics policy in its present form was fuelled by the reluctance of foreign capital to negotiate.[24]

In the intervening years, there appears to have been some pay-off to the policy of exclusion. Brazilian firms have built up a considerable technological capability not only in manufacturing but also in R&D (Cassiolato's Chapter 3 of this book; Hewitt 1988). They have also increased their bargaining strength with foreign capital. In the first international bid for licensed technology for microcomputers, in 1977, no leading foreign firm was interested in supplying technology to Brazil (Erber 1990a). In the second bid, for super-minicomputers in 1983, three large firms, DEC, Hewlett Packard and Data General, licensed technology to Brazilian firms. A recent survey of technological co-operation ventures between Brazilian IT firms and foreign suppliers listed 44 agreements involving 36 Brazilian and 34 foreign firms.[25]

Given the increased negotiating power which has resulted from the initial restrictions on foreign capital in Brazil, what do national firms and government want to negotiate over and what is the most effective means of going about it?

Many firms are interested in greater access to technology. There are various means which have been suggested in the past. These may include: licensing, joint ventures, screening by government, employment of foreign engineers and scientists, external training, and so on. The current debate in Brazil is centred on licensing and joint ventures. This is a controversial issue because, on the one hand, joint ventures carry with them the danger that foreign firms will capture markets already tested by local firms (Fung and Cassiolato 1976; Erber 1981) and, on the other, licensing is not a guarantee of access to technologies. In both cases, negotiating power is crucial.

The discussion in Chapter 3 about banking automation showed that the successful IT firms linked to the financial system are entering into licence agreements with foreign firms under the hypothesis that they are now better prepared to negotiate and have sufficient technical capabilities to reap real benefits from these arrangements.

On the other hand, the analysis of the introduction of electronics equipment in the Brazilian petrochemicals industry shows some of the shortcomings of licensing. Brazilian IT policy for internal production of process control equipment induced licensing as a means of a rapid access to frontier technology. However, foreign licensors were not willing to transfer the basic software of digital control equipment, nor have licensees invested significantly in mastering it. As a consequence, even if the quality of the nationally produced equipment seems to be reasonable, lack of capability in basic software has limited the chances of tailoring the configuration of the hardware to users' needs.

Of course the debate on licensing and joint ventures is not new. It has long been recognised that control over technology is an important competitive element and that it has a market value which the owners of technology will try to maintain (Cooper and Hoffman 1978:18). For developing countries, then, the value of access to technology through either means depends on the extent to which restrictive clauses can be avoided. In Luedde-Neurath's words,

> The dilemma facing all importers of technology is that if restrictive clauses are rejected outright, access to the desired technology may be refused. If on the other hand such clauses are too rapidly accepted, the value of the technology may be seriously undermined.
>
> (Luedde-Neurath 1988:92)

Such arguments were the rationale for early restrictions on licensing and joint ventures, as documented by Tigre (1983:164). The same arguments hold true today, but the balance of power has changed. The political conflict over the exclusion of foreign capital has been partially overtaken by the decision of the Collor government definitely to lift protective barriers by 1992. But even before this decision, some of the larger national firms with the capability to negotiate were keen to open up the protected market to gain greater access to technology.

Negotiations which are now occurring between national and foreign firms seem to be as much concerned with access to markets as to technology. For this reason, there is the possibility that internal technological capabilities will be of only limited advantage. It has been argued that some national firms are taking a quite different view of their position of bargaining strength (Banhos 1989). National firms and foreign corporations alike are looking ahead to the end of the reserved market with an eye to entering joint ventures where national firms have built up distribution and maintenance networks – both areas that foreign corporations would need a long time to develop.[26] The foreign firms, for their part, would bring superior technological capability, trade marks and export possibilities. In other words, the driving force behind national computer firms' strategy in the early 1990s may be more about access to foreign markets, via the established brand names and existing distribution channels of international firms than about continued strengthening of technological capability.[27] On the other hand, the rationale behind the behaviour of foreign firms in initiating agreements with national firms seems to lie in access to an internal market which even under severe crisis conditions showed a significant dynamism.

We should add finally that some of the chapters in this book show interesting evidence about the behaviour of user firms affiliated to multinational corporations. Chapter 5 by Carvalho suggests that at least one of the petrochemicals firms which exploited more effectively the dynamic potential of control system technology – locally produced by local firms – was the subsidiary of a major international enterprise. Chapter 6 by Silva on the automobile industry indicates that automation strategies of Ford's Brazilian subsidiary had to rely in some areas on locally produced equipment and that this did not harm its competitiveness.

These findings do not fit in with stereotype views according to which locally produced hi-tech goods would not meet the technical standards of such firms. On the contrary they support some of the main arguments of this book, namely that the effective use of hi-tech is positively correlated with the establishment of local production and innovation capabilities. Furthermore, the behaviour of foreign firms both in producer and user sectors suggests that the key policy issue is not whether they should be permitted to play a role in industrial development. More important is the issue of how a developing country induces multinational firms and their subsidiaries to contribute to building up technologically dynamic industries.

We can conclude that were it not for policy intervention, then the possibilities of bargaining with multinationals would not even be on the agenda. This is not to argue for a blind dismissal of foreign capital and the belief that developing countries can go it alone. On the contrary, foreign firms remain a most important source of technology. However, that technology is only useful in developmental terms if local firms, supported by the government, are able to find effective means of transferring it, and the know-how that goes with it, into local hands. From the above analysis, it appears that at least two prerequisites can assure this. The first is the build-up of a local capacity that knows what to do with the technology once it is obtained and the second concerns the timing of intervention and restriction on foreign capital in order to build up the capacity – the window of opportunity seems to be earlier not later.

9 LEARNING UNDER PRESSURE

This section seeks to draw out some political lessons from the Brazilian experience. As shown by Bastos in Chapter 9, the IT policy encountered increasing opposition in the course of the 1980s, from both outside and within the country. Indeed, internationally it has been one of the most heavily contested pieces of industrial policy. These pressures have contributed both to the relative success of this experience and to its limitations. Let us elaborate.

Suggesting that the opposition contributed to the success of the policy may surprise both supporters and critics. To explain, we refer to the earlier section on learning in government which concluded that the agencies in charge of the policy performed relatively well. As with other types of learning discussed in this chapter, learning in government is not automatic. That is, it does not happen simply as a function of time. We suggest that pressure from the opposition had – in this respect – a positive outcome: such pressure forced government agencies, in particular SEI, to address weaknesses (pointed out in Chapter 2). The fact that the policy was so severely and openly contested thus contributed to government performance, hence the heading for this section, 'Learning under pressure'.

Admittedly, this is a tricky argument since the pressure can become so strong that the policy disappears. As shown by Bastos in Chapter 9, it did not quite come to this. The pressure did however take its toll, particularly since the US threatened Brazil with trade war (*South*, November 1989). There was more to the conflict than access to the Brazilian markets for US computer firms, namely the possible demonstration effect for other LDCs. The US Trade Representative and the Department of State expressed concern that other LDCs would try to emulate the Brazilian model. There was also concern that the informatics example would spread in Brazil to other hi-tech sectors such as biotechnology and new materials (US Congress 1987:3 and 23). Indeed, it was not extended to these other sectors, largely owing to US pressure (Erber 1988). The compromises made in the software law, resulting in ineffective protection for national producers, also arose from this conflict (see Chapter 9).

What lessons of general relevance to LDCs emerge from this experience? Clearly national hi-tech policy which seeks to foster internal production and learning processes goes against the interest of leading international firms. The Brazilian case shows that it is hard to *sustain* such a policy against foreign opposition, particularly when the excluded foreign firms can muster support from their government. Most LDCs are more susceptible to foreign pressure than Brazil. Therefore, whatever the economic arguments for restricting imports or the entry of foreign producers, there are political limits to *sustaining* such a policy.

There are, however, ways of alleviating such pressure. By making the protection of national producers conditional upon performance, some of the wind can be taken out of the sails of the opposition. For example, export targets can be introduced which stipulate that a certain percentage of output has to be exported after a certain number of years and that otherwise the protection of the domestic market would cease. Such self-imposed conditionality can help industry to become more efficient and the policy to become more acceptable. In Brazil, government recognised by the end of the 1980s that exports of the informatics industry had to be raised but it stopped short of implementing measures which made protection conditional upon performance.

10 SHORT-TERM STABILISATION AND LONG-TERM INDUSTRIAL POLICY

In previous sections it was emphasised that a considerable amount of learning had occurred in industry and government by the end of the 1980s. This was achieved despite enormous difficulties: the policy was continuously under attack and the economy suffered various ups and downs.

We have argued above that the government and industry learnt to live with such adverse conditions and gained a certain resilience because of it. However, there are questions over whether the established platform is solid enough to cope with the changes to come. At the beginning of the 1990s, industry is forced to concentrate all its efforts on coping with galloping inflation and deepening recession, and government's main concern is the stabilisation of the economy.

These issues were on the agenda of industry and government throughout the 1980s, but were not then as acute as they are in the early 1990s. When all forces are absorbed by the need to survive in the short term, what space is there for long-run technological concerns? If in previous sections we have emphasised the importance of learning, we now have to address the issue of unlearning, that is the dismantling of capacity in industry and government owing to macro-economic crisis.

The conflict between short-run crisis management and the pursuit of long-run objectives is surprisingly little explored in the literature. The conflict is acute not just in Brazil, but has been the reality in a large number of countries. This is why it is taken up in this final section of the book even though we can do little more than show some direct and indirect ways in which long-run technology policy gets squeezed out by short-term crisis management.

In Chapter 9, Porteous investigates the differential impact which economic recession has on national and foreign firms' ability to produce computerised numerically controlled machine tools. His study concentrates on the recession which hit the sector in the first half of the 1980s, but his findings are of undiminished interest, not least because studies on the connection between economic crisis and technological capability are extremely rare. Porteous suggests that the recession shifts the competitive advantage towards foreign firms. National producers were severely hit by the recession whereas foreign subsidiaries were able to ride it out with continued support from the parent company. Government exacerbated the disadvantage of national producers by reducing – in some cases even withdrawing – long-run finance at low interest rates. These cuts occurred not by design, but by circumstance. In the wake of the country's foreign debt crisis capital flows from outside dried up. Internal tax revenue declined. Public sector borrowing was not allowed to increase (owing to pressure from, amongst others, the IMF and the World Bank). The resulting cuts in government expenditure hit disproportionately the availability of long-term loans for investment in technological development of national firms.

This conflict between short-term fiscal policy and long-term technology policy deserves to be explored further. Even if, as suggested in a previous section, implicit policy measures are more important for technical change than explicit ones, mechanisms for financing such activities have played an important role in firms' R&D investment decisions in Latin America. In Brazil the deficiencies of a long-term credit market and the almost complete lack of organised access to venture capital are seen as one of the main problems private firms have when they are faced with prospective investments in general and investments in technological development in particular (Erber 1981:22–3). Thus the existing financial system has inhibited the development of technological capabilities in Brazil. In this respect the availability of FINEP's funds for financing industrial technology since the mid-1970s has been an important exception since it provides resources with long pay-back periods and relatively low real interest rates. As a public company, FINEP's capacity to provide loans is dependent upon general monetary and fiscal policies. In the early 1980s, cuts in government expenditures meant decreasing resources for FINEP. The budget provision for FINEP (excluding grant-in-aid for public research institutions) fell from US$13.5 million in 1982 to US$1.8 million in 1985. A significant recovery in 1986 to US$10.3 million, after the Ministry of Science and Technology was created, was followed again by sharp decreases in the following years to US$5 million in 1987 and US$4.2 million in 1988 (CNPq 1988:156).

The erosion of FINEP's capacity to foster technological development in Brazilian industry since the early 1980s meant that the private sector was deprived of the most important mechanism to support long-run technological strategies. The situation which Porteous described, in Chapter 8, for the machine-tool sector is representative for the industrial sector as a whole: many technological plans had to be shelved and postponed. This is not to argue that no R&D took place in the informatics industry. It is remarkable that, in spite of the crisis, the total spent by national firms on R&D represented 8–10 per cent of total income from their sales for most of the 1980s (Proença and Caulliraux 1989; Erber 1990). It is hard to tell what the R&D intensity would have been if the state financing agencies had been able to provide more support. However, we can say that, because of the cuts, the links between the private sector and public R&D institutions diminished substantially. In the late 1970s and early 1980s, such links had been crucial to the emergence of design capabilities in Brazil (Tigre 1983:69–72). Let us elaborate.

In the course of the 1980s, government expenditures for scientific and technological infrastructure diminished as a consequence of short-run stabilisation measures. The three main sources of funding public science and technology institutions (the National Fund for Science and Technology Development of FINEP and the budgets for basic research of CNPq and CAPES) suffered sharp decreases. In 1985 they were allocated only 40 per cent of the 1979 amount (Bielchowsky 1985). Sagasti and Cook (1987) suggest that the Brazilian situation was similar to that of Argentina, Mexico, Venezuela, Peru and Chile where stabilisation measures were introduced in the same period.

Government expenditures for science and technology in Brazil not only decreased substantially in real terms, throughout the 1980s, but also as a share of total government expenditures. Figure 10.2 shows yearly initial provisions and final disbursements for science and technology as a share of government expenditures from 1982 to 1988. Apart from 1985, when the Ministry of Science and Technology was created and the political momentum was favouring internal technological development, a marked declining trend is observed. This suggests that the area has been suffering a more than proportional reduction in the overall government expenditure squeeze. One of the main victims of this squeeze was CTI (Technological Centre for Informatics), a government-owned R&D institute. One of the main functions designated to CTI was R&D in integrated circuits customised for the Brazilian market. CTI never took off since the necessary finance was not made available.

This occurred even though industry backed CTI, recognising that the success of their investment decisions could be furthered by the upgrading of Brazilian capabilities in basic and applied research in

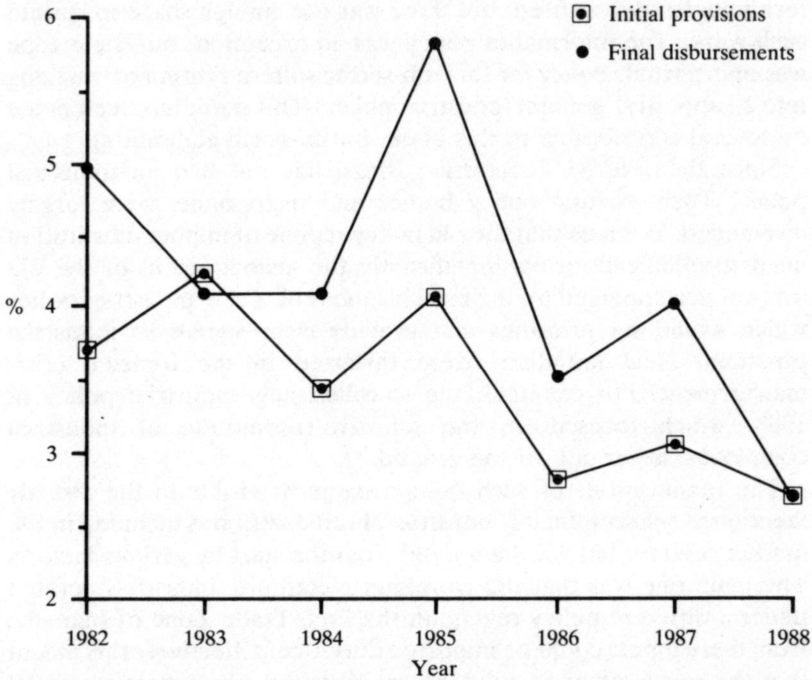

Figure 10.2 Share of S&T expenditures in total government expenditures 1982–8
Source: CNPq (1988)

micro-electronics (for details see Mammana 1989; Mendes and Borelli 1988; UNICAMP 1989). This withdrawal of support for CTI contributed to the relative failure in developing a Brazilian semiconductor industry. Indeed, this is the area where the informatics policy achieved least, and we return to it below.

So far we have emphasised the effect of government expenditure cuts on the availability of finance for technological development. The connection is obvious. There is another conflict between short-term stabilisation policy and technology policy which is less obvious but probably more severe. We use the case of Brazil to illustrate it, but it is widespread.

Economic policy in Brazil in the 1980s consisted of successive rounds of crisis management. The need to manage the external and internal debt and to control inflation dominated the government agenda. Pressing needs for short term stabilisation crowded out work on long-term economic strategy. Concern with industrial strategy and technology policy existed, but there was not enough space to sustain such work. The informatics policy was an exception, but the escape was only partial. Policy for this sub-sector suffered from not plugging into a supportive general industrial policy. This point has been made by several contributors to this book, but is worth elaborating.

Since the 1982/83 debt crisis, Brazil has not had an industrial policy. Then existing policy bodies and instruments were largely dismantled. It is true that the old policy regime of import substitution needed modification, but significantly the abandonment of the old was not accompanied by the establishment of a new industrial policy which would set priorities and provide clear signals to industrial investors. New initiatives were thwarted in the frenzied crisis management. For example, the so-called new industrial policy of 1988, which focused on the selective promotion of industrial complexes, never got off the ground.

The ramifications of such failure are most visible in the already mentioned semiconductor industry. This industry was included in the market reserve, but was hampered from the start by various factors. The main one was that the consumer electronics industry operated under a different policy regime in the Free Trade Zone of Manaus; from there inputs could be imported duty-free. Effectively this meant that the semiconductor industry was deprived of its main potential customers and from the learning opportunities that come with such linkages (Rosenthal 1987). These conflicting policies inhibited private firms from undertaking the heavy investments necessary in this industry (Mammana 1989).[28] There is no single reason for this blatant policy conflict,[29] but the fact that it was not put right in the course of the 1980s is in our view connected to the crisis conditions of the decade. As set out above, these conditions worked against a more comprehensive and strategic industrial policy.

It seems that the early 1990s bring no relief. On the contrary, the instruments of the market reserve are being dismantled and SEI, the institution which carried out the policy, has been demoted and marginalised. Worst of all, the previous cadre of experts within government has been disbanded, and a process of unlearning is under way. Again, this has to be seen in the context of the severe economic crisis and the new government's desperate attempt to get this crisis under control.

At the time of writing this final section of our book (February 1991), the changes are still under way and the outcome is hard to predict, but a number of points can be made, both on the changes in the informatics policy and the connection to the crisis. There was a recognition in both government and industry that the informatics policy needed some revision in liberalising certain imports and the participation of foreign firms in certain lines of production. Indeed, our argument was that both Brazilian industry and government had built up an ability to choose and to negotiate. Much of this ability is now being dismantled, not by design but by circumstance. To elaborate our argument it is necessary to mention the changes that have been introduced.

(1) SEI has been demoted to a department for informatics and automation in the Secretariat for Science and Technology. More generally, the government's cadre in informatics policy has dispersed – some have even left government.
(2) A number of products were taken out of the market reserve in an inconsistent manner. The choice had more to do with political pressure than with actual requirements.
(3) Import controls are to be eliminated in 1992 and replaced by tariffs.
(4) New projects for the development and manufacture of IT equipment and components will no longer need government approval. Foreign firms will be free to enter all sub-sectors of the informatics industry.[30]

Clearly modifications are necessary but it is questionable whether they are being made in such a way that they build on what has been achieved already. In our view this is not occurring. On the contrary, some of the gains from previous learning are being thrown away. Most seriously, by abandoning existing controls, Brazil loses much of the accumulated bargaining power *vis-à-vis* foreign firms. As shown in Section 8 of this Chapter, such power was crucial in changing attitudes of foreign firms and inducing them to license technology and collaborate with national firms. Mexico, for instance, opened its IT sector to foreign firms in 1986, but only did so after gaining concessions from such firms: in order to manufacture and sell in the local microcomputer market, IBM agreed to export 90 per cent of its local production, to maintain a positive trade balance, to invest 6 per cent of sales in internal R&D and to limit its share of the local market to 30 per cent (Tigre 1990:69). In contrast, Brazil is giving up all its controls without negotiation.

The apparent rationale for such a radical change is, as stated by government officials, to substitute 'positive' policies (i.e. based on stimulus) for the 'negative' ones (based on restrictions). Such 'positive' policies include an increase in government expenditures for R&D (both in the public and private sector). In fact the original budget provision for 1991 promised some increase in government R&D expenditures. However, the reality of crisis management asserts itself again: the new stabilisation plan of February 1991 cancelled such provision and the old tale is being repeated again. Indeed it is implausible that the planned government expenditures for supporting technological development (estimated to be US$3.2 billion over a four year period) will be made. In the end policies which needed revisions are being scrapped without any effective substitute.

To be sure, there are forces in government which recognise that a new industrial strategy is part and parcel of overcoming the crisis and advocate a strategy which involves step-by-step negotiations with national and foreign firms.[31] In practice, however, such efforts are relegated to the margin, because the need to stabilise the economy takes precedence.

Of course the lack of a coherent industrial policy and the dismantling of the informatics policy also has other causes, most notably vested interests. What is striking however, is that there is little space for debates in which such conflicts could be engaged in openly.

11 FINAL NOTE

In the introduction to this book we contrasted two approaches towards the fostering of hi-tech industries in LDCs. One says that the production and use of hi-tech is best left to market forces, particularly in the early stage of the product cycle. The other advocates that selective strategic state intervention yields superior results, particularly in the early stage of such new and pervasive technology. The former is most forcefully expressed by neo-liberals, the latter by neo-Schumpeterians. Indeed there are substantial differences in degree, timing and type of state intervention.

The Brazilian experience on which we have focused in this book does not provide unambiguous support for either of these positions. Neo-liberals are right in stressing the costs of intervention, evidence for which has been presented in several chapters. The weight of the empirical material, however, supports a more interventionist position. In fact, most contributors to this book would argue that key problems arose not because there was too much intervention but because there was too little. Previous sections in this chapter underlined how limited the policy was in terms of both instruments and sectoral coverage. We cannot be sure that a more comprehensive policy and more public investment would have been worthwhile. We can, however, conclude that the limited measures which were carried through had a significant impact and that both protected industry and protecting government agencies have shown a respectable performance.

To conclude that this performance vindicates the neo-Schumpeterian position would be somewhat misleading. Without other mechanisms to substitute for the initial import substitution measures and without a coherent industrial policy framework, the Brazilian IT experience was an isolated attempt at entering a new techno-economic paradigm, while government's and firms' behaviour remained largely wedded to the old paradigm. It is therefore amazing, also from a neo-Schumpeterian position, that so much was achieved.

In policy terms, the main achievement was that, by the end of the 1980s, a platform had been built up from which strategies for different sub-sectors could be pursued for the 1990s. In our view, such new strategies should include greater openness to imports and more collaboration with foreign firms in some areas but would not relinquish the key role of government in co-ordinating sub-sectoral strategies and structuring incentives. Indeed, the previous discussion on learning concluded that government had acquired the capacity to design and carry out more strategic policy for the electronics complex and that national industry had the capacity to respond. However, the signs are that this capacity will not be drawn upon, that many of the gains from learning will be wasted.

The severe crisis conditions of the early 1990s are a major cause discussed in the previous section. A further, even though related, reason lies in widespread disenchantment with the state. As a result of government failures in economic policy and provision of basic services, large sections of society are 'tired of the state'.[32] In such an anti-government climate it is hard to sustain the case for strategic intervention in the electronics complex, even if the performance of government and industry in some sub-sectors has been relatively favourable.

These wider contextual issues have to be mentioned because first, policies for the Brazilian hi-tech sectors in the 1990s will be influenced less by factors internal to these sectors than by external ones. Second, because in many LDCs, certainly in Latin America, there is a similar unavoidable constellation of discredited state and economic crisis. Third, because in Brazil and elsewhere, neo-liberal forces have gained strength as a result of such problems. President Collor of Brazil came into office in 1990 with an explicitly neo-liberal agenda.

Addressing this agenda has some positive effects. Long-dodged questions are faced, such as what the state should and should not do, and how its efficiency can be raised. In recent Brazilian practice this has meant that public spending was reduced but cuts were not necessarily made where most needed.[33]

While policy reform and expenditure review are required, pursuing these in the neo-liberal framework of 'less state' is not helpful. There are areas which may benefit from a retreat of the state but there are others in which more state is needed. Certainly the assumption of a generally over-blown Brazilian state is questionable. The share of gross tax revenue in GNP is small and has been falling from 26 per cent in 1970 to 22 per cent in 1988 (Alves 1989). By comparison, countries such as Italy and Spain increased their gross tax burden over the same period from 26 to 39 per cent of GNP (Dain and Afonso 1990). Furthermore, the net government tax revenue[34] in Brazil fell from 17 to 10 per cent of GNP during the last two decades (Alves 1989). This decline in net resources available to the state resulted in a sharp fall in public expenditure. In consequence public services deteriorated, in particular health, education and infrastructure.

The earlier mentioned popular mobilisation against government has to be seen in this context of state failure to deliver basic services. This failure is not just a result of inefficiency or corruption in the public sector; it is also a consequence of small and shrinking government revenue. A reversal is hard to imagine without a gradual increase in tax burden. It is equally hard to imagine how government can avoid getting absorbed in successive rounds of short-term crisis management without raising tax revenue. Our argument is not that such tax income should be used for hi-tech sectors as a priority but that increasing tax revenue is one condition for re-establishing the credibility and solvency of the state. Without these, strategic industrial policy can hardly be sustained.

Exploring these issues further is beyond the confines of this book, but ignoring them would mean escaping into fantasy. A discredited state and economic crisis are the unavoidable context for the industrial policy debate in a good number of LDCs. Once restored, successful policy for hi-tech industry does not necessarily require huge state investment. The message from the Brazilian case is that government can unleash an enormous indigenous learning process by relying primarily on its regulatory function. While not a clear-cut success story, it contrasts with the widespread gloom on what LDCs can do to foster hi-tech industry.

NOTES

1 We thank Afonso Fleury, Chris Freeman, John Humphrey and Carlota Perez for very helpful comments on a previous draft of this chapter.

2 The emphasis of this chapter is, nevertheless, on the lessons which can be drawn from the Brazilian experience for other LDCs and not on the lessons to be drawn from the latter for the former. A useful review of national policies in various countries can be found in Rushing and Ganz-Brown (1986). For a recent analysis of the international changes in information technologies and markets and their implications for industrial late-comers, see Ernst (1990).

3 In any case, the size of the internal market is not a given but is itself influenced by policies.

4 It is interesting to note that in Japan, in the early sixties, the same complaint was made about the Japanese computer industry (Anchordugui 1988).

5 A detailed study compared mark-ups in 1980 in Brazilian and OECD countries (Furtado 1990). One of the main findings is that the average figure for Brazil is 50 per cent compared with 24 per cent for OECD countries (ranging from 8 per cent in the Netherlands to 36 per cent in Sweden). Another recent study analysed the behaviour of leading firms in all sectors of Brazilian industry during the 1980s (123 national firms and 58 subsidiaries of multinationals were selected) (Ferreira *et al.* 1990). It was found that in spite of the crisis of the Brazilian economy these firms increased their gross profit margin by 23 per cent from 1978–80 to 1989. National firms increased their mark-up from an average of 54 per cent in 1984 to 102 per cent in 1989 and subsidiaries of multinationals from 36 per cent in 1984 to 51 per cent in 1989.

6 Imported softwares do not carry duties in Brazil since they are considered services and not products (Tigre 1990:64).

7 Ideally value data need to be supplemented with data on physical quantities. This could be done either with absolute figures (such as number of microcomputers for each country) or with an alternative relative indicator which is not GDP-dependent (for example number of computers per employee). Unfortunately lack of comparable data on numbers of microcomputers prevented us from doing so. However, size of the market (expressed in monetary terms) relative to industrial product is a reasonable proxy which shows that Brazil achieved high diffusion levels in spite of the crisis; nevertheless it is important to bear in mind that this kind of figure tends to underestimate the market size of countries with lower prices, notably the US.

8 Equally interesting are studies which show what happens in the absence of user–producer interactions. When producers dominated users (or when users had a limited technical competence) there was a tendency towards 'hyper automation', i.e., users were faced with design values different from their needs and with overly complex and costly capital goods. In such cases, instead of attaining productivity gains, automation led to the upsurge of diseconomies (Lundvall 1989:16–17). In the case of waste-water technology and office technology in Denmark, the lack of local user competence had a negative effect upon the systems developed.

9 The mismatch between user and producer of new information technologies seems also to be present in the petrochemical sector in more advanced economies. Hagedoorn and Kalff (1985) suggested as explanations for the slow diffusion of programmable logic controllers that 'the process industry has a long and successful tradition of its own in automation [and] this obstructs the entrance of computer-technology to the market; there is no standard solution for all process control automation problems as each system is tailor-made' (p. 138).

10 While we would argue that some electronics production is no longer at an early stage of technological development, generic technologies grouped around information systems are still in their infancy.

11 The innovation–diffusion dynamics of information technology has a unique characteristic: the information requirements of the various processes shape technologies and do not just result from them as with previous vintages of capital goods. This is made possible by the soft component that supplements the hardware. The user is particularly important to the innovation–diffusion process since the software component of IT is flexible and appropriate for user's intervention. Such intervention allows users to insert the knowledge specific to their own processes into the IT system (Aksoy 1990).

12 This survey was organised by Rogerio Viana of the Special Secretariat for Informatics (SEI).

13 The firms interviewed were four leading microcomputer manufacturers (Microlab, Itautec, Monydata and Scopus), two successful peripheral producers (Conpart and Elebra Periféricos) and two multinational subsidiaries (Unysis and Texas).

14 Not untypically, this firm was established by four engineers who left other IT firms to start their own venture.

15 'Original Equipment Manufacture'.

16 This literature has been subsequently criticised for being over-simple in the emphasis on global sourcing for cheap labour and for ignoring a more complex set of global investment decisions by corporations which were driven as much by market access criteria (Scott 1987; Jenkins 1987:6; Elson 1988:20).

17 Fleury (1990) and discussions with the authors.

18 See Lim and Fong (1986) for the case of Singapore; Tchoi and Sanyal (1985) for South Korea; Mody (1985) for Taiwan and South Korea.

19 For some products, however, the definition and management of the technological gap is more controversial. The gap might be huge, but – as emphasised by Erber (1990a) – the extent to which this matters depends on the 'margins of tolerance' of local users. This simply means that what local users demand may be influenced more by what international technology offers (performance parameters) than by the range of functions actually required.

20 In the case of Brazil, Evans (1979) makes an important contribution to the debate.

21 This finding is corroborated by other studies. For South Korea see, for example, Kim, Westphal and Dahlman (1985:185–7).

22 'The method of assimilating and improving upon imported technologies was mainly some form of "reverse engineering". This involved trying to

manufacture a product similar to one already available on the world market but without direct foreign investment or transfer of blueprints for product and process design. The widespread use of reverse engineering in the 1950s and 1960s had several major consequences for the Japanese system of innovation' (Freeman 1987:40–4).

23 Direct foreign investment was not permitted at this time since, it was argued, national firms would not be able to compete on the internal market even with licensed technology from other sources. More importantly, the explicit objective of policy intervention in the computer industry was to build up the competitiveness of national firms by fostering the technological capability of these firms.

24 The inflexibility of foreign capital may have reinforced and strengthened the glue between the national alliance, particularly with respect to the military's involvement in the formulation of informatics policy.

25 Including four agreements with IBM, two with Hitachi, Mitsubishi and Fujitsu and one with Data General, BASF, Nixdorf, DEC, Hewlett Packard, Sun, ATT, Bull and Philips (Meyer-Stamer 1990:90).

26 It is no coincidence that there are already a number of foreign corporations which have started some form of distribution but not production in Brazil.

27 Access to the European market and brand name was, for example, the main reason behind the acquisition by Sistema (the leading Brazilian firm producing printers) of an Italian printer manufacturer in late 1990 (personal communication by one of Sistema's directors in February 1991).

28 As a result, at the end of the 1980s only two national firms maintained manufacturing operations which contained R&D in the design of integrated circuits. Both these firms, Sid and Itaucom, are subsidiaries of the big banks which are behind the success of banking automation in Brazil. Owing to the needs of their upstream partners, Sid and Itaucom have focused on niche markets. In contrast to South Korean producers they do not want to compete in standard integrated circuits where minimum scales are high. Instead they concentrate on custom and semi-custom ICs where 'technological advance may have lowered barriers to chip design . . . and . . . human ingenuity rather than massive capital is the key to successful design entry', a strategy that was followed successfully by some Taiwanese firms (Hobday 1990:viii–ix).

29 For further discussion of the policy conflict, see Erber (1990a), Meyer-Stamer (1990) and Chapter 9 of this book.

30 This change still requires approval of Congress.

31 The industrial policy launched in 1990 is set out in IPEA (1990). For a critique see Erber (1990b).

32 For a comparison with previous anti-government campaigns, see Fiori (1989).

33 The new government, for instance, abolished practically all subsidies but bowed to political pressures and eventually maintained the incentives and subsidies to the North-Eastern and Amazon regions, which – in the way they are applied – are hard to defend on efficiency and equity grounds and which have historically been a source of malpractice and corruption.

34 This net (or disposable) tax revenue amounts to the sum of all direct and indirect taxes minus interest payments on the external and internal debt and (non-fungible) national insurance payments.

BIBLIOGRAPHY

Aksoy, A. (1990) 'Innovation and diffusion dynamics of the information technology paradigm', Working Paper No. 11, Centre for Information and Communications Technologies, Brighton: Science Policy Research Unit, University of Sussex.

Alves, S.R. (1989) 'Tratamento da correção, monetária dos juros da divida interna', *Revista de Economia Politica*, Vol. 9, No. 3.

Alves Filho, A.G., Marx, R. and Zilbovicius, M. (1989) 'Fordism and New Best Practice: some issues on the transition in Brazil', *IDS Bulletin*, Vol. 20, No. 4, October, Brighton: Institute of Development Studies, University of Sussex.

Anchordugui, M. (1988) 'Mastering the market: Japanese government targeting of the computer industry', *International Organisation*, Vol. 42, No. 3.

Andersen, E., Dalum, B. and Villumsen, G. (1981) 'The importance of the home market for the technological development and the international specialisation of manufacturing industry', *Technical Innovation and National Economic Performance*, IKE Seminar, Aalborg: Aalborg University Press.

Azpiazu, D., Basualdo, E. and Nochteff, H. (1990) 'Industrial policies and industrial restructuring: lessons from the Argentine informatics sector', paper presented at the Workshop 'Hi-tech for industrial development', Brighton: Institute of Development Studies, University of Sussex.

Banhos, S. (1989) 'Corporate strategy in the computer industry in Brazil', M.Sc. dissertation, Brighton: Science Policy Research Unit, University of Sussex.

Baptista, M., Caulliraux, H., Possas, M. and Tauile, J. (1990) 'A indústria de informática no Brasil', paper prepared for the research project *Desenvolvimento Tecnológico da Indústria e a Constituicão de um Sistema Nacional de Inovação no Brasil*, Instituto de Economia, Universidade Estadual de Campinas, Campinas.

Bell, M., Ross-Larson, B. and Westphal, L.E. (1984) 'Assessing the performance of infant industries', *World Bank Staff Working Paper* No. 666, Washington, DC: World Bank.

Bielchowsky, R. (1985) 'Situação do apoio do Governo Federal à pesquisa fundamental no Brasil', mimeo, Rio de Janeiro, FINEP.

Castaño, A., Katz, J. and Javajas, F. (1980) 'Etapas históricas y conductas tecnológicas en una planta metalmecánica argentina', mimeo, IDB/ECLA/UNDP Research Program on Scientific and Technological Development in Latin America, Buenos Aires.

Cline, W.R. (1987) *Informatics and Development: Trade and Industrial Policy in Argentina, Brazil and Mexico*, Washington, DC: Economics International Inc., February.

CNPq (1988) *Ciência e Tecnologia: Recursos do Tesouro da União – 1980 a 1988*, Brasilia: CNPq, Secretaria Especial de Ciência e Tecnologia, Presidência da República.

Cooper, C. and Hoffman, K. (1978) 'Developing countries and international markets in industrial technology', Brighton: Science Policy Research Unit, University of Sussex, January.

Coutinho, L. and Suzigan, W. (1990) *Desenvolvimento Tecnológico da Indústria e a Constituição de um Sistema Nacional de Inovação no Brasil*, Campinas: Instituto de Economia, Universidade Estadual de Campinas.

Dahlman, C. and Westphal, L. (1982) 'Technological effort in industrial development: a survey', in F. Stewart and J. James (eds) *The Economics of New Technology in Developing Countries*, London: Frances Pinter.

Dain, S. and Afonso, J.R.R. (1990) 'O funcionamento do setor público no Plano Collor', in M.C. Tavares *et al.* (eds) *Aquarela do Brasil – Ensaios Políticos e Econômicos sobre o Governo Collor*, Rio de Janeiro: Rio Fundo Editora.

Eicher, H. (1989) 'A contribuição da empresa estrangeira', in Seminário de Avaliação da Politica Nacional de Informática, Instituto de Economia, Universidade Estadual de Campinas, Campinas: UNICAMP.

Elson, D. (1988) 'Transnational corporations in the new international division of labour: a critique of the 'cheap labour' hypothesis', *Manchester Papers on Development*, Vol. 4, No. 3.

Erber, F.S. (1981) 'Science and technology policy in Brazil: a survey of the literature', *Latin American Research Review*, Vol. 16, No. 1.

—— (1988) 'A transformação dos regimes de regulação: desenvolvimento tecnologico e intervenção do Estado nos países industrializados e no Brasil', tese de Professor Titular, Faculdade de Economia e Administração, Universidade Federal do Rio de Janeiro.

—— (1990a) 'Politica de informática e politica industrial', Rio de Janeiro: IPEA.

—— (1990b) 'A politica industrial – paradigmas teóricos e modernidade', in M.C. Tavares *et al.* (eds) *Aquarela do Brasil – Eusaios Politicos e Economicos sobre o Governo Collor*, Rio de Janeiro: Rio Fundo Editora.

Ernst, D. (1985) *Automation, Employment and the Third World: The Case of the Electronics Industry*, IDPAD Project on Microelectronics, Institute of Social Studies, The Hague.

—— (1990) 'Global competition, new information technologies and international technology diffusion – implications for industrial latecomers', mimeo, Paris: OECD.

Ernst, D. and O'Connor, D. (1990) *New Technologies, Global Competition and Latecomer Industrialisation – an Agenda for the 1990s*, Paris: OECD.

Evans, P. (1979) *Dependent Development: The Alliance of Multinational, State and Local Capital in Brazil*, Princeton: Princeton University Press.

—— (1985) 'Varieties of nationalism: the politics of the Brazilian computer industry', paper presented at the MIT Symposium on The Computer Question in Brazil, Cambridge, Massachusetts, April.

—— (1986) 'State, capital and the transformation of dependence: the Brazilian computer case', *World Development*, Vol. 14, No. 7.

Ferraz, J.C. (1989) 'A heterogeneidade tecnológica da indústria brasileira: perspectivas e implicações para política', *Revista Brasileira de Economia*, Vol. 43, No. 3.

Ferreira, C., Almeida, J. and Novais, L. (1990) 'O ajuste na grande empresa privada nos anos 80', paper prepared for the research project *Desenvolvimento Tecnológico da Indústria e a Constituição de um Sistema Nacional de Inovação no Brasil*, Instituto de Economia, Universidade Estadual de Campinas, Campinas.

Fiori, J.L. (1989) 'Sobre a Crise do Estado Brasileiro', *Revista de Economia Politica*, Vol. 9, No. 3.

Fleury, M.T. (1990) 'Mudanças e persistências nos modelos de gestão de pessoal em setores de tecnologia de ponta – o caso brasileiro em contraponto com o japones', mimeo, Departamento de Economia e Administração, Universidade de São Paulo.

Freeman, C. (1982) *The Economics of Industrial Innovation*, London: Frances Pinter.

—— (1986) 'The challenge of new technologies', paper presented to the OECD Symposium on Opportunities and the Risks for the World Economy: the Challenge of Increasing Complexity, Paris: OECD.

—— (1987) *Technology Policy and Economic Performance – Lessons from Japan*, London: Frances Pinter.

Frischtak, C. (1989) 'Specialisation, technical change and competitiveness in the Brazilian electronics industry', paper presented at the OECD Workshop on Technological Change and the Electronics Sector – Perspective and Policy Options for Newly Industrialising Countries, Paris: OECD.

Fröbel, F., Heinrichs, J. and Kreye, O. (1980) *The New International Division of Labour*, Cambridge: Cambridge University Press.

Fung, S. and Cassiolato, J. (1976) *The International Transfer of Technology to Brazil Through License Agreements: Characteristics of Government Control System and Commercial Transactions*, Center for Policy Alternatives, Massachusetts Institute of Technology, Cambridge, Massachusetts: MIT.

Furtado, J. (1990) *Produtividade na Indústria Brasileira: Padrões Setoriais e Evolução – 1975/80*, M.Sc. dissertation, Instituto de Economia, Unicamp, Campinas.

Gaio, F.J. (1990) 'The development of computer software technological capabilities in developing countries – a case study of Brazil', Ph.D. thesis, University of Sussex, April.

Georghiou, L. *et al.* (1986) *Post-Innovation Performance: Technological Development and Competition*, London: Macmillan Press.

Grieco, J. (1984) *Between Dependency and Autonomy: India's Experience with the International Computer Industry*, Berkeley, California: University of California Press.

Hagedoorn, J. and Kalff, P. (1985) 'Sectoral and technological development: information technology and process control equipment', Interim Report, Amsterdam: STB.

Heeks, R.B. (1991) 'State policy and the development of the Indian software industry', Ph.D. thesis, Open University, Milton Keynes.

Hewitt, T. (1988) 'Employment and skills in the electronics industry: the case of Brazil', D.Phil. thesis, University of Sussex.

Hobday, M. (1990) 'The world semiconductor industry: industrial organisation, technological change and prospects for developing countries', paper prepared for the FECAMP/UNDP Project on Technological Development of Industry and the Constitution of a National System of Innovation in Brazil, Brighton: Science Policy Research Unit, University of Sussex.

IDRC (1978) 'Science and technology for development', Main Comparative

Report of the STPI Project, International Development Research Council, Ottawa.
IPEA (1990) 'Política industrial e de comércio exterior: apoio à capacitaçâo tecnológica da industria', Documento de Politica No. 1, Instituto de Pesquisa Economica Aplicada, Brasilia.
Israel, A. (1988) *Institutional Development – Incentives to Performance*, Baltimore and London: Johns Hopkins University Press, for the World Bank.
Jenkins, R. (1987) *Transnational Corporations and Uneven Development: The Internationalisation of Capital and the Third World*, London: Methuen.
Justman, M. and Teubal, M. (1989) 'A structuralist perspective on the role of technology in economic growth and development', *Discussion Paper* No. 13, The Jerusalem Institute for Israel Studies, Jerusalem.
Kaplinsky, R. (1988) 'Industrial restructuring in LDCs: the role of information technology', paper prepared for Conference on Technology Policy in the Americas, Stanford University, December.
—— (1989a) 'Technological revolution and the international division of labour in manufacturing: a place for the third world?', in R. Kaplinsky and C. Cooper (eds) *Technology and Development in the Third Industrial Revolution*, London: Frank Cass.
—— (1989b) 'Industrial intellectual property rights in the Uruguay round and beyond', *Journal of Development Studies*, Vol. 25, No. 3.
Karlsson, C. (1988) 'Corporate families to handle galloping technologies', in N. Hood and J. Vahlne (eds) *Strategies in Global Competition*, Beckenham: Croom Helm.
Katz, J. (1987) 'Domestic technology generation in LDCs: a review of research findings', in J. Katz (ed.) *Technology Generation in Latin American manufacturing Industries: Theories and Case-Studies Concerning its National Magnitude and Consequences*, London: Macmillan Press.
—— (ed.) (1987) *Technology Generation in Latin American Manufacturing Industries*, London: Macmillan Press.
Kim, L., Westphal, L. and Dahlman, C. (1985) 'Reflections on the Republic of Korea's acquisition of technological capability', in N. Rosenberg and C. Frischtak (eds) *International Technology Transfer: Concepts, Measures and Comparisons*, New York: Praeger Press.
Lim, L. and Fong, P.E. (1986) 'Trade, employment and industrialisation in Singapore', *Employment, Adjustment and Industrialisation 2*, Geneva: ILO.
Luedde-Neurath, R. (1988) 'State intervention in "outward-looking" development: in South Korea', in G. White (ed.) *Developmental States in East Asia*, London: Macmillan.
Lundvall, B.-Å. (1983) *Technology, Competitiveness and Small Countries*, Smaskrift No. 37, Aalborg: Aalborg University Press.
—— (1985) 'Product innovation and user–producer interaction', *Industrial Development Research Series*, No. 31, Aalborg: Aalborg University Press.
—— (1986) 'Technological revolutions and the international division of labour', paper presented at the Workshop on Innovation and Regional Development, University of Architecture, Venice.
—— (1988) 'Innovation as an interactive process: from user–producer

interaction to the national system of innovation', in G. Dosi *et al.* (eds) *Technical Change and Economic Theory*, London: Pinter Publishers.

—— (1989) 'Innovation, the organised market and the productivity slow-down', paper presented at the International Seminar on Science, Technology and Economic Growth, Paris: OECD.

Mammana, C. (1989) *Considerações sobre a Microeletrônica no Brasil*, Instituto de Economia, Universidade Estadual de Campinas, Campinas.

MCT (1986) 'Relatório final dos grupos de trabalho de recursos humanos em informática', Ministério de Ciência e Tecnologia, Brasilia, October.

Mendes, M. and Borelli, M. (1988) 'The semiconductor policy and ICs design in Brazil', in C. Piragibe (coord.) *Electronics Industry in Brazil: Current Status, Perspectives and Policy Options*, Brasilia: Center of Studies in Scientific and Technological Policy, Ministry of Science and Technology.

Metcalfe, J. (1986) 'Technological innovation and the competitive process', in P. Hall (ed.) *Technology, Innovation and Economic Growth*, Southampton: Camelot Press.

Meyer-Stamer, J. (1990) 'From import substitution to international competitiveness – Brazil's informatics industry at the crossroads', German Development Institute (GDI), Berlin, mimeo.

Microelectronics Monitor, (1989) Vienna: UNIDO, October.

Mody, A. (1985) 'Recent evolution of microelectronics in Korea and Taiwan: an institutional approach to comparative advantage', Boston University, mimeo.

—— (1987) 'Information industries: the changing role of newly industrialising countries', *Technology and Government Policy in Telecommunications and Computers*, Washington, DC: Brookings Institute.

Morgan, K. and Sayer, A. (1988) *Microcircuits of Capital, 'Sunrise Industry' and Uneven Development*, Cambridge: Polity Press.

O'Brien, R. (1985) 'Technology and industrial development: the Irish electronics industry in an international context', in J. Fitzpatrick and J. Kelly (eds) *Perspectives on Irish Industry*, Dublin: Irish Management Institute.

O'Connor, D. (1985) 'The computer industry in the third world: policy options and constraints', *World Development*, Vol. 13, No. 3.

OECD (1980) *Impact of Multinational Enterprises on National Scientific and Technical Capacities: Analytical Report*, Paris: OECD.

Pack, H. and Westphal, L. (1986) 'Industrial strategy and technological change: theory versus reality', *Journal of Development Economics*, Vol. 22.

Pavitt, K. (1988) 'International patterns of technological accumulation', in N. Hood and J. Vahine (eds) *Strategies in Global Competition*, Beckenham: Croom Helm.

Pavitt, K., Robson, M. and Townsend, J. (1989) 'Technological accumulation, diversification and organisation in U.K. companies, 1945–1983', *Management Science*, Vol. 35, No. 1.

Perez, C. (1983) 'Structural change and the assimilation of new technologies in the economic and social systems', *Futures*, Vol. 15, No. 5.

—— (1986) 'Las nuevas tecnologias: una vision de conjunto', in C. Ominami (ed.) *La Tercera Revolucion Industrial: Impactos Internacionales del Actual Viraje Tecnologico*, Buenos Aires: Grupo Editor Latino-Americano.

—— (1988) 'New technologies and development', in C. Freeman and B. Lundvall (eds) *Small Countries Facing the Technological Revolution*, London: Pinter Publishers.

—— (1989) 'The present wave of technical change: implications for competitive restructuring and for institutional reform in developing countries', paper prepared for the Strategic Planning Department of the World Bank, Washington, DC.

—— (1990) 'Electronics and development in Venezuela – a user oriented strategy and its policy implications', OECD Development Centre, *Technical Papers* No. 25, October, Paris: OECD.

Piragibe, C. (1985) *Indústria de Informática: Desenvolvimento Brasileiro e Mundial*, Rio de Janeiro: Editora Campus.

Porto, J., Silva, A. and Laplane, M. (1990) *Avaliação da Politica Nacional de Informática*, Instituto de Economia, Universidade Estadual de Campinas, Campinas.

Proença, A. and Caulliraux, H. (1989) *Criação e Evolução de Empresas de Base Tecnológica – Avaliação Setorial Indústria de Informática*, final report, Rio de Janeiro: FINEP/OEA.

Robinson, R. (1976) *National Control of Foreign Business Entry: a Survey of Fifteen Countries*, New York: Praeger Publishers.

Rosenthal, D. (1987) 'Microelectronics and industrial policies in developing countries: the case of the semiconductor industry', Ph.D. thesis, University College London.

Rushing, F. and Ganz-Brown, C. (eds) (1986) *National Policies for Developing High Technology Industries: International Comparisons*, Boulder and London: Westview Press.

Sagasti, F. and Cook, C. (1987) 'La ciencia y la tecnologia en América Latina durante el decenio de los ochenta', *Comercio Exterior*, Vol. 37, No. 12.

Schware, R. (1987) 'Software industry development in the third world: policy guidelines, institutional options and constraints', *World Development*, Vol. 15, Nos. 10/11.

Scott, A.J. (1987) 'The semiconductor industry in South East Asia: organisation, location and the international division of labour', *Regional Studies*, Vol. 21, No. 2.

SEI (1989) *Panorama do Setor de informática*, Séries Estatistícas, Vol. 2, No. 1, 20–1.

Soete, L. (1985) 'International diffusion of technology, industrial development and technological leapfrogging', *World Development*, Vol. 13, No. 3.

South (1989) November.

Tchoi, C.I. and Sanyal, B.C. (1985) 'Development of the computer and microelectronics industry and of automation and the consequences for education and training in the Republic of Korea', paper presented in November at the IIEP International Seminar on Education, Industrialisation and Technical Progress, Monastir, Tunisia.

Teixeira, A.P. (1990) 'Attaining technological maturity: protection and liberalisation in the informatics sector – the segment of PC compatibles in Brazil, 1980 to 1990', M.Phil. dissertation, Brighton: Institute of Development Studies, University of Sussex.

Teubal, M. (1979) 'On user needs and need determination: aspects of the theory of technological innovation', in M. Baker (ed.) *Industrial*

Innovation: Technology, Policy and Diffusion, London: Macmillan Press.
Tigre, P.B. (1983) *Technology and Competition in the Brazilian Computer Industry*, London: Frances Pinter.
—— (1988) 'How does Latin America fit hi-technology?', paper prepared for the International Symposium, 1–3 December: Technology Policy in the Americas, Stanford.
Tigre, P. (1990) 'Análise do complexo eletrônico brasileiro', paper prepared for the research project *Desenvolvimento Tecnológico da Indústria e a Constituição de um Sistema Nacional de Inovação no Brasil*, Instituto de Economia, Universidade Estadual de Campinas, Campinas.
UNICAMP (1989) *Seminário de Avaliação da Politica Nacional de Informática*, Instituto de Economia, Universidade Estadual de Campinas, Campinas.
US Congress (1987) 'Informatics trade problems with Brazil', Subcommittee on Commerce, Consumer Protection and Competitiveness of the Committee on Energy and Commerce, House of Representatives, US Government, 15 July.
US Department of Commerce (1987) *A Competitive Assessment of U.S. Microcomputers*, Washington, DC: US Government Printing Office.
Walsh, V. (1988) 'Technology and the competitiveness of small countries: a review', in C. Freeman and B.-Å. Lundvall (eds) *Small Countries Facing the Technological Revolution*, London: Pinter Publishers.
World Bank (1989) *World Development Report*, Washington, DC: World Bank.

Index